Building Energy-Saving Technology—2nd Edition

Building Energy-Saving Technology—2nd Edition

Guest Editors

Yaolin Lin
Wei Yang

MDPI

Basel • Beijing • Wuhan • Barcelona • Belgrade • Novi Sad • Cluj • Manchester

Guest Editors

Yaolin Lin
Built Environment and Energy Engineering
University of Shanghai for Science and Technology
Shanghai
China

Wei Yang
Faculty of Architecture, Building and Planning
The University of Melbourne
Melbourne
Australia

Editorial Office
MDPI AG
Grosspeteranlage 5
4052 Basel, Switzerland

This is a reprint of the Special Issue, published open access by the journal *Buildings* (ISSN 2075-5309), freely accessible at: https://www.mdpi.com/journal/buildings/special_issues/8S2EGC5MGI.

For citation purposes, cite each article independently as indicated on the article page online and as indicated below:

Lastname, A.A.; Lastname, B.B. Article Title. *Journal Name* **Year**, *Volume Number*, Page Range.

ISBN 978-3-7258-3193-7 (Hbk)
ISBN 978-3-7258-3194-4 (PDF)
https://doi.org/10.3390/books978-3-7258-3194-4

Contents

About the Editors

Yaolin Lin

Dr. Yaolin Lin is a Shanghai University of Science and Technology associate professor. Before that, he worked as an associate professor at Wuhan University of Technology and as a principal engineer at a leading building energy efficiency consulting company in California, USA. He received his PhD degree in building engineering from Concordia University in Canada. He has over 7 years of industrial experience in building HVAC controls, building energy management, retro-commissioning of building systems, and advanced controller development, and over 10 years of research experience on green and sustainable buildings, HVAC system control optimization, and building energy system modeling and simulation, leading over 50 industrial projects across five different countries, including the USA, Canada, France, Singapore, and China. He has published over 90 journal papers, book chapters, and peer-reviewed conference papers on building energy efficiency and simulation. Dr. Lin is an associate editor of the SCI Journal of *Energy Exploration and Exploitation*. He serves as an editorial member for the SCI journals *Buildings* and *Advances in Meteorology* and the EI journal *Energy Engineering*. He is a reviewer for over 40 renowned international journals, including *Energy, Applied Energy, Energy and Buildings, Building and Environment*, and *Renewable Energy*.

Wei Yang

Dr. Wei Yang is a senior lecturer in the Faculty of Architecture, Building and Planning at The University of Melbourne. She holds a PhD in building science, awarded by the National University of Singapore (NUS). She has broad research expertise and interest in building energy efficiency, sustainable building design, building performance evaluation, and life cycle assessment. She has over 80 peer-reviewed journal publications, and her Google Scholar h-index is 26. Currently, she serves as the associate editor for *Frontiers in Built Environment*. She is also on the topical advisory panel for the Journal of *Atmosphere*.

Preface

Building energy conservation is particularly important in achieving carbon peak and carbon neutrality. The building energy-saving technologies include passive technologies focusing on building and environmental design and active technologies focusing on mechanical system control. This Special Issue presents state-of-the-art progress and trends in advanced building energy-saving technologies. This Special Issue aims to guide building design and operation for the building designer and facility management and identify future directions for the researchers. As the world is going towards a low carbon economy, we feel it is critical to know the carbon reduction technology in the build section as it consumes 40% of global energy and a similar amount of carbon emission. A large audience, including building energy researchers, designers, facility managers, policymakers, etc., can benefit from this Special Issue.

Yaolin Lin and Wei Yang
Guest Editors

buildings

MDPI

Review

Scopes for Improvements in Energy Conservation and Thermomechanical Performance of Building Blocks in the Kingdom of Bahrain: A Literature Review

Payal Ashish Modi [1,*], Abdelgadir Mohamed Mahmoud [1] and Yousif Abdalla Abakr [2]

[1] Razak Faculty of Technology and Informatics, University of Technology, Kuala Lumpur 54100, Malaysia; abdgadir.kl@utm.my
[2] Department of Mechanical Engineering, University of Nottingham, Semenyih 43500, Malaysia; yousif.abakr@nottingham.edu.my
* Correspondence: ashish@graduate.utm.my

Abstract: In regions with hot climates such as Bahrain, the utilization of air conditioning is indispensable in both public and private buildings to attain thermally comfortable indoor environments. External walls, constructed with building blocks, play a crucial role in the heat penetration into the building system. Despite extensive research on the cavity designs of building blocks, there has been a lack of comparison between individual block systems and integrated multi-block systems, considering both thermal and mechanical performance criteria simultaneously. Therefore, it is imperative to gather and review information on key parameters influencing the thermomechanical performance of building blocks, along with investigating techniques used to evaluate these parameters according to international standards. This review primarily focuses on these aspects. Additionally, it presents the historical evolution of housing types and the standard procedures followed by the Bahraini Government and block manufacturing companies in alignment with energy efficiency policies in Bahrain. Ultimately, this review aims to inspire researchers to explore other viable and innovative designs for enhancing the thermal insulation of building walls. By doing so, this work will contribute to Bahrain's 2030 goals of fostering sustainability and mitigating environmental impact at a local level, while also aligning with the United Nations' Sustainable Development Goals (SDGs) for 2030, specifically SDG 11, which aims to "make cities and human settlements sustainable".

Keywords: energy efficient buildings; thermal and structural performance of building blocks; improved building blocks; traditional buildings; insulation materials; in-lab and in-situ methods; building standards and regulations; SDGs

Citation: Modi, P.A.; Mahmoud, A.M.; Abakr, Y.A. Scopes for Improvements in Energy Conservation and Thermomechanical Performance of Building Blocks in the Kingdom of Bahrain: A Literature Review. *Buildings* **2024**, *14*, 861. https://doi.org/10.3390/buildings14040861

Academic Editors: Chi Yan Tso, Wei Yang and Yaolin Lin

Received: 30 January 2024
Revised: 12 March 2024
Accepted: 16 March 2024
Published: 22 March 2024

1. Introduction

At the turn of the 21st century, the high level of consumption of fossil fuels and dramatic shifts in the global climate patterns left no economic sectors untouched, including the building industry. Buildings, responsible for nearly 40% of energy consumption and approximately one third of greenhouse gas emissions, primarily devote this energy to heating and cooling purposes [1]. The majority of this energy is lost or gained through the building envelope, which resembles a container punctuated by doors, windows, roofs, external walls, and ventilations. Consequently, the design of each of these elements significantly impacts the energy efficiency of building envelopes. With rapid urbanization, architectural design as a whole plays a pivotal role in energy consumption [2]. Additionally, human comfort within buildings is influenced by building-related characteristics [3,4]. Implementing energy efficiency measures in buildings holds long-term implications, given their extended lifespan. If established energy efficiency practices are adopted, global building energy demand could decrease by one-third by 2050 [5]. Efforts towards this end are underway globally, resulting in the development of sustainable materials and technologies, such as phase change materials, green building materials, and advanced insulation materials [6]. Energy-efficient and renewable energy technologies focusing on HVAC systems, lighting, and building walls [7,8], and building-integrated photovoltaic systems are also being developed [9]. The latest advancements include the integration of Industry 4.0 technologies like [10] deep learning [11] to forecast building energy consumption. Moreover, smart solutions like photovoltaic green roofs [12] and building automation systems, employing machine learning and Internet of Things (IoT) technologies, are enhancing energy efficiency in buildings [13]. Sustainable energy technologies, such as nanofluid-based solar collectors, are being explored to achieve net-zero energy goals [14].

In hot climates, notably in the Gulf countries, the high energy demand and diminishing fossil fuel reserves are major concerns. Consequently, enhancing the energy efficiency of building walls has become a critical research area, particularly in regions like the Gulf, where high atmospheric temperatures persist for a significant portion of the year. Thus, the measurement and investigation of Thermal Resistance [15] alongside the Mechanical Strength of building walls [16] are crucial, providing real-time insights and serving as a basis for more effective insulation techniques [17,18], especially when combined with numerical modeling applications. This review paper primarily focuses on such investigations in the Kingdom of Bahrain.

Need for the Investigation of Building Blocks in Bahrain

The Kingdom of Bahrain, being one of the Gulf countries, experiences a hot and humid climate for most months of the year. Over the past twenty years, Bahrain's total primary energy supply has grown by 4.2%, total final energy consumption has grown by 5.3%, and non-industrial electricity consumption has grown by 6.6% per year [19]. The primary energy consumption for Bahrain increased from 0.38 (1.11×10^8 MWh) quadrillion Btu in 2000 to 0.67 quadrillion Btu (1.96×10^8 MWh) in 2019, growing at an average annual rate of 3.14% [20]. Per capita energy consumption reached 9.7 toe (112.81 MWh) in 2019 (three times higher than the Middle East average and five times the global average), while electricity consumption per capita was 21 MWh, which was five times the Middle East average and seven times the global average [21]. In the Kingdom of Bahrain, the residential sector used most of the grid electricity, accounting for 49.5% of total grid consumption during the baseline period. The commercial sector was the second-largest user, accounting for 36.6% during the same period. Around 60% of the residential sector's annual electricity use is related to air conditioning, which is similar for the commercial sector, at 55% annual electricity use [19]. The residential and commercial sectors therefore have high potential for significant energy efficiency improvements. In accordance with the available data from the Ministry of Electricity and Water (MEW-2021) [22] of the Kingdom of Bahrain, there has been an increase of 5% in the population during the period from 2017 to 2021. Eventually, the per capita consumption of electrical energy has significantly increased, thereby increasing the requirements to meet the national energy needs. The total energy consumption in all sectors has increased from 16,559 to 18,091 kWh (9.25%), with the total per capita energy consumption increasing from 11,083 to 11,871 kWh (7.10%) for the period from 2017 to 2021 [22], as shown in Figure 1, while the electricity consumption in different sectors with respect to per capita consumption is shown in Figure 2. It is interesting to note that 2019 and 2020 were the years of COVID-19 during which the majority of industrial and commercial services were under lockdown conditions. Therefore, in Figure 2, it can be observed that for this time period, there exists a drop in energy consumption in industrial and commercial sectors while there is a clear increase in the domestic sector.

Bahrain has five electrical power stations with a combined capacity of 4794 MW. Additionally, with other private power links (Alba link and GCC link), the net capacity turns out to be 5994 MW (MEW-2021) [22], as shown in Figure 3.

Figure 4 depicts Bahrain's sizable generating capacity, having serviced a peak load of 3751 MW in 2021. Based on consistent increases in population and energy consumption trends and no significant energy improvement plans in Bahrain, it is estimated that the annual peak demand is likely to be over 5400 MW by 2030 [23]. The linear trendlines of Figures 1 and 4 also support this outcome.

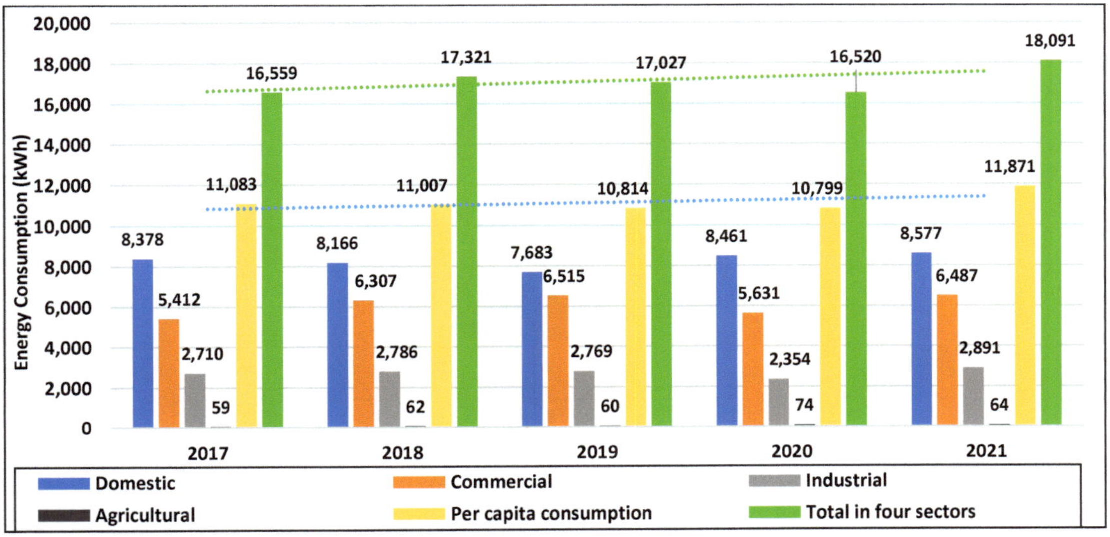

Figure 1. Increase in per capita electrical energy consumption in Bahrain: 2017–2021.

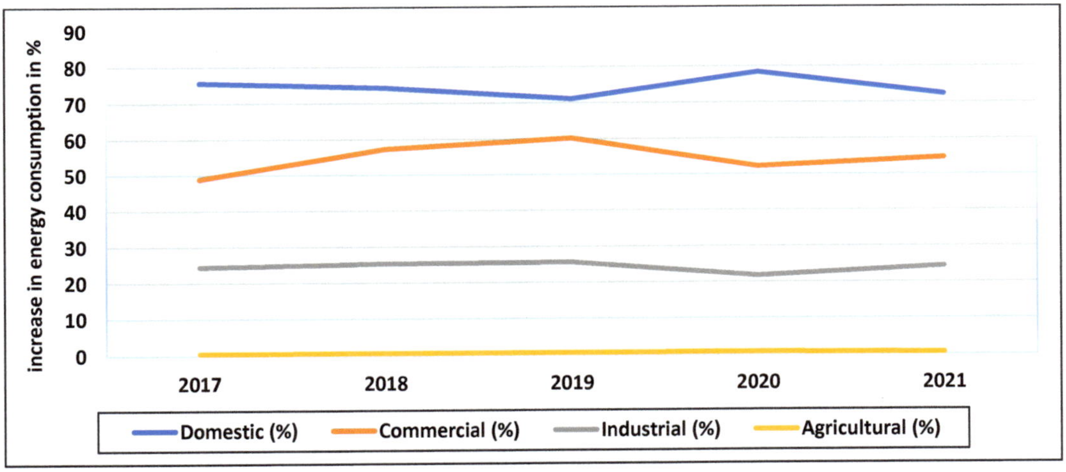

Figure 2. Increase in energy consumption in different sectors with respect to per capita electrical energy consumption in Bahrain: 2017–2021.

The Economic Development Board (EDB) expects 56,208 households to qualify for social (subsidized) housing by 2030. The report shows that non-Bahraini housing demand is expected to increase from an estimated 61,117 housing units today to 121,581 in 2030. Total housing demand (Bahraini and non-Bahraini) is expected to increase to 346,718 in 2030, up from a current stock of 145,181. Additionally, due to the geographical location of Bahrain, it experiences a hot climate for most months of the year, i.e., from April to November, and the warmest month is August, with an average temperature of 39 °C, as shown in Figure 5. The average annual maximum temperature is 31 °C, and the average annual minimum temperature is 23.5 °C. However, the average annual percentage of humidity is high, up to 73% [24], due to which the effective heat felt is much higher than the actual temperature.

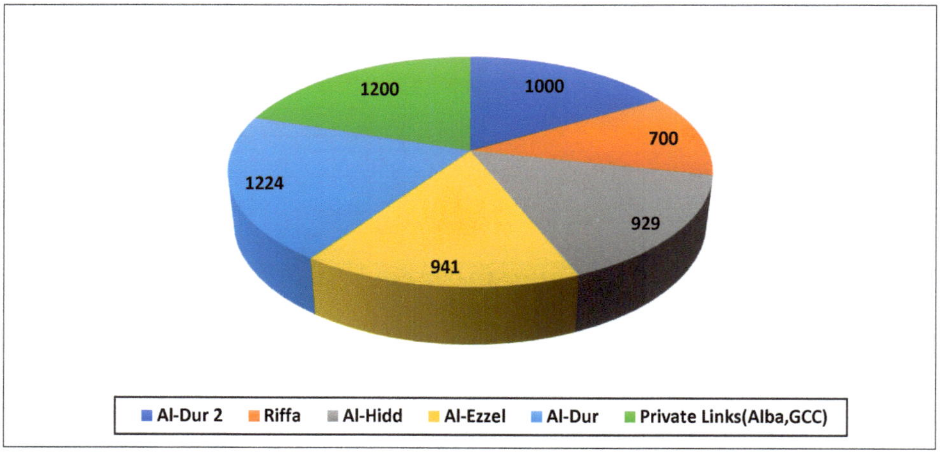

Figure 3. Bahrain's power stations and their capacities in MW.

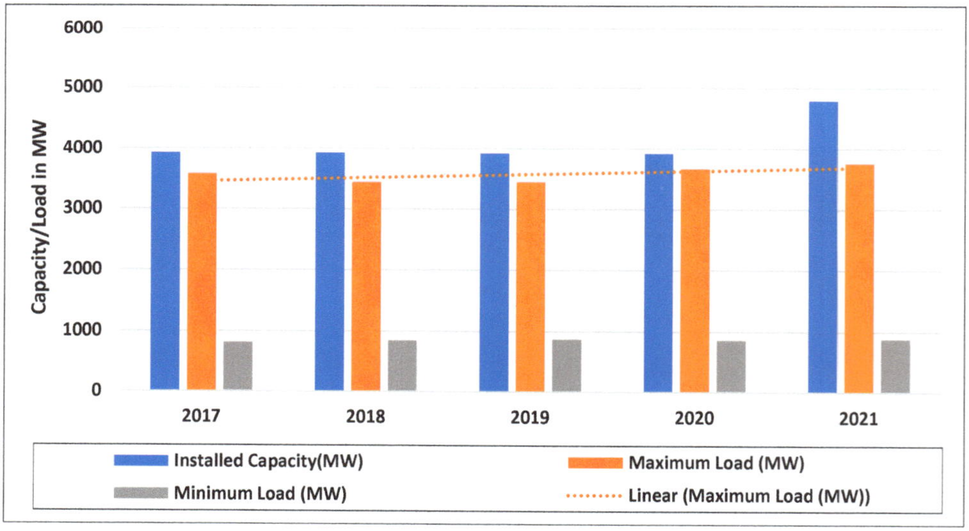

Figure 4. Comparison of maximum and minimum loads with the installed electrical energy generating capacity in Bahrain.

These reasons contribute to the fact that residents rely heavily on air conditioners to obtain a comfortable indoor temperature. Around 47–57% of the electricity demand in summer is related to the usage of air conditioners [25].

It was estimated that 6.34 TWh of electrical energy was used for cooling during May to October 2016, most of which was utilized during the months of June to September, and these results were translated into a total electricity cost of 507.443 million USD [26]. There are two major energy efficiency opportunities for the residential and commercial sectors in Bahrain: first, to significantly reduce the electricity used in buildings, and second, to improve the thermal efficiency of buildings.

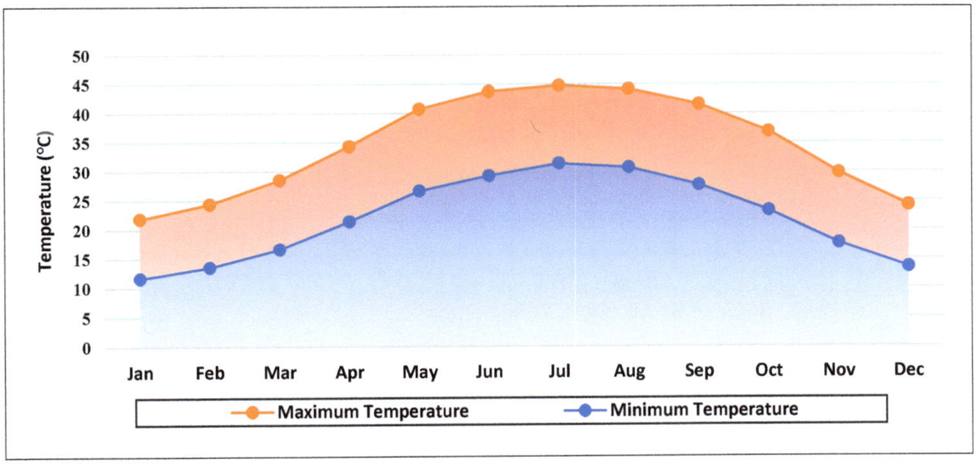

Figure 5. Average minimum and maximum temperatures in Bahrain [24].

2. Motivation

In the Kingdom of Bahrain, only three research studies in this direction have been undertaken. Hasan Radhi in 2016 [27] investigated the cool roof strategy and its potential in Bahrain. Thermal simulations of five standard roofs were performed during summer days. These included lightweight concrete screed, bituminous roofing felt, tile-light, tile-dark, and metal decking. Results show that by using light tile roofs and metal decking roofs, almost 80% of the sunlight can be reflected by the surface, and only 20% of the sunlight heats the building. However, this study mainly focused on the roof systems of buildings in Bahrain. Modi Payal in 2017 [28] designed and developed a mini-scale hot box to evaluate the thermal resistance of an insulation building block prototype using experimental and analytical methods. However, this study mainly focused on affirmation and validation of the experimental results with the thermal performance results of the block from the manufacturing companies. The analysis of Moncef Kranti in 2018 [23] indicated that the development and enforcement of a more stringent building energy efficiency code improves the energy productivity of buildings, as well as the country, with a reduction of over 87 MW in peak demand. However, this study mainly dealt with the implementation of new energy efficiency codes and PV systems in buildings. Eventually, it is clear that not much research is conducted on the technicalities of the walls, which are mainly responsible for the majority of energy transfers through a building. These technicalities of a wall include their thermal and mechanical performances, aggregate materials, insulation materials, dimensions, cavity designs, mechanical strength, thermal resistance, and so on. On one hand, very limited work is presented by the researchers, while on the other hand, the government of Bahrain has been very active and agile.

The Bahrain Economic Vision 2030 [29] states, "Bahraini nationals and residents enjoy a sustainable and attractive living environment," and some major steps are undertaken in this direction. To improve the efficiency of AC units available for retail, the government enacted a Minimum Energy Performance Standard (MEPS) and energy efficiency labeling of small AC units (Ministerial Order No. 70/2015) [30]. Thermal Insulation regulations were introduced in 1999 (Ministerial Order No. 8/1999) [31], which mandated all buildings over four stories to be insulated and stipulated minimum requirements for the efficiency of the envelope of residential and commercial buildings. In 2012, the regulations were extended to cover all buildings (Ministerial Order No. 63/2012) [32]. The current review can be considered as one step forward in the research area, as it provides appropriate information with regard to the influential factors for the performance of building blocks, in general. While this review is mainly focused on Bahrain, it is simultaneously important to mention the history of the houses and their building blocks. Along with that, the local standard procedures followed for thermal and mechanical testing and government approvals with allowable technical parameters of blocks are clearly presented. The information from the present work will be helpful in the future for extended numerical/analytical/experimental investigations of building blocks in Bahrain. On a global level, this study is also aligned with one of the 17 sustainable goals (SDGS) [33] set by the United Nations' 2030 Agenda. Sustainable Development Goal 11 aims to "make cities and human settlements sustainable," which promotes inclusive, safe, resilient, and sustainable construction industry activities.

3. Influential Factors for the Performance of Building Blocks

The performance of building blocks is mainly affected by three main parameters, as discussed in Sections 3.1–3.3.

3.1. Aggregate Materials

Various interesting analyses and proposals regarding modified building blocks' aggregate materials have been presented by researchers. This includes changes in the aggregates of the concrete and material concentrations to improve the thermal and mechanical performance of masonry blocks. A study shows that special masonry mortar made up of sintered coal gangue, obtained as an industrial waste, can save 65% of energy compared to normal cement mortar, which can save only 50% of energy. Keeping the same conditions, numerical simulation shows that the maximum heat flow of sintered coal gangue insulation block is 23.48 W/m^2 °C, while that of the ordinary clay block is 66.48 W/m^2 °C [34]. The special masonry mortar can provide better thermal performance compared to traditional clay brick. It has been observed that the manufacturing procedure of cement used in mortar involves huge emissions of CO_2 (approximately 1 ton of CO_2 for 1 ton of cement production). Recent research in this direction shows that Geopolymerization Technology is an effective method of converting wastes consisting of alumina and silica into useful products. The structural performance of geopolymers is similar to that of Ordinary Portland Cement (OPC), and therefore, geopolymers can easily replace OPC, thereby reducing energy consumption and CO_2 emissions. So, geopolymers can replace ordinary Portland cement [35]. However, the use of geopolymers in building blocks has not been investigated. In another study, customized mortar blocks were cast as a replacement for Ordinary Portland Cement (OPC) with varying amounts of Date Palm Ash (DPA) in the range of 10–30%. Experiments and simulation results depicted that thermal resistance (R-value) of the block loaded with 10%, 20%, and 30% DPA, in comparison to the control block (0% DPA), increased by 9.6%, 8.1%, and 47.4%, respectively. As per Noman Ashraf [36], DPA is a promising insulation material as it can reduce the average monthly indoor temperature by 2 °C without compromising the physical, chemical, and mechanical properties of the block, and cost analysis shows that the block with 30% DPA is the most effective, with a unit cost of $2.76. However, although, in this study, ASTM C90 standard [37] was followed to maintain the density and hence the mechanical strength of the blocks, appropriate analysis to measure the mechanical strength of the blocks was not presented. Another study shows that a new concrete material with red mud aggregate is preferred over natural basalt aggregate, as the red mud has higher wettability to the cement paste, strong adhesion, and enhanced mechanical properties compared to the basalt aggregate, which is a mixture of basalt and limestone. Results show that it increases the compressive strength of the modified block by over 10% and flexural strength by over 20% [38]. It seems that red mud can be a good option for use in concrete aggregate. The latest study shows that researchers are exploring the possible use of Recycled Concrete Aggregate (RCA) as another option for an aggregate material [39,40] or Recycled Expanded Polystyrene (EPS) [41] to manufacture the blocks. With this, the compressive strength of the block will be slightly lowered; however, its manufacturing process is environmentally friendly). These studies show good mechanical analysis; however, thermal analysis was not conducted.

Waleed A. Al-Awash [42] conducted research on different blocks by implementing changes in the aggregates (partially replacing sand with insulation materials—expanded polystyrene EPS, low-density polyethylene LDPB, vermiculate VL, volcanic scoria VS, crushed crumb-rubber RU) and produced a control block with three-row rectangular cavities to reduce the effective thermal conductivity. Experimental tests were conducted considering the selected control block design and varied insulation materials. Results showed that the VS (volcanic scoria)-treated block had optimum performance with a 15% increment in thermal resistance over the control block and 318% over the market block. This study presents good experimental and analytical (Ansys) results for the blocks of Saudi Arabia. However, optimum performance was obtained by considering only the thermal and economic criteria; however, mechanical analysis was not conducted. In a similar study [43], insulation materials (perlite, rubber, and polyethylene) were added to concrete mixtures to produce proposed optimized hollow blocks followed by an experimentation process. The insulation materials were ranked in increasing order of their performance as perlite, rubber, and polyethylene. The thermal conductivity of the proposed optimized blocks was reduced up-to 40% as compared to the conventional ones available in the market. The study shows good analytical and experimental investigation, including thermal and mechanical parameters, to prove that perlite, rubber, and polyethylene are good aggregate materials.

3.2. Insulation Materials

While following the recommendations presented by the researchers with regard to insulation materials in building blocks, it was observed that traditional thermal insulation materials like polystyrene and mineral wool are still in demand [44,45]. However, varieties of nature-based thermal insulation materials have been rediscovered as less amounts of fossil fuels are required for their production. Nature-based thermal insulation materials can be grouped as follows: rediscovered materials—cork, wood wool, fiberboard, etc.; recently introduced materials—cellulose insulation, cotton, sheep wool, etc.; and insulation materials in an experimental stage—cornstalk block, corncob board, palm fibers, etc. From market research, as shown in Figure 6, it is found that currently, all over the world, 57% of the rediscovered materials are used, with 43% of recently introduced materials, out of which cellulose, comprising 32%, is used as thermal insulation materials, while the remaining is still in the experimental stage [46]. Eventually, palm fiber is a good insulation material with a low thermal conductivity of 0.041 W/m K and can be a good option for Gulf countries.

A good comparison between two specific insulation materials like wood fiber and mineral wool has been presented by Lars Gullbrekken [47], which shows that the air permeability of wood fiber insulation is nearly 50% higher than that of mineral wool despite its high density, and consequently, in order to achieve the same U-value, 20 mm and 30 mm less insulation layers are required for the 250 mm and 400 mm walls, respectively. This study shows a good comparison between conventional insulation—mineral wool— and the newly developed wood fiber. Currently, the most widely accepted insulation materials are Phase Change Materials (PCM) that are integrated into the hollow bricks of building walls, thereby improving the thermal performance of external building walls [48,49]. The high latent heat of fusion of PCMs reduces the heat flow through the building envelope, which further stabilizes and reduces indoor temperature fluctuations [48]. However, this study does not clearly specify the placements of PCMs in the hollow blocks that can provide optimum thermal performance. Refs. [50,51] conducted an extensive numerical investigation on the integration of PCM into building walls to establish the key conditions required for effective utilization of PCM in reducing heat gains in the cooling season and heat losses in the hot season. Results show that PCM-integrated walls are advantageous mostly in moderate climates and that employing PCMs in building walls does not always lead to improvement. In fact, as shown in Figure 7, incorrect placements of PCMs can substantially increase energy use in buildings. The heat gain is higher for $x = 0.0$ and 1.0 while it is lowest at $x = 0.5$ [50].

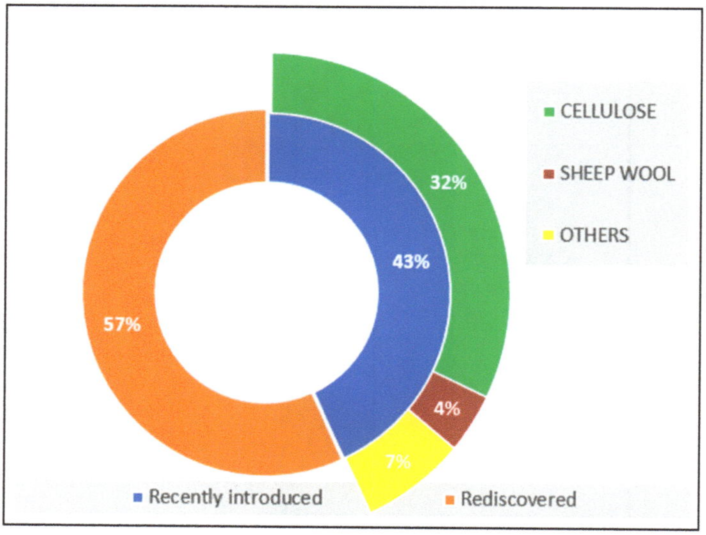

Figure 6. Market distribution of nature-based thermal insulation materials.

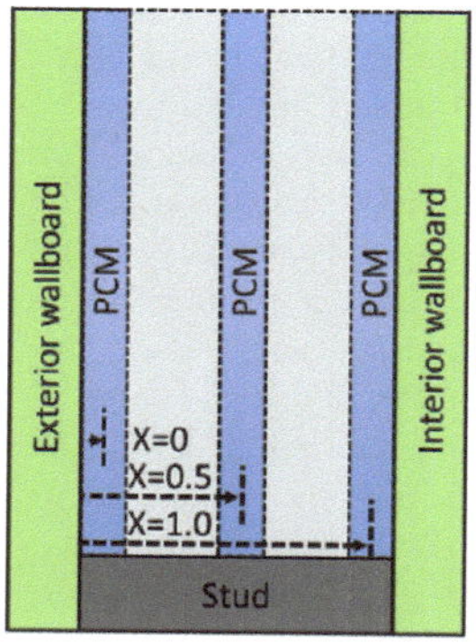

Figure 7. Variable location of PCM in the wall and its effect on cumulative heat gain [50].

Analysts are looking forward to investigating effective insulation materials and techniques to propose newer methods to reduce energy consumption. A good approach using plant fibers, recycled wastes, photochromic glass [52], agricultural wastes [53], or coconut fibers [54] as applications for green construction materials on building walls to contribute to sustainable development has been proposed.

However, Raad Homod [55] preferred to undergo a comparative study of building walls with traditional ones. Hybrid vernacular buildings made of hard external mud wall with finishing layers of a composite of clay and straw on the interior and exterior sides were compared with hybrid vernacular buildings made of AAC (Autoclaved Aerated Concrete). It was reported that AAC buildings consume 47.83% more energy compared to Vernacular buildings. In this study, the traditional approach of using mud and straw has been used in developing hybrid buildings, and they have turned out to be more effective. However, the moisture gain of the walls was not discussed. It is a general assumption that the application of insulation material layers on the exterior walls plays a vital role in enhancing the energy efficiency of the buildings. Iffa Emishaw [56] implemented this technique and used Polyisocyanurate-PIR rigid boards, as they have high thermal resistance compared to other commonly used insulation materials applied on the exterior wall. The results showed that PIR accumulated a small amount of moisture compared to other insulation boards, which can work well in colder climates; however, for other climatic conditions, due to lower permeance, its use for exterior walls may not be very desirable. Eventually, PIR is not suitable for hot countries like Bahrain. This means that insulation materials need to be selected based on climatic conditions [57,58]. As such, research depicts that the application of a small thickness of aerogel-based rendering on the exterior surface of insulated or non-insulated walls significantly reduces heat losses and removes the risk of moisture, thereby improving the energy efficiency of the buildings [59,60].

Various studies have been carried out to utilize industrial waste in the insulation industry as well. Over 60 million tons of Aluminum (Al) are produced annually, requiring sustainable and eco-friendly recycling methods for Al waste. In the research done by Thenappa [61], Al metal waste is utilized in the fabrication of Al hydroxide aerogels using a cost-effective and environmentally friendly process. Thus, the aerogels manufactured. with changes in the contents of Al and Polyvinyl Alcohol (PVA) as a binder, exhibit low density (0.06–0.108 g/cm^3), high porosity (92.3–95.5%), low electrical conductivity ([1.8–5.2] $\times$ 10^{-8} S/m), ultra-low thermal conductivity 0.028–0.032 W/mK (as shown in Table 1), and are capable of bearing high temperatures of 800 °C with less than 50% decomposition. Therefore, synthesized aerogels can be promising candidates for engineering applications such as thermal insulation of buildings and pipes to expand the usage of recycled Al [61]. The implementation of the usage of aerogel and hence aluminum in building insulation can be helpful to Bahrain as it comprises a company named ALBA, which is the world's largest aluminum smelter with a production capacity of 1.651 M mtpa [62].

Table 1. Thermal conductivity of Aluminum Hydroxide Aerogels with different contents of Al and PVA [61].

Sample Name	Aluminum (Al) Content wt. (%)	PVA Content wt. (%)	Thermal Conductivity Kavg (W/m K)
AA1	0.25	2.0	0.029 ± 0.003
AA2	0.50	2.0	0.031 ± 0.001
AA3	0.75	2.0	0.032 ± 0.001
AA4	0.50	1.0	0.028 ± 0.002
AA5	0.50	1.5	0.030 ± 0.001

The most important point to note is that although new insulation materials are being developed and proposed, block manufacturing companies still use polymer-based materials like polystyrene and polyurethane foam. As such, these materials have huge environmental impacts during their production process. Therefore, researchers should keep in mind that the new materials they develop should not only have excellent properties but should also be environment friendly, like composite materials [63,64] or industrial by-products [65,66].

3.3. Geometrical Configurations of Cavities of Blocks

Various studies have been conducted to investigate the effects of varied geometrical configurations of hollow concrete building blocks on their thermal resistance, thereby considering different modes of heat transfer: conduction, convection, and radiation [67,68]. In one such study conducted by M.P. Morales [69], it was observed that a rhomboid layout of voids with a longer diagonal at right angles to the heat flux, alongside rhomboidal cavities at grooves and tongues in alternate rows, breaks the thermal bridge and can be a good option to improve the thermal resistance of blocks (as seen in Figure 8b). With this type of arrangement, the direct thermal bridge in the block was broken, and around a 16% improvement was observed in the thermal properties of the block compared to the original one with rectangular cavities (as seen in Figure 8a), which was used in Spain, thereby keeping the U value below 0.57 W/m^{2}K, which was well within the range of Spanish regulations. However, convection and radiation modes of heat transfers were neglected, and mechanical analysis of the blocks was not conducted.

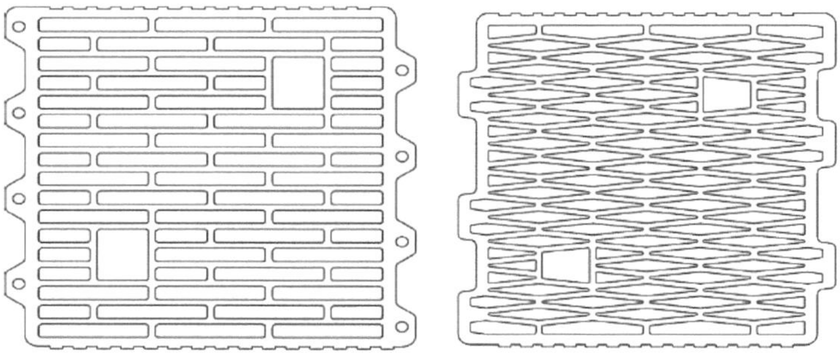

(**a**) (**b**)

Figure 8. Sections of block with rectangular cavities (**a**) and rhomboidal cavities (**b**) [69].

In a similar study by Yuan Zhang [70], it was found that for the same block width, the thermal performance can be improved remarkably by increasing the number of hole rows compared increasing the hole thickness, as shown in Figure 9. So, it is advisable to increase the hole rows with a staggered arrangement over increasing the hole thickness if the hole thickness is larger than 0.02 m. Factors influencing the thermal performance of the blocks are ranked in descending order as follows: (a) raising block thickness (with an increase in hole spacing), (b) increasing hole rows with small hole thickness, (c) raising block thickness (with a rise in hole thickness), (d) increasing hole rows with large hole thickness, (e) reducing the hole number of each row, (f) staggered arrangement of holes [70,71]. This study provides good insights into the various structural factors that affect the thermal resistance of the block; however, it is limited to only rectangular-shaped cavities within the block. An important factor to be noted is that changes in the number and dimensions of cavities of the block significantly affect the mechanical strength of the block, although this lack related analysis in the study [70].

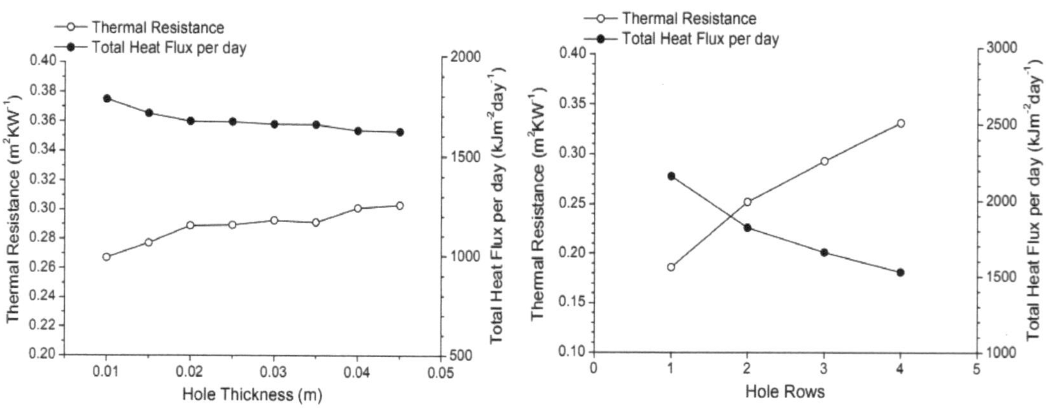

Figure 9. Rectangular cavities and comparison of the effects of hole thickness and hole rows on the thermal resistance of the block while keeping the block width constant [70].

H. Ghailane [72] investigated the reduction of heat flow through hollow clay bricks with cylindrical voids (as per the sample shown in Figure 10) of varying numbers, diameters, and cavity filling materials (air, wool, and expanded foam), relative to solid blocks. The findings presented a reduction in energy consumption through the usage of hollow blocks compared to solid blocks. It was found that effective reduction in the thermal conductivity of hollow blocks was obtained when the volume of the voids and the reduction in the cross-sectional area for heat transfer was more than 10%, as shown in Figure 10. The figure provides results for cavities filled with air; however, similar results were obtained for wool and expanded foam. Nevertheless, this study was limited to only cylindrical-shaped cavities, and a comparative study of cylindrical cavities with widely used rectangular cavities was not conducted. Also, changes in the structural configuration might have significantly affected the mechanical properties of the block, which was not discussed. Additionally, only the conduction mode of heat transfer was considered across the thickness of the blocks, while the convection and radiation modes of heat transfer at either side of the block and within the air cavities were neglected.

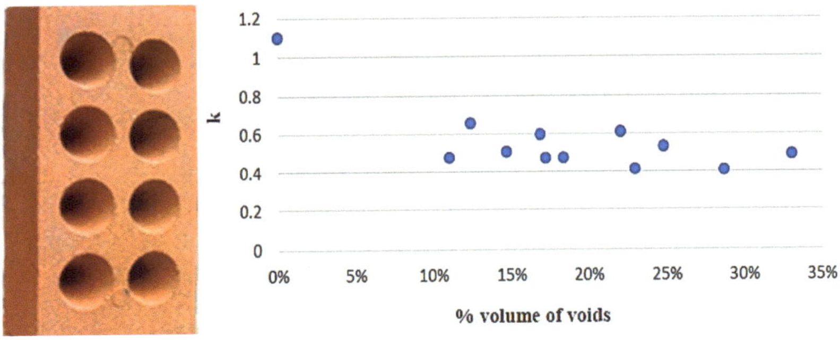

Figure 10. Cylindrical cavities and effects of % volume of voids on thermal conductivity [72].

In 2020, a numerical two-dimensional steady state analysis was conducted by Marcelo Fogiatto [73] to study the effect of baffles within the rectangular cavities of blocks as shown in Figure 11 was tested. It was observed that the rate of heat transfer in rectangular cavities were reduced by 22.2% when baffles were introduced within these cavities. In 2020, Ahmed S. Al-Tamimi [43] developed a 3D FEM model to propose an optimal geometry of hollow blocks with rectangular cavities (as shown in Figure 12) with higher thermal efficiency compared to the hollow blocks available in the market. From Ahmed's studies, it was concluded that thermal flow of heat can be reduced by increasing the Hollow Ratio (the total area of holes in a block/the cross-sectional area of the block), Aspect Ratio (the width perpendicular to heat flow/height of the cavity) and reducing the cavity width (parallel to heat flux) [43]. Further, results showed that, the baffles in rectangular cavities had a greater influence compared to rectangular cavities [73].

However, the tested blocks had one particular shapes of cavities; either had rectangular or circular cavities, while no block was tested that had both rectangular and circular cavities in one model. Also, for the staggered arrangement, it was observed that the complexity of the staggered arrangement of cavities on the extreme sides of the blocks could further investigate the thermal performance of the blocks. Moreover, the mechanical strength of the blocks was not investigated, and no experimental analysis was conducted.

In [43], an excellent analysis was conducted to show that for the same Hollow Ratios, rectangular cavities were better than the cylindrical ones, including appropriate experimental, thermal, and mechanical analyses. However, again, no block having both types of cavities was tested. Secondly, the optimal design was proposed by considering thermal analysis, without simultaneously considering both thermal and mechanical analyses. Thirdly, the complexity of staggered arrangement on extreme sides of the blocks was less.

Apparently, a three-dimensional CFD-based numerical study, considering the conduction method in the solid part of the block and the convection method in the cavities and wall surfaces (inner and outer), was carried out by Erdem Cuce [74] in 2020, in which rectangular cavities were arranged in a staggered pattern at the sides of the blocks. The results showed that the depth of cavities within the hollow enclosure, with continuous hollow cavities from top to bottom, provided a maximum improvement of 53% with U value of 0.43 W/m^2K as compared to the commercially used hollow block with a U-value of 0.916 W/m^2K. However, increasing the depth of cavities notably reduced the weight of the blocks. While thermal analysis of the design of the cavities and the joints of the blocks were performed to propose some optimized blocks, later, the mechanical strength of the blocks was investigated only for these optimized blocks. Therefore, the simultaneous consideration of thermal and mechanical parameters was missing.

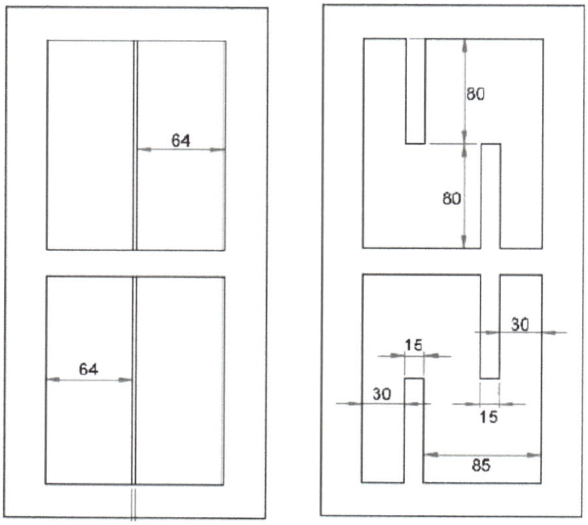

Figure 11. Comparison of heat transfer through models of hollow blocks with rectangular cavities without baffles and with baffles [73].

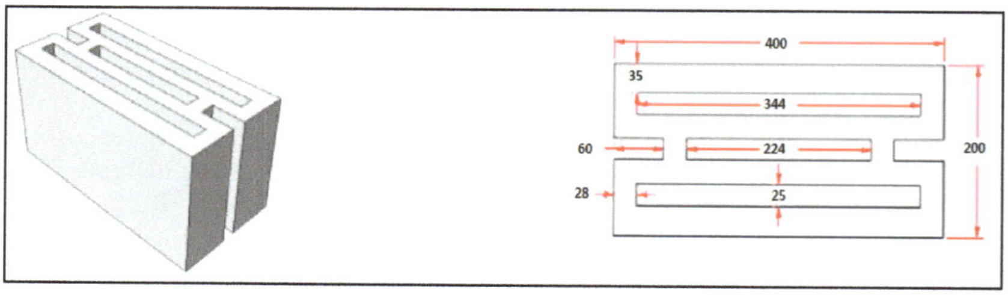

Figure 12. Proposed model of hollow block with optimal geometry [43].

Out of the three study possibilities discussed in the above sections, it can be observed that to increase the thermal performance of the building blocks, the first two possibilities of proposing new insulation materials/techniques and new aggregate materials may not be cost-effective, as they may increase the material cost, and hence, the overall cost of the block may increase. Conversely, the third possibility involves suggestions for modifications of the cavities of the block, which would contribute to enhancing the internal design without increasing the price of the block. It is very clear that the performance of the block depends on various parameters of the internal design, including the Aspect Ratio of the cavity (the ratio of width perpendicular to heat flux to the height), Hollow Ratio (the ratio of the total area of the block to the cross-sectional area of the block), thermal bridges, number of cavities, number of rows of cavities, dimension of cavities parallel and perpendicular to heat flux, and linear or staggered arrangement of the cavities. These parameters are summarized in Table 2. These parameters not only contribute to changes in thermal performance but the mechanical performance of the block as well. Therefore, there is a requirement to introduce a single non-dimensional criterion that can evaluate the effectiveness of building blocks in terms of thermal and structural parameters for better comparison and selection purposes.

Table 2. Summary of the effect of variations in geometrical parameters of cavities on the thermal performance of blocks from the literature [43,67–74].

Effective Geometrical Parameters	Parameter	Thermal Resistance (R)
Number of cavities	Decreases	Increases
Number of rows of cavities	Increases	Increases
Width of cavity (parallel to heat flux)	Decreases	Increases
Thickness of cavity (perpendicular to heat flux)	Increases	Increases
Aspect Ratio (width perpendicular to flow of heat/cavity height)	Increases	Increases
Hollow Ratio (total area of holes in a block/the cross-sectional area of the block)	Increases	Increases
Thermal bridges	Decreases	Increases
Staggered arrangement	Increases	Increases

4. Influential Technical Parameters for the Performance of Building Blocks

The main influential technical parameters responsible for the performance of building blocks are discussed in Sections 4.1 and 4.2.

4.1. Thermal Transmittance (U-Value)

The thermal behavior of building walls and their energy consumption can be determined by evaluating the thermal and mechanical (thermo–mechanical) properties of such systems. Apparently, thermal transmittance (U-value) and mechanical strength of the wall system become the most important parameters in the overall thermo-mechanical performance of a building.

4.1.1. Evaluation Techniques for Thermal Transmittance (U-Value)

There are two main methods to evaluate the thermal transmittance of building wall systems:

(i) Theoretical method
(ii) Experimental method

Theoretical Method

As per the BRE publication 'Conventions for U-value calculations', the different U-value theoretical calculation methods are:

a. Calculation by simplified method using a BRE U-value calculator.
b. Calculation by 'numerical methods' using different computer simulation software.
c. Combination of calculation and numerical methods.
    - However, the theoretical evaluation of thermal transmittance (U-value) becomes difficult in existing building walls. Two main difficulties that are encountered are:
    - Traditional buildings are built using non-homogenous materials whose thermal conductivity values are not available.
    - Destructive methods are required to be adopted in order to establish exact dimensions of layers of the wall.

Experimental Method

Various experimental methods for measuring the thermal transmittance (U-value) or thermal resistance (R-value) of a building wall element have been developed. These methods further include measurements and analyses of wall systems within controlled environment (in-lab) or in real environments (on-site). In 2013, Abdulaziz Almujahid [75] constructed a test room measuring 600 × 730 cm^2 with a height of 220 cm to evaluate the thermal performance of building wall systems under real conditions (on-site method). The surface area of each wall system was 120 × 120 cm^2, thereby accumulating most varieties of blocks manufactured in the country. Each module of the block was separated by 20 cm-thick siporex insulation material on all sides, allowing heat flow only in the normal direction to the wall surface. The roof consisted of corrugated metal sheet insulated with 5 cm fiber glass insulation. The room inside was air conditioned using two split ac units, each with a capacity of 1.5 tons. The purpose of air conditioning the room was to simulate real conditions. In another experimental study measuring the thermo–mechanical properties of building blocks used in hot desert arid climates, Miloud Hatia conducted research in 2021 [76], for which different models of wall systems (on-site method) were constructed, as shown in Figure 13. Transient heat transfer conditions, along with the effects of radiation heat transfer, were studied. It was noticed that during the first hours, the temperature inside was higher than outside because during the daytime, the slab made of concrete had absorbed and stored heat. In the early hours, with minimal solar radiation, the wall lost heat (the heat was transferred from the hot medium to the cold medium). But, from the middle of the day untill 17:00, the external temperature remained higher than the interior (due to solar radiation). After 17:00, the opposite became true, and the system behaved as a storehouse of building elements (slab and walls).

Figure 13. Wall construction and testing with varieties of block materials [76].

Waleed A. Al-Awsh [42] conducted experimental and numerical investigations on innovative masonry walls for industrial and residential buildings in Saudi Arabia in 2020. Compared to the previous two studies, the experimental work in this study was conducted in the controlled environment of a laboratory room (in-lab method). The samples of blocks (Figure 14) were tested using a specific device called a guarded hot plate instrument (Figure 15). The temperature difference across the blocks' thickness was maintained from 30° C to 40 °C, while the measurements were noted on an hourly basis until they reach steady-state values to eventually evaluate the thermal conductivity of the blocks. In one such study, a mathematical model was developed to evaluate the R-value, heat flux, and thermal diffusivity of individual masonry blocks based on experimental results in a controlled environment [77], while in other studies, the experimental results were compared with numerical results for a multilayered wall [78], or the experimental results were compared with modeled data [79].

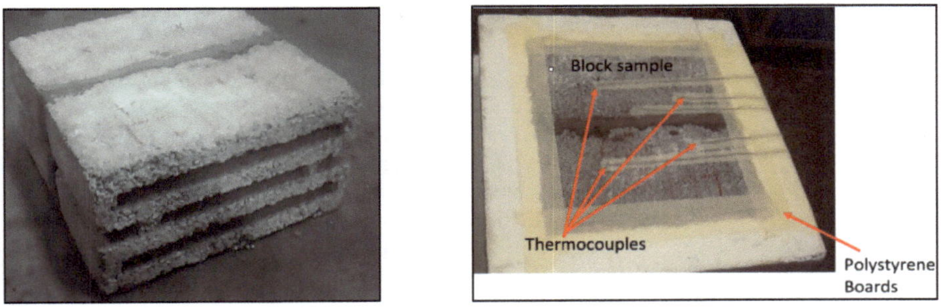

Figure 14. Block and sample of hot guarded plate (insulated by polysterene) [42].

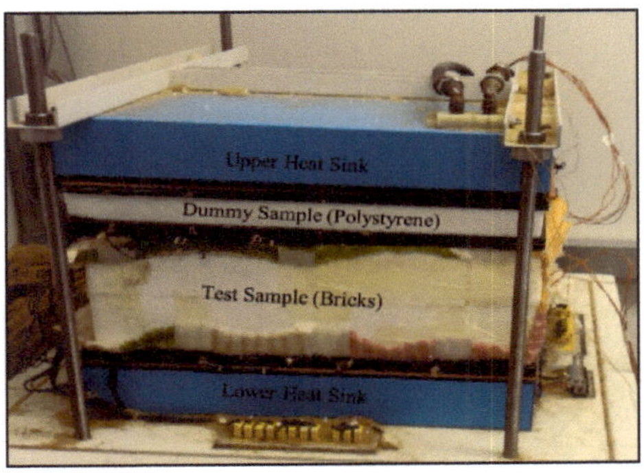

Figure 15. Guarded hot plate instrument [42].

To summarize, a good attempt was made by Abdulaziz Almujahid [75] to consider real transient conditions by building a test room and conducting experiments in real environmental conditions (on-site method). Miloud Hatia [76] effectively discussed the effects of radiation on the thermal behaviour of walls by conducting real experiments in open environmental conditons (on-site method). However, in both studies, readings were noted after 24 h based on the assumption that the system reaches a steady-state condition within that time period. On the other hand, Waleed A. Al-Awsh [42] validated the results by comparing experimental measurements in a controlled environment (in-lab method) with the computational fluid dynamics (CFD) model implemented by the Ansys Fluent program. However, since on-site (uncontrolled environment) analysis was not conducted, it is not clear as to how the results would be affected by changes in ambient conditions.

Since, in most nations, the topic of measuring the thermal performance of building walls is of high importance in energy conservation campaigns, a common trusted procedure is required to be followed as a reference. Therefore, standards for experimental methods have been developed and are widely used [75,76,80,81] by researchers in this field. These standards include the International standard ISO 9869 [82] and the American standard ASTM 1046 and 1155 [83]. The standard procedures, descriptions, measuring methods, measurement equipment, conditions, etc., in the on-site method, which is also named the 'In-Situ method', are well-discussed in both ISO and ASTM standards. These two standards were compared by I. A. Atsonios [84] in 2017, and it was found that other than small differences in details, the principles of the two standards remain the same. The In-Situ method evaluates the U-value of a wall system under steady-state conditions using temperature sensors on both sides of the wall element and a heat flow meter on one side of the wall, preferably the inner side of the wall, which has higher stability in temperature. As per the International Standard ISO 9869-1 [82], it is always difficult to achieve steady-state conditions in on-site problems. Therefore, a few methods have been developed to overcome this difficulty:

- Achieving steady-state conditions using a hot and cold box, which is commonly used in laboratory experimentation methods (ISO 8990) [85].
- Assuming that the mean values of temperatures and heat flow rates over a long period of time provide an estimation of steady-state conditions (Average Method)
- Implementing dynamic theory to consider variations in temperatures and heat flow rates in the analysis of the recorded data.

When observing the difference between in-lab and on-site methods (as seen in Table 3) for the thermal performance of walls, the In-Situ technique is highly recommended. The most popular method of the In-Situ technique is the Average Method, as suggested in ISO 9869 standard [82], using one heat flux sensor and two thermocouples [81].

Table 3. Summary of factors under consideration for in-lab and in-situ methods.

Sr. No.	Factors under Consideration	In-Lab Method	In-Situ Method
1	Evaluation of thermal performance of walls considering the actual climatic conditions	No	Yes
2	Consideration of large temperature drifts	No	Yes
3	Consideration of wind velocity	No	Yes
4	Consideration of moisture content	No	Yes
5	Consideration of material aging	No	Yes
6	Consideration of 1D heat transfer conditions	Yes	Yes
7	Consideration of transient heat transfer conditions	Yes	Yes
8	Realistic actual heat flux measurements	No	Yes
9	Realistic actual temperature measurements	No	Yes
10	Short measurement times	Yes	No
11	Accuracy of results if conditions are followed appropriately	No	Yes
13	Preferred method on international level	No	Yes
14	Accuracy with low temperature gradients	Yes	No
15	Stability of boundary conditions	Yes	No

However, there are two main problems that are encountered when using the In-Situ average method. Firstly, the issue of the long duration of measurements due to unstable boundary conditions, and secondly, the problem with the precision of the U-value. As per ISO 9869 [82], considering that all conditions are satisfied to report an acceptable U-value, the main criteria to fulfill and stop the measurement include the following:

- The measurement period should exceed 72 h with a certain range of fixed intervals.
- The R-value obtained at the end of the test should not deviate by more than 5% from the value obtained 24 h before.
- The R-value obtained by analyzing the data from the first time-period during INT ($2 \times DT/3$) should not deviate by more than 5% from the values obtained from the data the last time period of the same duration. DT is the duration of test in days and INT is the integer part.

In 2018, Arash Rasooli [81] demonstrated a minor modification in the standard In-Situ average method to improve the results with regard to the duration and precision. As shown in Figure 16, an additional heat flux sensor, opposite to the first one, was incorporated into the wall system. The two R-values differ for each heat flux sensor and eventually converge, thereby obtaining an average R-value with higher precision in a shorter time period.

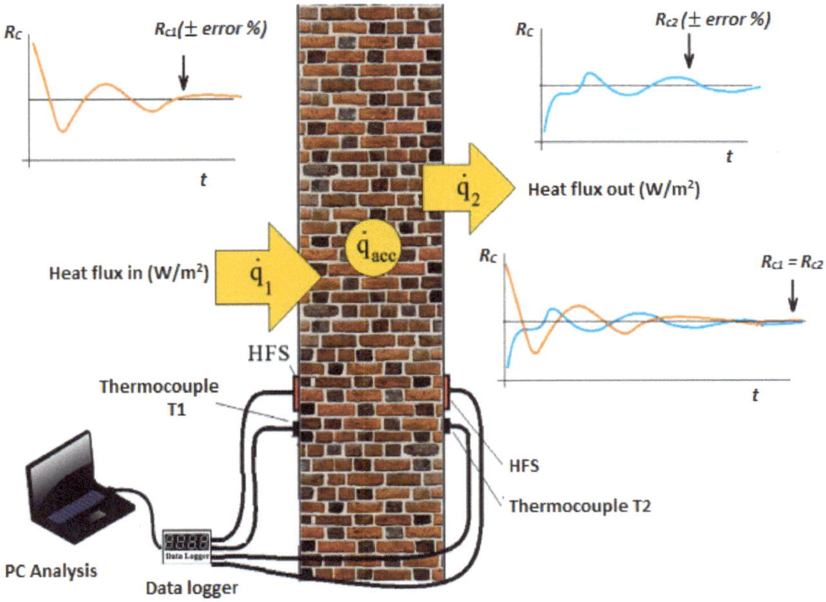

Figure 16. General configuration of ISO 9869 standard measurement with one extra heat flux sensor [81].

4.2. Compressive Strength of Insulation Building Block

Different tests are required to measure the strength of concrete before using it in a construction project. These strength measurement tests include compressive strength [86,87], tensile strength [88,89], and flexural strength [90]. However, in the case of building blocks, the compressive strength of concrete blocks is a crucial parameter to determine suitability in building construction work, including high-rise buildings. Concrete masonry blocks are generally made of cement, aggregate, and water. The compressive strength test is of high importance as walls are constructed by placing blocks one above the other, and the compression on the blocks increases gradually from the top to bottom layers, with the lowest layer bearing the maximum load for the entire wall. In accordance with historical practice, the compressive strength of concrete is usually determined from standard procedures that include sampling, making, curing, and testing of specimens [91]. ASTM Standard C 140-08 [92] provides various testing procedures commonly used for evaluating the characteristics of concrete masonry units and related concrete units. Methods are provided for sampling, measurement of dimensions, compressive strength, density, absorption, moisture content, flexural load, and ballast weight. All these methods are not used for all types of concrete units; however, specific testing and reporting procedures are included in this standard [93–95]. Further, the testing of physical quantities of sample blocks includes the following procedures, as shown in Figure 17.

In addition to unit compressive strength testing, specimens are tested in a compression testing machine wherein the load is applied gradually at a rate of 140 kg/sq cm/min until the specimen fails. The load at failure, when divided by the area of the specimen, yields the compressive strength of the concrete. The strength of concrete increases with age; for example, the strength is 1% on day 1 and 99% on day 28 [96].

The compressive strength test of the blocks is conducted as per ASTM C140 [92], according to which 3.45 MPa is the minimum strength requirement of an individual block. In 2020, Waleed A. Al-Awsh [42] conducted compressive tests on individually controlled designs. Sample blocks were first dried out, weighed, and capped using a special high-strength cement mortar for uniform pressure distribution. Rigid plates were placed on the top and bottom surfaces of the blocks for uniform load distribution over the cross-sectional area. Later, the load was applied gradually at a rate of 0.3 kN per second until the specimen failed. The failure load was divided by the net normal area of the block to calculate the compressive strength of the block. Concrete masonry blocks of two types, solid and hollow, using four different mortars were constructed and tested by Tatheer Zahra in 2021 [94] for compressive tests using the prism method. Results showed that the compressive strengths of hollow blocks varied from 12.5 to 15.6 MPa, while for solid blocks, they varied from 15.4 to 21.0 MPa for different mortars used. The material behavior of the blocks and mortar were simulated using the 'Concrete Damage Plasticity Model' (CDP), which includes the failure mechanisms of tensile and compressive crushing. Apparently, both thermal transmittance and compressive strength are equally important for the performance of building blocks. Therefore, there is a requirement to establish a single parameter to measure the thermal and mechanical behavior of building blocks.

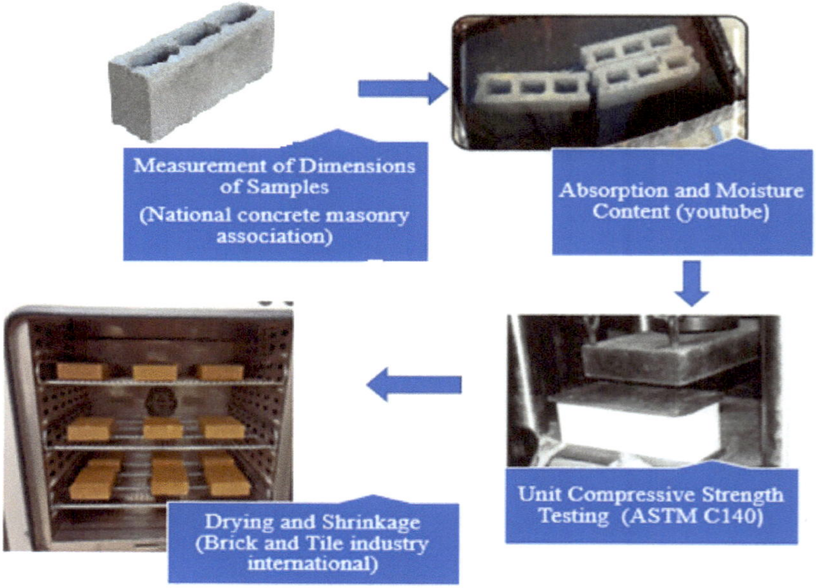

Figure 17. Testing physical quantities of sample blocks.

5. Houses and Building Blocks in the Kingdom of Bahrain

5.1. History of Houses in Bahrain

The Kingdom of Bahrain is an island country and is one of the smallest in the Middle East. Traces depict that the island was inhabited in 3000 B.C. Initially, people lived in straw, stone, and mud houses, as shown in Figure 18.

In the past 200 years, the country started to gain its architectural identity when people started to move in to houses made up of bearing walls. These house structures consisted mainly of 60 cm-wide walls, which were grounded 50 cm deep to form the foundation. The thickness in the foundation was 1 m. Irregular desert stones of different sizes were used to build the walls, joined together using a mortar mixture of sand, gypsum/limestone, and water in particular ratios. The gaps between the big stones were filled by smaller stones and gravel, followed by a layer of mortar to make the walls smooth. Approximately five layers of stone above ground level were arranged, upon which a layer of timber wooden round beams was applied around the perimeter of the house as shown in Figure 19. These beams were strengthened by diagonal connections using 1 cm diameter rope. The space between the timber beams was again filled with stones, followed by a mortar mixture.

Figure 18. Early Bahraini houses made of straw, stone, and mud.

Figure 19. Bahraini house around 200 years ago.

One such traditional house, built in 1834 and located in Muharraq, the traditional city of Bahrain, is shown in Figure 20. This house was selected for renovation and rehabilitation purposes, and the images below depict its condition prior to the renovation work [97].

When modern industries came to Bahrain in the 1950s, the materials used for constructing buildings were more industrialized than traditional methods. The most common material used for construction was concrete. Bahraini houses are now built with walls made of concrete blocks, followed by a layer of cement plaster and paint, as shown in Figure 21.

Figure 20. Bahraini house in the year 1834: front view, roof, external wall, and internal wall [97].

Figure 21. Modern-day Bahraini houses.

5.2. History of Building Blocks in Bahrain

The leading companies that began manufacturing concrete blocks during the 1950s were Haji Hasan group, 'Al Manaratain', and 'Bahrain Blocks [98,99]. To this day, they remain the largest manufacturers in Bahrain, with a variety of block types. Al Manaratain's [98] history extends back to 1959, when it started as a manufacturer of local manual blocks. Over the last fifty years, the company has evolved to become an industry leaders in the production of concrete blocks and other products. The rising demand for blocks encouraged the group to acquire additional machinery throughout its existence, resulting in a large industrial expansion in 1985 in the form of another company named Bahrain Blocks.

In 1995, the company introduced a fully automated block with the capability to manufacture 45,000 concrete blocks per day to increase its manufacturing capability and quality standards. Bahrain Blocks presently manufactures various varieties of concrete blocks, including the insulation blocks that are included in the present study. Other companies such as Middle East Block Factory [100], Aljalahma Blocks [101], RedX company [102], Ahmed Isa Block factory [103], etc., that came into production later in the late 1990s and 2000s have a lower scale in concrete block production, mainly focused on hollow blocks.

Varieties of blocks are manufactured in the Kingdom. Some simple blocks, as shown in Figure 22, are used to build compound walls of a house or for the partition walls between two rooms of a house or for the pavement. However, there are some other blocks that are specifically manufactured to build the outer walls of a house or a building, as shown in Figure 23. These blocks are called 'Insulation block'.

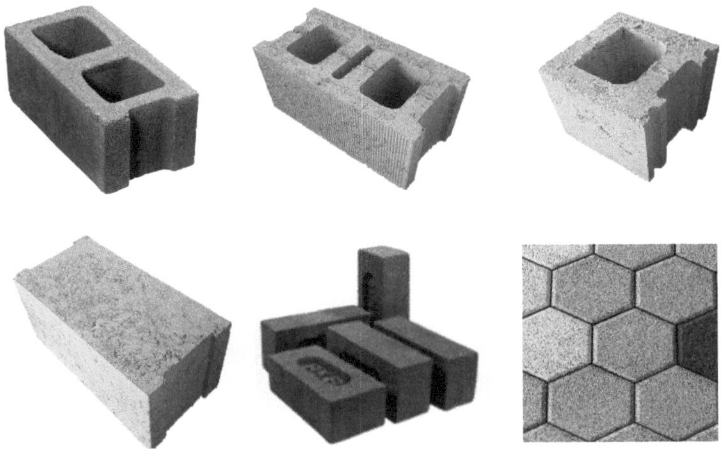

Figure 22. Simple building blocks of Bahrain.

Figure 23. Insulation building blocks of Bahrain.

5.3. Initial Data Collection and Standard Procedures for Building Blocks in Bahrain

In order to propose an improvement in the performance of the building blocks, it is important to review the existing scenario in the Kingdom. For this, the following activities were conducted:

1. Collect information from the Ministry of Electricity and Water Authorities (EWA) of the Kingdom of Bahrain.
2. Collect information from the leading insulation block manufacturing companies in Bahrain, particularly related to the internal and external designs of blocks, cavities, materials, strength, and application parameters.
3. Collect information from the block testing laboratories in Bahrain related to the thermal and mechanical properties of the building blocks.

The collected data were reviewed for further analysis.

The government body, Electricity and Water Authority (EWA) of Bahrain, has set a certain value of thermal transmittance (U-value) for the insulation blocks. As per ministerial order 63/2012 [32], the maximum allowable U-value of the block is 0.57 W/m^2 °C. After testing the samples, the test report issued by the Ministry of Works includes all the thermal and mechanical characteristics of the blocks. However, it mainly focuses on the thermal resistance (R-value) and thermal transmittance (U-value) of the block. Eventually, it is clear that the most important parameter of an insulation block is the U-value, followed by its mechanical strength.

All the companies in the Kingdom more or less manufacture blocks with identical designs, with almost similar thermal and mechanical characteristics required to fulfill the conditions laid down by the government of the Kingdom. Companies cannot merely manufacture and sell the blocks directly in the market. Instead, a standard procedure, as shown in Figure 24, is required to be followed by each company for each variety of block they manufacture.

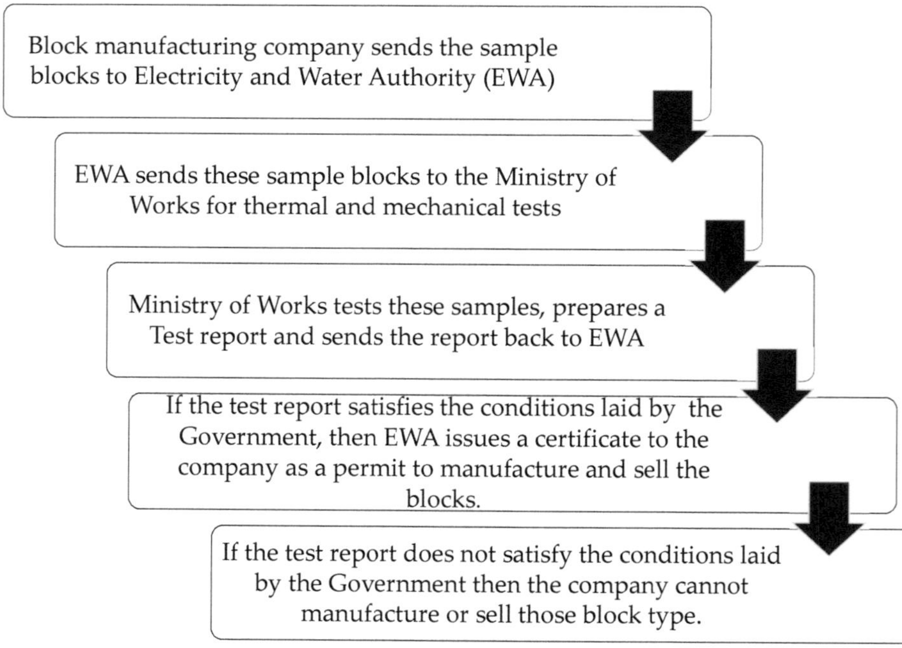

Figure 24. Flowchart representing the steps followed by the company for legal certification of insulation blocks.

5.4. Comparison of Widely Used Insulation Building Blocks in Bahrain

In accordance with discussions with leading manufacturing companies in Bahrain, namely Al Manaratain and Bahrain Blocks, it is clear that two types of insulation blocks are widely used for the construction of buildings in Bahrain, as shown in Figure 25, due to their desirable thermal and mechanical parameters, which are compared in Table 4.

(**a**) Sandwich block (**b**) IMSI block

Figure 25. Sandwich block and IMSI block.

Table 4. Comparison table of Sandwich and IMSI insulation blocks.

Block Type	8″ Sandwich Insulation	8″ IMSI Insulation
Overall dimensions	400 mm × 200 mm × 200 mm	400 mm × 200 mm × 200 mm
Thermal Conductivity (k)	0.1251 W/m K	0.2330 W/m K
Thermal Transmittance (U)	0.6255 W/m^2 K (Non-standalone)	1.165 W/m^2 K (Non-standalone)
Compressive Strength	7.9 N/mm^2	12.2 N/mm^2
Weight	19 g	18 kg

It was also clear that after a few years, the Sandwich block had issue with the formation of cracks on the walls, while no such issues were observed in the IMSI block. Eventually, in today's context, the IMSI block can be considered one of the preferred options for detailed investigation in the near future.

6. Conclusions

The Kingdom of Bahrain is grappling with a significant challenge of escalating energy consumption, particularly driven by air conditioning systems in both residential and commercial buildings. Recognizing the potential for substantial reductions in building energy demand through energy efficiency initiatives, the government has formulated ministerial orders and aligned with Bahrain Economic Vision 2030 [29] to foster sustainable construction practices. These efforts are in line with SDG 11 of the United Nations' 2030 agenda [5]. This paper delves into the influential factors affecting building block performance, including aggregate materials, insulation, and cavity geometries. It also outlines the two major technical parameters to predict the performance of building blocks: thermal transmittance and compressive strength. Thermal transmittance can be evaluated using theoretical and experimental methods, while the later has two types—in-lab and In-Situ methods. The need for both these testing methods, in particular, under real environmental conditions as per ISO [82,85] and ASTM [37,83,92], is well-emphasized in this paper. The study also presents the standard procedure to measure the compressive strength of the building blocks under ASTM Standards.

Despite limited research on improving the thermal and mechanical performance of Bahrain's building blocks, there is a pressing need for investigation. The paper provides insights into the historical evolution of building materials in Bahrain and highlights the shift towards more efficient blocks with insulation. It was observed that the type of materials used for building walls has evolved to adapt to changes in environmental conditions. Due to modernization, mud, straw, and clay in walls of the houses were replaced with hollow blocks; however, it was found that these houses were not comfortable. So, to find thermal comfort in present times, people are looking for more efficient building blocks with insulation inserted in their cavities. Eventually, the varieties of blocks currently manufactured in Bahrain were discussed in this paper, among which the IMSI block is the most popular one. This paper also depicts the standard procedure followed by block manufacturing companies in the country to obtain approval from the government to sell their products in the market. However, the test reports presented by manufacturers to the government include testing done under controlled environments (in-lab) and for a limited number of blocks, which varies from one to three blocks. Nevertheless, the results of a wall may vary in real environmental conditions, particularly when there exists a wall consisting of a multi-block system with plaster. Hence, the paper concludes by proposing various paths for future research, between individual and multi-block systems using both in-lab and In-Situ methods. A few of these future scopes are listed below:

i. Propose improved aggregate materials.
ii. Propose improved insulation materials.
iii. Propose improved cavity designs.
iv. Evaluate the difference in performance for an individual block system compared to a multi-block system in controlled environmental (in-lab) conditions.
v. Evaluate the difference in performance for an individual block system compared to a multi-block system in real environmental (In-Situ) conditions.
vi. Evaluate the difference in performance of both systems for the two methods: in-lab and In-situ methods.
vii. Propose a single non-dimensional factor to simultaneously measure the thermal and mechanical performance of both individual and multi-block systems.

It is understood that on a global level, a lot of research is consistently ongoing in the field of energy conservation for buildings and their materials. However, a considerable amount of work needs to be done in Bahrain. Overall, this review aims to lay the groundwork for technical analyses and testing procedures to optimize Bahrain's building block performance in energy conservation.

Author Contributions: Conceptualization, P.A.M.; methodology, P.A.M.; data curation, P.A.M.; validation, A.M.M.; writing-original draft preparation, P.A.M.; writing-review and editing, A.M.M.; visualization, Y.A.A.; supervision, Y.A.A. All authors have read and agreed to the published version of the manuscript.

Funding: This research received no external funding.

Data Availability Statement: Not applicable.

Conflicts of Interest: The authors declare no conflict of interest.

References

1. Cao, X.; Dai, X.; Liu, J. Building energy-consumption status worldwide and the state-of-the-art technologies for zero-energy buildings during the past decade. *Energy Build.* **2016**, *128*, 198–213. [CrossRef]
2. Sansaniwal, S.K.; Mathur, J.; Mathur, S. Review of practices for human thermal comfort in buildings: Present and future perspectives. *Int. J. Ambient. Energy* **2022**, *43*, 2097–2123. [CrossRef]
3. Andargie, M.S.; Touchie, M.; O'Brien, W. A review of factors affecting occupant comfort in multi-unit residential buildings. *J. Affect. Disord.* **2019**, *160*, 106182. [CrossRef]
4. Cascone, S.; Leuzzo, A. Thermal Comfort in the Built Environment: A Digital Workflow for the Comparison of Different Green Infrastructure Strategies. *Atmosphere* **2023**, *14*, 685. [CrossRef]
5. Sustainable Energy for All, Building Efficiency Accelerator. 2015. Available online: https://www.seforall.org/partners/building-efficiency-accelerator (accessed on 15 March 2024).
6. Rashid, F.L.; Al-Obaidi, M.A.; Dulaimi, A.; Mahmood, D.M.N.; Sopian, K. A Review of Recent Improvements, Developments, and Effects of Using Phase-Change Materials in Buildings to Store Thermal Energy. *Designs* **2023**, *7*, 90. [CrossRef]

7. Wani, A.A.; Shahadat, M.; Ali, S.W.; Ahammad, S.Z.; Uddin, M.K. Recent advances and future perspectives of polymer-based magnetic nanomaterials for detection and removal of radionuclides: A review. *J. Mol. Liq.* **2022**, *365*, 119976. [CrossRef]
8. Christopher, S.; Vikram, M.; Bakli, C.; Thakur, A.K.; Ma, Y.; Ma, Z.; Xu, H.; Cuce, P.M.; Cuce, E.; Singh, P. Renewable energy potential towards attainment of net-zero energy buildings status—A critical review. *J. Clean. Prod.* **2023**, *405*, 136942. [CrossRef]
9. Maghrabie, H.M.; Abdelkareem, M.A.; Al-Alami, A.H.; Ramadan, M.; Mushtaha, E.; Wilberforce, T.; Olabi, A.G. State-of-the-Art Technologies for Building-Integrated Photovoltaic Systems. *Buildings* **2021**, *11*, 383. [CrossRef]
10. Bilal, M.; Afzal, M.; Irfan, M.; Musarat, M.A. Energy Efficient Buildings in the Industry 4.0 Era: A Review. *Preprints* **2023**, 2023100301. [CrossRef]
11. Olu-Ajayi, R.; Alaka, H.; Sulaimon, I.; Sunmola, F.; Ajayi, S. Building energy consumption prediction for residential buildings using deep learning and other machine learning techniques. *J. Build. Eng.* **2022**, *45*, 103406. [CrossRef]
12. Shafique, M.; Luo, X.; Zuo, J. Photovoltaic-green roofs: A review of benefits, limitations, and trends. *Sol. Energy* **2020**, *202*, 485–497. [CrossRef]
13. Metallidou, C.K.; Psannis, K.E.; Egyptiadou, E.A. Energy Efficiency in Smart Buildings: IoT Approaches. *IEEE Access* **2020**, *8*, 63679–63699. [CrossRef]
14. Said, Z.; Iqbal, M.; Mehmood, A.; Le, T.T.; Ali, H.M.; Cao, D.N.; Nguyen, P.Q.P.; Pham, N.D.K. Nanofluids-based solar collectors as sustainable energy technology towards net-zero goal: Recent advances, environmental impact, challenges, and perspectives. *Chem. Eng. Process.-Process. Intensif.* **2023**, *191*, 109477. [CrossRef]
15. Na Talang, R.P.; Sirivithayapakorn, S. Comparing environmental burdens, economic costs and thermal resistance of different materials for exterior building walls. *J. Clean. Prod.* **2018**, *197*, 1508–1520. [CrossRef]
16. Ruiz, G.; Zhang, X.; Edris, W.F.; Cañas, I.; Garijo, L. A comprehensive study of mechanical properties of compressed earth blocks. *Constr. Build. Mater.* **2018**, *176*, 566–572. [CrossRef]
17. Vrachopoulos, M.G.; Koukou, M.K.; Kotsiovelos, G.; Kravvaritis, E.; Ioannidou, P. Experimental evaluation of the performance of reflective insulation for improvement of indoor thermal environmental conditions. *WSEAS Trans. Environ. Dev.* **2006**, *2*, 173–177.
18. Shi, T.; Zhang, X.; Hao, H.; Chen, C. Experimental and numerical investigation on the compressive properties of interlocking blocks. *Eng. Struct.* **2021**, *228*, 111561. [CrossRef]
19. World Bank. *Kingdom of Bahrain Energy Efficiency Program (KEEP)*; World Bank: Washington, DC, USA, 2016.
20. World Data Atlas, Bahrain Energy. 2019. Available online: https://knoema.com/atlas/Bahrain/Primary-energy-consumption#:~:text=In%202019%2C%20primary%20energy%20consumption,average%20annual%20rate%20of%203.14%25 (accessed on 15 March 2024).
21. Alnaser, N. Building integrated renewable energy to achieve zero emission in Bahrain. *Energy Build.* **2015**, *93*, 32–39. [CrossRef]
22. MEW (Ministry of Electricity and Water). *Annual Report 2021 Bahrain*; EWAStatistics2021.pdf; MEW (Ministry of Electricity and Water): Manama, Bahrain, 2021.
23. Krarti, M.; Dubey, K. Benefits of energy efficiency programs for residential buildings in Bahrain. *J. Build. Eng.* **2018**, *18*, 40–50. [CrossRef]
24. Bahrain Weather. (n.d.). Climate. Available online: https://www.bing.com/ck/a?!&&p=5ef09116abced6edJmltdHM9MTcxMDk3OTIwMCZpZ3VpZD0xYTJhZjA4NS0wYTM2LTY5OWMtMjllMC1lMTVlMGIzZjY4NmUmaW5zaWQ9NTE5NQ&ptn=3&ver=2&hsh=3&fclid=1a2af085-0a36-699c-29e0-e15e0b3f686e&psq=Climate.+Available+online%3a+https%3a%2f%2fweather-and-climate.com%2faverage-monthly-min-max-Temperature%2cBahrain%2cBahrain&u=a1aHR0cHM6Ly93d3cuYmFocmFpbndlYXRoZXIuZ292LmJoL2NsaW1hdGU&ntb=1 (accessed on 15 March 2024).
25. Ma-Investment. *Thermal Insulation in Buildings*; Ma-Investment: Houston, TX, USA, 2016.
26. Bayram, I.S.; Al-Qahtani, M.; Saffouri, F.; Koc, M. Estimating the Cost of Summer Cooling in Bahrain. In Proceedings of the 2017 9th IEEE-GCC Conference and Exhibition (GCCCE), Manama, Bahrain, 8–11 May 2017; pp. 1–9.
27. Radhi, H.; Sharples, S.; Taleb, H.; Fahmy, M. Will cool roofs improve the thermal performance of our built environment? A study assessing roof systems in Bahrain. *Energy Build.* **2017**, *135*, 324–337. [CrossRef]
28. Modi, P.; Bushehri, R.; Georgantopoulou, C.; Mavromatidis, L. Design and development of a mini scale hot box for thermal efficiency evaluation of an insulation building block prototype used in Bahrain. *Adv. Build. Energy Res.* **2017**, *11*, 130–151. [CrossRef]
29. Bahrain Economic Vision 2030. Available online: https://www.bing.com/ck/a?!&&p=5b16bfc9398feee6JmltdHM9MTcxMDk3OTIwMCZpZ3VpZD0xYTJhZjA4NS0wYTM2LTY5OWMtMjllMC1lMTVlMGIzZjY4NmUmaW5zaWQ9NTIxNQ&ptn=3&ver=2&hsh=3&fclid=1a2af085-0a36-699c-29e0-e15e0b3f686e&psq=Bahrain+Economic+Vision+2030&u=a1aHR0cDovL2JhaHJhaW4uYmgvYmFocmFpbjIwMzA&ntb=1 (accessed on 15 March 2024).
30. Ministerial Order 70/2015, Energy Efficiency Labelling of AC Units. Electricity Conservation Tips. Available online: https://www.ewa.bh/en/Conservation/Electricity/electricity-conservation-tips (accessed on 15 March 2024).
31. Ministerial Order 08/1999. Available online: https://www.ewa.bh/en/Conservation/Electricity/Documents/GuidelinesThermalInsulationImp.pdf (accessed on 15 March 2024).

32. Ministerial Order 63/2012. Available online: https://www.bing.com/ck/a?!&&p=caf68c760227bd38JmltdHM9MTcxMDk3OTIwMCZpZ3VpZD0xYTJhZjA4NS0wYTM2LTY5OWMtMjllMC1lMTVlMGIzZjY4NmUmaW5zaWQ9NTE5Ng&ptn=3&ver=2&hsh=3&fclid=1a2af085-0a36-699c-29e0-e15e0b3f686e&psq=%3a+https%3a%2f%2fwww.ewa.bh%2fen%2fConservation%2fElectricity%2fbuilding-regulations&u=a1aHR0cHM6Ly93d3cuZXdhLmJoL2VuL0NvbnNlcnZhdGlvbi9FbGVjdHJpY2l0eS9idWlsZGluZy1yZWd1bGF0aW9ucw&ntb=1 (accessed on 15 March 2024).
33. United Nations, Sustainable Goals 2030 Agenda. Available online: https://sdgs.un.org/topics/sustainable-cities-and-human-settlements (accessed on 15 March 2024).
34. Hongxia, Z.; Kang, P.; Baoquan, C.; Xuanyi, L. Numerical Simulation of Thermal Theory of Sintered Coal Gangue Self-Insulation Block and Wall. *MATEC Web Conf.* **2018**, *175*, 01021. [CrossRef]
35. Shehata, N.; Sayed, E.T.; Abdelkareem, M.A. Recent progress in environmentally friendly geopolymers: A review. *Sci. Total. Environ.* **2021**, *762*, 143166. [CrossRef]
36. Ashraf, N.; Nasir, M.; Al-Kutti, W.; Al-Maziad, F.A. Assessment of thermal and energy performance of masonry blocks pre-pared with date palm ash. *Mater. Renew. Sustain. Energy* **2020**, *9*, 17. [CrossRef]
37. *ASTM C90*; Standard Specification for Load Bearing Concrete Masonry Units. Available online: https://www.uop.edu.jo/PDF%20File/uop%20%5BArchitecture_Ebook%5D_Masonry_Design_and_Detailing_For_Architects_and_Contractors-20615_Part279.pdf (accessed on 15 March 2024).
38. Li, X.; Zhang, Q.; Mao, S. Investigation of the bond strength and microstructure of the interfacial transition zone between cement paste and aggregate modified by Bayer red mud. *J. Hazard. Mater.* **2021**, *403*, 123482. [CrossRef]
39. Guo, Z.; Tu, A.; Chen, C.; Lehman, D.E. Mechanical properties, durability, and life-cycle assessment of concrete building blocks incorporating recycled concrete aggregates. *J. Clean. Prod.* **2018**, *199*, 136–149. [CrossRef]
40. Guo, Z.; Kong, X.; Zhang, J.; Sun, Y.; Tu, A.; Jiang, T. Life-cycle assessment of concrete building blocks incorporating recycled concrete aggregates—A case study in China. In *Woodhead Publishing Series in Civil and Structural Engineering, Advances in Construction and Demolition Waste Recycling*; Woodhead Publishing: Sawston, UK, 2020; pp. 515–535. ISBN 9780128190555. [CrossRef]
41. Pavlu, T.; Fortova, K.; Divis, J.; Hajek, P. The Utilization of Recycled Masonry Aggregate and Recycled EPS for Concrete Blocks for Mortarless Masonry. *Materials* **2019**, *12*, 1923. [CrossRef] [PubMed]
42. Al-Awsh, W.A.; Qasem, N.A.; Al-Amoudi, O.S.B.; Al-Osta, M.A. Experimental and numerical investigation on innovative masonry walls for industrial and residential buildings. *Appl. Energy* **2020**, *276*, 115496. [CrossRef]
43. Al-Tamimi, A.S.; Al-Amoudi, O.S.B.; Al-Osta, M.A.; Ali, M.R.; Ahmad, A. Effect of insulation materials and cavity layout on heat transfer of concrete masonry hollow blocks. *Constr. Build. Mater.* **2020**, *254*, 119300. [CrossRef]
44. Schiavoni, S.; D'Alessandro, F.; Bianchi, F.; Asdrubali, F. Insulation materials for the building sector: A review and comparative analysis. *Renew. Sustain. Energy Rev.* **2016**, *62*, 988–1011. [CrossRef]
45. Awoyera, P.O.; Akinrinade, A.D.; Galdino, A.G.D.S.; Althoey, F.; Kirgiz, M.S.; Tayeh, B.A. Thermal insulation and mechanical characteristics of cement mortar reinforced with mineral wool and rice straw fibers. *J. Build. Eng.* **2022**, *53*, 104568. [CrossRef]
46. Bozsaky, D. Nature-Based Thermal Insulation Materials from Renewable Resources—A State-Of-The-Art Review. *Slovak J. Civ. Eng.* **2019**, *27*, 52–59. [CrossRef]
47. Gullbrekken, L.; Grynning, S.; Gaarder, J.E. Thermal Performance of Insulated Constructions—Experimental Studies. *Buildings* **2019**, *9*, 49. [CrossRef]
48. Mahdaoui, M.; Hamdaoui, S.; Msaad, A.A.; Kousksou, T.; El Rhafiki, T.; Jamil, A.; Ahachad, M. Building bricks with phase change material (PCM): Thermal performances. *Constr. Build. Mater.* **2020**, *269*, 121315. [CrossRef]
49. Hasan, A.; Al-Sallal, K.A.; Alnoman, H.; Rashid, Y.; Abdelbaqi, S. Effect of Phase Change Materials (PCMs) Integrated into a Concrete Block on Heat Gain Prevention in a Hot Climate. *Sustainability* **2016**, *8*, 1009. [CrossRef]
50. Kishore, R.A.; Bianchi, M.V.; Booten, C.; Vidal, J.; Jackson, R. Optimizing PCM-integrated walls for potential energy savings in U.S. Buildings. *Energy Build.* **2020**, *226*, 110355. [CrossRef]
51. Patel, B.; Rathore, P.K.S.; Gupta, N.K.; Sikarwar, B.S.; Sharma, R.; Kumar, R.; Pandey, A. Location optimization of phase change material for thermal energy storage in concrete block for development of energy efficient buildings. *Renew. Energy* **2023**, *218*, 119306. [CrossRef]
52. Wang, H.; Chiang, P.-C.; Cai, Y.; Li, C.; Wang, X.; Chen, T.-L.; Wei, S.; Huang, Q. Application of Wall and Insulation Materials on Green Building: A Review. *Sustainability* **2018**, *10*, 3331. [CrossRef]
53. Souza, A.B.; Ferreira, H.S.; Vilela, A.P.; Viana, Q.S.; Mendes, J.F.; Mendes, R.F. Study on the feasibility of using agricultural waste in the production of concrete blocks. *J. Build. Eng.* **2021**, *42*, 102491. [CrossRef]
54. Iwaro, J.; Mwasha, A. Effects of Using Coconut Fiber–Insulated Masonry Walls to Achieve Energy Efficiency and Thermal Comfort in Residential Dwellings. *J. Arch. Eng.* **2019**, *25*, 04019001. [CrossRef]
55. Homod, R.Z.; Almusaed, A.; Almssad, A.; Jaafar, M.K.; Goodarzi, M.; Sahari, K.S. Effect of different building envelope materials on thermal comfort and air-conditioning energy savings: A case study in Basra city, Iraq. *J. Energy Storage* **2020**, *34*, 101975. [CrossRef]
56. Iffa, E.; Tariku, F.; Simpson, W.Y. Highly Insulated Wall Systems with Exterior Insulation of Polyisocyanurate under Different Facer Materials: Material Characterization and Long-Term Hygrothermal Performance Assessment. *Materials* **2020**, *13*, 3373. [CrossRef]

57. Dong, Y.; Kong, J.; Mousavi, S.; Rismanchi, B.; Yap, P.-S. Wall Insulation Materials in Different Climate Zones: A Review on Challenges and Opportunities of Available Alternatives. *Thermo* **2023**, *3*, 38–65. [CrossRef]
58. Rehman, H.U. Experimental performance evaluation of solid concrete and dry insulation materials for passive buildings in hot and humid climatic conditions. *Appl. Energy* **2017**, *185*, 1585–1594. [CrossRef]
59. Yang, J.; Wu, H.; Xu, X.; Huang, G.; Xu, T.; Guo, S.; Liang, Y. Numerical and experimental study on the thermal performance of aerogel insulating panels for building energy efficiency. *Renew. Energy* **2019**, *138*, 445–457. [CrossRef]
60. Golder, S.; Narayanan, R.; Hossain, R.; Islam, M.R. Experimental and CFD Investigation on the Application for Aerogel Insulation in Buildings. *Energies* **2021**, *14*, 3310. [CrossRef]
61. Thenappa, S.; Nguyen, P.T.; Do, N.H.; Le, D.K.; Thai, Q.B.; Le, P.K.; Phan-Thien, N.; Duong, H.M. Correction to: Advanced fabrication and multi-properties of aluminium hydroxide aerogels from aluminium wastes. *J. Mater. Cycles Waste Manag.* **2021**, *23*, 1276.
62. ALBA-2021. Available online: https://www.albasmelter.com/en/article/alba-discloses-its-financial-results-for-the-fourth-quarter-and-full-year-of-2021 (accessed on 15 March 2024).
63. Abu-Jdayil, B.; Mourad, A.-H.; Hittini, W.; Hassan, M.; Hameedi, S. Traditional, state-of-the-art and renewable thermal building insulation materials: An overview. *Constr. Build. Mater.* **2019**, *214*, 709–735. [CrossRef]
64. Reboul, N.; Larbi, A.S.; Ferrier, E. Two-way bending behaviour of hollow concrete block masonry walls reinforced by composite materials. *Compos. Part B Eng.* **2018**, *137*, 163–177. [CrossRef]
65. Asdrubali, F.; D'Alessandro, F.; Schiavoni, S. A review of unconventional sustainable building insulation materials. *Sustain. Mater. Technol.* **2015**, *4*, 1–17. [CrossRef]
66. Martinez-Amariz, A.; Muñoz, L.P.; Posso, F.; Bellon, D. Compressive evaluation of mansory bricks produced with industrial byproduct aggregates. *Int. J. Appl. Ceram. Technol.* **2020**, *17*, 2681–2689. [CrossRef]
67. Antar, M.A. Thermal radiation role in conjugate heat transfer across a multiple-cavity building block. *Energy* **2010**, *35*, 3508–3516. [CrossRef]
68. Huelsz, G.; Barrios, G.; Rojas, J. Evaluation of heat transfer models for hollow blocks in whole-building energy simulations. *Energy Build.* **2019**, *202*, 109338. [CrossRef]
69. Morales, M.; Juárez, M.; Muñoz, P.; Gómez, J. Study of the geometry of a voided clay brick using non-rectangular perforations to optimise its thermal properties. *Energy Build.* **2011**, *43*, 2494–2498. [CrossRef]
70. Zhang, Y.; Wang, Q. Influence of Hollow Block's Structural Configuration on the Thermal Characteristics of Hollow Block Wall. *Procedia Eng.* **2017**, *205*, 2341–2348. [CrossRef]
71. Ismaiel, M.; Chen, Y.; Cruz-Noguez, C.; Hagel, M. Thermal resistance of masonry walls: A literature review on influence factors, evaluation, and improvement. *J. Build. Phys.* **2022**, *45*, 528–567. [CrossRef]
72. Ghailane, H.; Ahamat, M.A.; Padzi, M.M.; Beddu, S. Steady-state heat flow through hollow clay bricks. *IOP Conf. Series Mater. Sci. Eng.* **2020**, *834*, 012021. [CrossRef]
73. Fogiatto, M.A.; Santos, G.H.d.; Catelan, J.V.R. Numerical Two-Dimensional Steady-State Evaluation of the Thermal Transmittance Reduction in Hollow Blocks. *Energies* **2019**, *12*, 449. [CrossRef]
74. Erdem, C.U.; Cuce, P.M.; Besir, A.B. Improving thermal resistance of lightweight concrete hollow bricks: A numerical optimi-sation research for a typical masonry unit. *J. Energy Syst.* **2020**, *4*, 121–144.
75. Almujahid, A.; Kaneesamkandi, Z. Construction of a test room for evaluating thermal performance of building wall systems under real conditions. *Measurement* **2013**, *2*.
76. Hatia, M.; Kriker, A.; Belatrache, D. Experimental Study of the Thermo-Mechanical Properties of Building Blocks Used in Hot Desert Arid Climate. *EAI Endorsed Trans. Energy Web* **2020**, *8*. [CrossRef]
77. Vivancos, J.-L.; Soto, J.; Perez, I.; Ros-Lis, J.V.; Martínez-Máñez, R. A new model based on experimental results for the thermal characterization of bricks. *Build. Environ.* **2009**, *44*, 1047–1052. [CrossRef]
78. Mavromatidis, L.E.; Michel, P.; El Mankibi, M.; Santamouris, M. Study on transient heat transfer through multilayer thermal insulation: Numerical analysis and experimental investigation. *Build. Simul.* **2010**, *3*, 279–294. [CrossRef]
79. Bruno, R.; Ferraro, V.; Bevilacqua, P.; Arcuri, N. On the assessment of the heat transfer coefficients on building components: A comparison between modeled and experimental data. *Build. Environ.* **2022**, *216*, 108995. [CrossRef]
80. Yang, Y.; Chen, Z.; Wu, T.V.; Sempey, A.; Batsale, J.-C. In situ methodology for thermal performance evaluation of building wall: A review. *Int. J. Therm. Sci.* **2022**, *181*, 107687. [CrossRef]
81. Rasooli, A.; Itard, L. In-situ characterization of walls' thermal resistance: An extension to the ISO 9869 standard method. *Energy Build.* **2018**, *179*, 374–383. [CrossRef]
82. *ISO 9869-1*; Thermal Insulation, Building Elements, In-Situ Measurement of Thermal Resistance and Thermal Transmittance-Part 1: Heat Flow Meter Method. BSI: London, UK, 2014. Available online: https://www.bing.com/ck/a?!&&p=dd21a8746fb356c7JmltdHM9MTcxMDk3OTIwMCZpZ3VpZD0xYTJhZjA4NS0wYTM2LTY5OWMtMjllMC1lMTVlMGIzZjY4NmUmaW5zaWQ9NTE5Nw&ptn=3&ver=2&hsh=3&fclid=1a2af085-0a36-699c-29e0-e15e0b3f686e&psq=ISO+9869-1%3b&u=a1aHR0cHM6Ly93d3cuaXNvLm9yZy9zdGFuZGFyZC81OTY5Ny5odG1s&ntb=1 (accessed on 15 March 2024).
83. ASTM 1046 and 1155. Available online: www.astm.org (accessed on 15 March 2024).
84. Atsonios, I.A.; Mandilaras, I.D.; Kontogeorgos, D.A.; Founti, M.A. A comparative assessment of the standardized methods for the in–situ measurement of the thermal resistance of building walls. *Energy Build.* **2017**, *154*, 198–206. [CrossRef]

85. *ISO 8990*; Thermal Insulation-Determination of Steady-State Thermal Transmission Properties-Calibrated and Guarded Hot Box. 2019. Available online: https://www.iso.org/standard/16519.html (accessed on 15 March 2024).
86. Fortes, E.S.; Parsekian, G.A.; Camacho, J.S.; Fonseca, F.S. Compressive strength of masonry constructed with high strength concrete blocks. *Rev. Ibracon Estruturas Mater.* **2017**, *10*, 1273–1319. [CrossRef]
87. Aubert, J.E.; Maillard, P.; Morel, J.-C.; Al Rafii, M. Towards a simple compressive strength test for earth bricks? *Mater. Struct.* **2016**, *49*, 1641–1654. [CrossRef]
88. Álvarez-Pérez, J.; Mesa-Lavista, M.; Chávez-Gómez, J.H.; Miguel, G.F.-S. Experimental investigation on tensile strength of hollow concrete blocks. *Mater. Struct.* **2021**, *54*, 164. [CrossRef]
89. Haach Vladimir, G. Development of a New Test for Determination of Tensile Strength of Concrete Blocks. 2013. Available online: https://hdl.handle.net/1822/26309 (accessed on 15 March 2024).
90. Bae, B.-I.; Choi, H.-K.; Choi, C.-S. Flexural Strength Evaluation of Reinforced Concrete Members with Ultra High Performance Concrete. *Adv. Mater. Sci. Eng.* **2016**, *2016*, 2815247. [CrossRef]
91. Chen, W.F. The Civil Engineering Handbook. 2002. Available online: https://books.google.com.bh/books?id=9Q3OBQAAQBAJ&printsec=frontcover&dq=civil+engineering+material+handbook&hl=en&sa=X&ved=2ahUKEwi1qrbw-dz6AhXC_rsIHW8pC4EQuwV6BAgKEAY#v=onepage&q=civil%20engineering%20material%20handbook&f=false (accessed on 15 March 2024).
92. ASTM C140-99b. 2003. Available online: https://standards.globalspec.com/std/3806313/astm-c140-99b (accessed on 15 March 2024).
93. Sitton, J.D.; Zeinali, Y.; Heidarian, W.H.; Story, B.A. Effect of mix design on compressed earth block strength. *Constr. Build. Mater.* **2018**, *158*, 124–131. [CrossRef]
94. Zahra, T.; Thamboo, J.; Asad, M. Compressive strength and deformation characteristics of concrete block masonry made with different mortars, blocks and mortar beddings types. *J. Build. Eng.* **2021**, *38*, 102213. [CrossRef]
95. Holmes, N.; O'Malley, H.; Cribbin, P.; Mullen, H.; Keane, G. Performance of masonry blocks containing different proportions of incinator bottom ash. *Sustain. Mater. Technol.* **2016**, *8*, 14–19. [CrossRef]
96. The Constructor, Building Ideas, Compressive Strength of Concrete-Cube Test [PDF], Procedure, Results. Available online: https://theconstructor.org/concrete/compressive-strength-concrete-cube-test/1561/ (accessed on 15 March 2024).
97. Sirriyeh, M.M. *The House of Bahrain Press Heritage*; Sustainable Development and Planning II; WIT Press: Southampton, UK, 2005; Volume 2.
98. Al-Manratain Company. *Bahrain Building Material Supplier Since 1959—Al Manaratain*; Al-Manratain Company: Jidhaf, Bahrain, 1959.
99. Bahrain Blocks, Company, Home | Bahrain Blocks. Available online: https://www.bing.com/ck/a?!&&p=f42ef0deeeed66acJmltdHM9MTcxMDk3OTIwMCZpZ3VpZD0xYTJhZjA4NS0wYTM2LTY5OWMtMjllMC1lMTVlMGIzZjY4NmUmaW5zaWQ9NTIwNA&ptn=3&ver=2&hsh=3&fclid=1a2af085-0a36-699c-29e0-e15e0b3f686e&psq=bahrain+blocks+company+bahrain+official+website&u=a1aHR0cHM6Ly9iYWhyYWluYmxvY2tzLmNvbS8&ntb=1 (accessed on 15 March 2024).
100. Middle East Block Company, Al-Ali Group. AMA GROUP | Building Materials. Available online: https://al-aali.com/ (accessed on 15 March 2024).
101. Aljalahma Blocks. Ahmed Rashid Aljalahma & Brothers Co. Available online: https://aljalahmacompany.com (accessed on 15 March 2024).
102. RedX Company. RedX Industries. Available online: https://www.bing.com/ck/a?!&&p=3b1b3014f4975d98JmltdHM9MTcxMDk3OTIwMCZpZ3VpZD0xYTJhZjA4NS0wYTM2LTY5OWMtMjllMC1lMTVlMGIzZjY4NmUmaW5zaWQ9NTIxMA&ptn=3&ver=2&hsh=3&fclid=1a2af085-0a36-699c-29e0-e15e0b3f686e&psq=redex+industries+bahrain+official+website&u=a1aHR0cHM6Ly93d3cucmVkeGluZHVzdHJpZXMuY29tLw&ntb=1 (accessed on 15 March 2024).
103. Ahmed Isa Block Factor; Ahmed Isa Block Factory S.P.C—Manufacturing and Supplying Concrete Masonry Block. Ahmed Isa Block Factory | Buildeey. Available online: https://buildeey.com/profile/ahmed-isa-block-factory (accessed on 15 March 2024).

buildings

MDPI

Article

Energy Monitoring and Analysis of a Residential House in China

Yanzhi Wang [1], Shaotong Han [2], Qiuqi Zhang [1], Jing Sun [1,3,*], Zhibao Cheng [1,3] and An Chen [1,3]

1 School of Civil Engineering, Beijing Jiaotong University, Beijing 100044, China; 20115042@bjtu.edu.cn (Y.W.); 22125952@bjtu.edu.cn (Q.Z.); chengzb@bjtu.edu.cn (Z.C.); anchen@bjtu.edu.cn (A.C.)
2 United World College Changshu China, Changshu 215556, China; sthan22@uwcchina.org
3 Beijing's Key Laboratory of Structural Wind Engineering and Urban Wind Environment, Beijing Jiaotong University, Beijing 100044, China
* Correspondence: jsun@bjtu.edu.cn

Abstract: The energy consumption of residential buildings plays a crucial role in overall energy consumption and environmental sustainability. This paper aims to conduct an energy analysis of a residential house located in China, with a focus on comparing the accuracy of the model, identifying areas for improvement, and proposing energy-efficient solutions. Four sets of temperature sensors were placed to monitor the ambient temperature at which the building is located and the indoor temperature of the residential building during a heating season. The energy consumption of keeping the building running at a low temperature was recorded and compared with the simulation results to verify the accuracy of the model. The monitoring results give the weekly average temperature of each zone on each floor, and the door and window positions, room layouts, and orientations are discussed to analyze the thermal response of the building. In addition, the effect of the heat transfer coefficient of the exterior walls, the heat transfer coefficient of the roof, and the solar heat gain coefficient (SHGC) of the exterior windows on the heating energy consumption of the building are further analyzed through simulations. The results show that, after adding a certain thickness of insulation to the exterior walls and roofs of a building, increasing the thickness of the insulation layer produces little extra energy saving. The use of building windows with high SHGC can effectively reduce building heating energy consumption.

Keywords: residential building; energy monitoring; energy simulation; parametric study; optimization analysis

Citation: Wang, Y.; Han, S.; Zhang, Q.; Sun, J.; Cheng, Z.; Chen, A. Energy Monitoring and Analysis of a Residential House in China. *Buildings* **2024**, *14*, 2930. https://doi.org/10.3390/buildings14092930

Academic Editors: Gerardo Maria Mauro and Apple L. S. Chan

Received: 10 July 2024
Revised: 6 September 2024
Accepted: 13 September 2024
Published: 16 September 2024

1. Introduction

With the progress and development of society, issues such as climate change and energy consumption have attracted widespread attention and discussion. Globally, the building sector constitutes approximately 36% of total energy consumption and over 40% of total carbon emissions. Among all sectors, the building industry possesses the greatest potential to implement effective energy saving measures [1,2]. In recent years, China has made significant efforts to address its energy consumption and environmental challenges. The residential sector, accounting for a substantial portion of the country's energy consumption, presents a unique opportunity for energy optimization [3]. As the country undergoes rapid urbanization and experiences a growing demand for housing, it becomes imperative to evaluate and analyze the energy performance of residential buildings. By analyzing the energy use of representative dwellings and proposing ideas for energy-optimized design, policymakers, architects, and owners can be guided to adopt more sustainable and energy-efficient building practices.

For the energy use of buildings, researchers' primary research methods include model simulation and field monitoring. Vartholomaio [4] used the dynamic building energy simulation software EnergyPlus 8.3 to perform a parametric simulation analysis of the heating and cooling energy consumption of residential buildings in three urban forms, focusing on the Mediterranean city of Thessaloniki, which has a heating and cooling demand. The study concluded that there are synergies between a highly compact urban form and passive solar design. Additionally, Al-Saggaf et al. [5] integrated 3D building information modeling (BIM) models of three different design options for a residential building in Saudi Arabia using Ecotect analysis 2010 software for energy simulation. The results demonstrated the effectiveness of the Architectural-Based Energy Impact Scoring System (AEISS). Comprising seven building design features and over 40 design options, the AEISS exhibited a positive response. It effectively delivers energy-efficient and low-energy-consuming design options for decision-makers. Wang et al. [6] selected several typical buildings in Beijing as the research objects. They utilized eQUEST software to investigate the buildings' energy consumption and derived corresponding building energy consumption prediction model. These models can predict the overall energy consumption of various buildings types during urban regional planning stage, offering insights for urban energy conservation planning. Santana et al. [7] conducted a similar study, monitoring and comparing two dwellings in Barcelona, Spain, during the winter. Their analysis aimed to understand the impact of the envelope's thermal performance on indoor temperature, alongside other factors such as outdoor temperature and solar gain. This analysis suggested that a building's thermal performance involves multiple variables beyond envelope thermal transmittance, such as thermal inertia, solar gain building orientation, and an urban environment. Charai et al. [8] employed a validated PMV-PPD model in EnergyPlus to assess the benefits of using biosourced earth in building envelopes during the summer, based on real-time monitoring of an existing building. Results indicated that the proposed wall improved thermal comfort by regulating indoor temperature and humidity, with 24.6% fewer dissatisfied occupants compared to a typical wall. Shan et al. and Yu et al. [9,10] designed and constructed a test building to investigate the thermal performance of a new type of bamboo–steel composite wall. They measured the heat transfer coefficient of the composite wall and validated the simulation model through the actual measurement of the energy consumption of the test building. Simulation analyses were then carried out to investigate the energy saving effect of composite walls in different climatic regions. Hassan et al. [11] investigated the effects of courtyard layout, aspect ratio, and courtyard orientation on energy consumption in residential buildings using nine case studies. The results showed that the highest energy savings of 18.73% were achieved when the optimal location of the courtyard layout, i.e., southwest elevation, was adopted.

Further research on energy design optimization of buildings, based on building energy performance studies, can provide valuable insights for policymakers, architects, and building owners, fostering sustainable building development. In the realm of energy design optimization, many researchers have focused on reducing building energy consumption and enhanced indoor thermal comfort through modifications to the thermal performance of the building envelopes. Cabeza [12] conducted experiments on four small rooms constructed according to the traditional Mediterranean building system and compared the thermal performance of three typical insulation materials, polyurethane, polystyrene, and mineral wool, over an extended period. Tunçbilek et al. [13] indicated that the doping of highly conductive nanoparticles decreased the energy saving performance of phase change materials. The results showed that increasing the nanoparticle concentration to 3 vol% could increase the negative effect to 1.7%. Alyami et al. [14] investigated the effects of location and insulation materials on the energy performance of residential buildings across five climatic regions in the Kingdom of Saudi Arabia. Five commonly used insulation materials with distinct thermal properties, namely, polyurethane panels (PU), expanded polystyrene (EPS), glass wool (GW), urea–formaldehyde foam (UFF), and expanded perlite (EP), were analyzed in different climatic regions, revealing significant variations in energy performance of the insulation materials based on location. Guo et al. [15] constructed a comparative study between two identical model buildings, one insulated with 38 mm expanded polystyrene (EPS) panels as exterior insulation. The dynamic indoor thermal environment and heating energy consumption of both buildings were monitored comparatively at the same heating temperature. The results showed that the energy consumption of the two buildings was similar in the initial 20 min of heating, and the energy saving rate of the EPS building increased slowly with the heating duration. Amani et al. [16] modelled and analyzed the thermal behavior of buildings and identified design options to optimize building energy consumption and reduce global warming potential through an insulation system. The model incorporated a four-layer insulation system with the ability to select the material and thickness of each layer that offers 130 different design combinations. Other researchers have also proposed energy-efficient building designs focusing on cooling and heating loads and types of cooling and heating sources. Shibuya et al. [17] tested multistory office buildings at three different locations in Japan and investigated the potential impacts of climate change on the energy demand for office cooling and heating by simulating three periods of reference weather data through thermal analysis. The results emphasized the potential of office buildings in Japan for energy saving and CO_2 emissions reduction without relying on nuclear power generation. Li et al. [18] utilized DOE-2 software to simulate the cooling energy consumption and electricity demand in an office building in Hong Kong under various envelope performance and lighting systems, pointing out the importance of rational daylight use in the peak electricity consumption and peak cooling demand, and that appropriate lighting design could reduce the cooling energy consumption and electricity demand of buildings.

The aim of this paper Is to conduct energy monitoring and analysis of a residential building located in China, including the monitoring of outdoor/indoor temperatures and energy consumption of the building during a heating season, and analyze the thermal behavior of the building. The results from the energy analysis model are compared with the monitoring results to verify the validity of the model. On this basis, the heat transfer coefficients of the exterior wall and roof and the SHGC of the exterior window are taken as the key parameters to simulate and analyze the influence of the key parameters on the energy consumption of the building, providing a reference for the optimal design of the building's energy systems.

2. Building Information

2.1. Building Geometry

The demo house is a typical three-story reinforced concrete house in Haituo Valley, Zhangjiakou City, Hebei Province, China. Zhangjiakou is located at 39°30′–42°10′ N latitude and 113°50′–116°30′ E longitude, with a population of about 4.05 million in 2023, at an elevation of about 760 m above sea level, and has a temperate continental climate [19]. The photo of the building is shown in Figure 1. It is 7.7 m wide, 12.3 m long, and 9.9 m high. The main living room extends in height over all three floors. The total area of the demo house, including both heated and non-heated spaces, encompasses around 180 m^2. Of this, the heated area comprises approximately 144 m^2, providing comfortable living conditions during the colder months. The design and layout of the house were planned according to the codes in China, striking a balance between thermal comfort and energy consumption.

Figure 2 shows the location of the house and surrounding buildings, which are similar to the house in this study. Additionally, the house uses a shear wall structural system [20], which is also common. Furthermore, according to the *China Statistical Yearbook* [21], the per capita housing area in this region is 38.08 m^2. The house currently accommodates five people, resulting in a per capita housing area of approximately 36 m^2. Therefore, this house can be considered representative of the common housing typology in the region, and this study provides a reference for the energy consumption of the buildings in this region.

Overall, despite its elegance and functionality, the house's energy efficiency could be further improved. An analysis of its energy consumption can yield valuable insights into optimizing energy usage, enhancing sustainability, and promoting eco-friendly practices in residential construction. By identifying areas for improvement, such as insulation, the house can serve as a model for environmentally conscious residential design in the future.

Figure 1. Building photo.

Figure 2. Geographical location of the house and surrounding buildings.

2.2. Building Construction

2.2.1. Wallboards

The exterior walls of the demo house are multi-layered, consisting of (from the exterior to the interior) a layer of 120 mm rock wool board, 10 mm cement mortar, 8 mm polymer mortar, and 200 mm reinforced concrete. The composite structure of the exterior wall and the interior wall are different. The interior walls that connect each room consist of a layer of 200 mm reinforced concrete and a 10 mm panel of cement mortar. The details of the structures are shown in Figure 3 and Table 1.

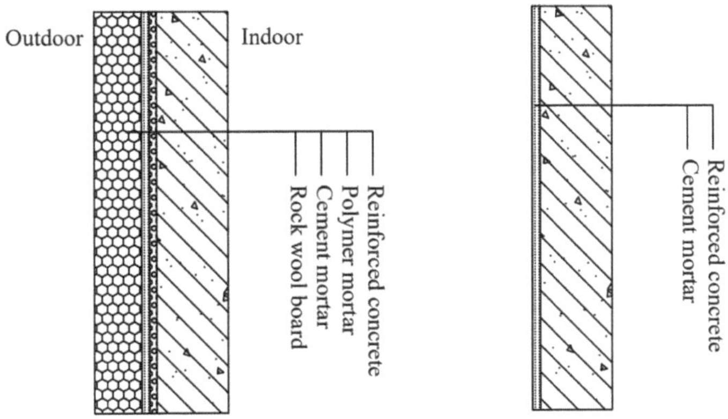

(**a**) Exterior wall (**b**) Interior wall

Figure 3. Wallboard configuration.

Table 1. Building construction.

Building Component	Configuration		Thickness (mm)
Wallboards	Exterior wall	Reinforced concrete	200
		Polymer mortar	8
		Cement mortar	10
		Rock wool board	120
	Interior wall	Reinforced concrete	200
		Cement mortar	10
Roof	Fine stone concrete		40
	Plastic benzoic board		110
	SBS		4
	Cement mortar		15
	Reinforced concrete		150
Floor	Cement mortar		20
	Pisolite concrete		50
	Plastic benzoic board		20
	Cement mortar		20
	Reinforced concrete		120

2.2.2. Roof

The roof of the demo house consists of (from exterior to the interior) a layer of 40 mm fine stone concrete, 110 mm plastic benzoic board, 4 mm SBS (styrene–butadiene–styrene), 15 mm of cement mortar, and 150 mm reinforced concrete. The angle of the roof is 30° of elevation. The details are shown in Figure 4 and Table 1.

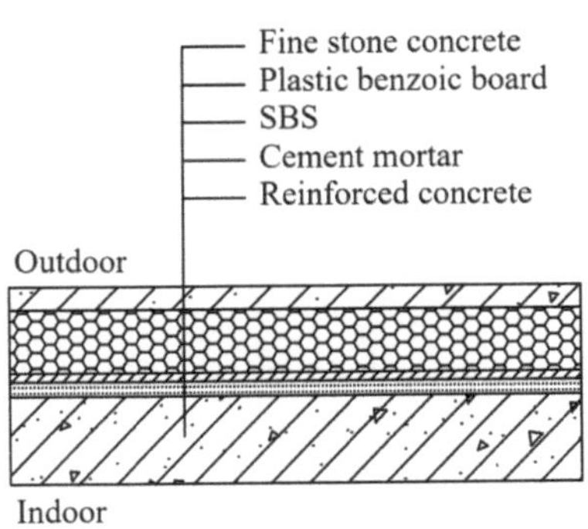

Figure 4. Roof configuration.

2.2.3. Floor

The floor of the demo house, in contact with the ground, is constructed of 120 mm reinforced concrete, 20 mm cement mortar, 20 mm plastic benzoic board, 50 mm pea gravel concrete, and 20 mm cement mortar. On the first floor and the second floor, they have the same structure as the floor that contacts with the ground. The details are shown in Figure 5 and Table 1.

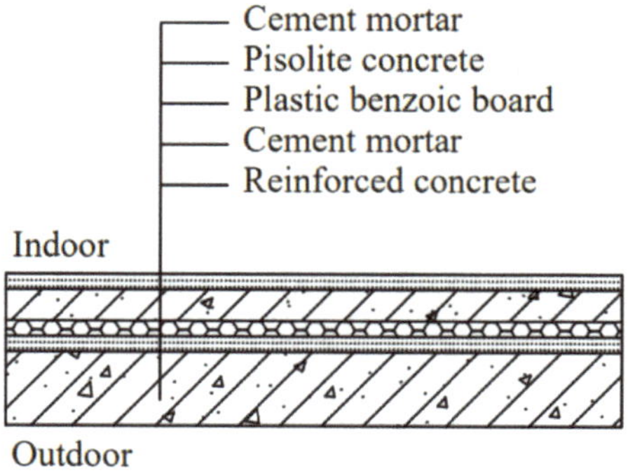

Figure 5. Floor configuration.

3. Methodology

3.1. Monitoring Methodology

To analyze the thermal response of the building as well as to carry out 3D heat transfer model simulations, the average indoor temperatures of each zone on each floor of the building and the outdoor temperatures at the location of the building were monitored.

Temperature and humidity are important factors in building energy consumption. Studies have shown that buildings consume more energy in areas with higher humidity compared to those in dry environments [22–24]. The climate of this residence is temperate continental, which is relatively dry compared to areas such as the coast. Moreover, the building remains closed and unoccupied for most of the monitoring period, with only three days occupied. Therefore, humidity has a negligible effect on this building's energy consumption, and this study mainly focuses on temperature.

3.1.1. Indoor Average Temperature

A device for temperature monitoring of houses in low-temperature operations, the G9-TH instrument, was used, with data readings taken every 15 min. When selecting the sensor placement, several principles were followed: Firstly, representative rooms were chosen based on different orientations and floors. Secondly, the sensors were positioned to avoid interference from radiators and direct sunlight, ensuring a relatively stable temperature state. Additionally, the sensors were kept away from windows, considering that windows had higher heat transfer coefficients, which could result in localized temperature drops due to cold air infiltration. Furthermore, the sensors were placed at a horizontal distance of more than 0.5 m from the wall, with a recommended height between 0.6 m and 1.1 m.

On each floor, we installed four high-precision temperature sensors, and in addition, there was one main unit per floor, responsible for connecting and logging data from the four high-precision temperature sensors. The main unit was utilized to record the temperature measurements taken by the temperature sensors. The specific locations for sensor placement can be referred to in Figure 6, where the pentagram represents the main unit's position, and the circle denotes the position of the temperature sensor.

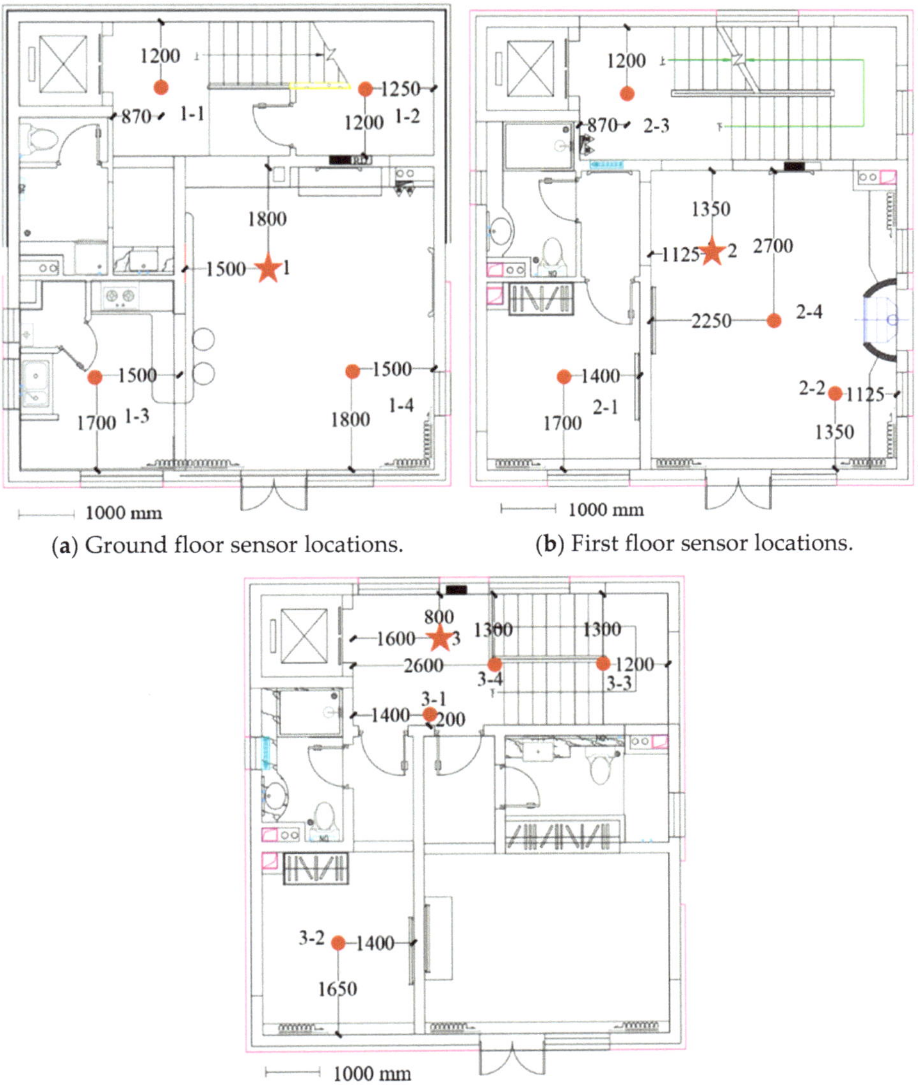

(**a**) Ground floor sensor locations. (**b**) First floor sensor locations.

(**c**) Second floor sensor locations.

Figure 6. Locations of indoor temperature sensors (unit: mm): red color denotes sensors' locations, where the pentagram and circle indicate locations of the main unit and temperature sensor, respectively.

3.1.2. Outdoor Average Temperature

The sensors were placed in a well-ventilated area of the second-floor balcony to monitor the outdoor air temperature. Waterproofing was implemented to protect the sensors. The data recording interval was set to 15 min. The main unit 4 and 4-2 measurement points were positioned on the inner and outer sides of the living room's exterior wall, which were immediately adjacent to the windows, as shown in Figure 7. The pentagram and circle indicate the locations of the main unit and temperature sensor, respectively.

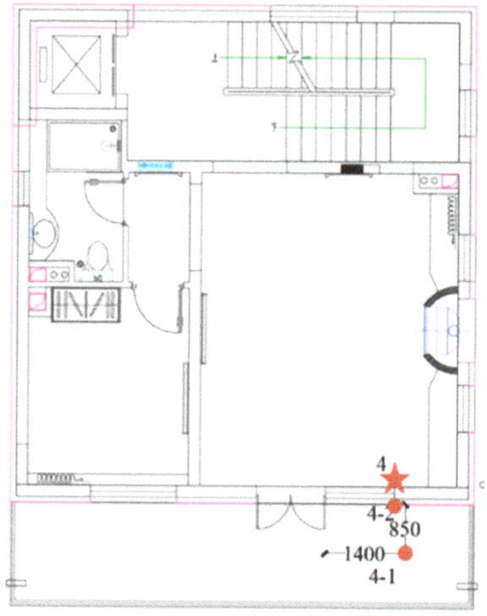

Figure 7. Locations of outdoor temperature sensors (unit: mm): red color denotes sensors' locations, where the pentagram and circle indicate locations of the main unit and temperature sensor, respectively.

3.1.3. Temperature Control

The house remained closed and unoccupied for the vast majority of the monitoring period, occupied for only a few days in mid-January. The indoor temperature of the building was controlled at 12 °C from November 1st to January 15th, 15 °C from January 16th to January 23rd, and 13 °C from January 24th to March 31st.

3.2. Modeling

We used SketchUp to model the house for accurate energy simulation in EnergyPlus, with the building model shown in Figure 8. The SketchUp model reflected the actual conditions of the base, which consisted of the three-story building located in Zhangjiakou. Due to significant temperature fluctuations between day and night in Zhangjiakou region, indoor temperatures were influenced by solar radiation during the daytime, resulting in substantial diurnal temperature variations. The temperature control settings in the model were consistent with the actual conditions. SketchUp enables zoning and supports zone-based modelling. In this study, different zones are divided, but the parameters are the same in each zone. Therefore, an in-depth discussion is not carried out.

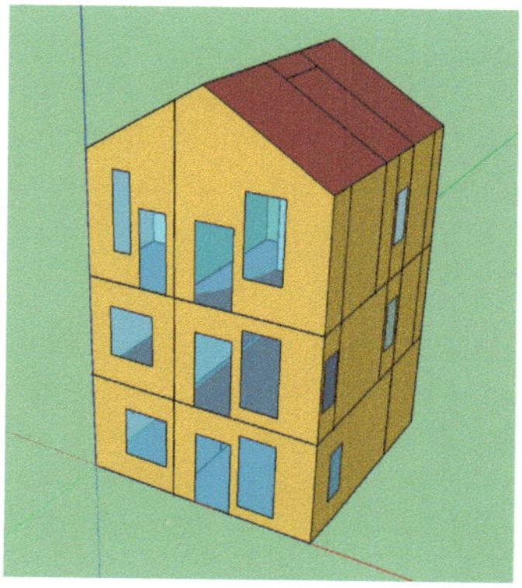

Figure 8. Building model (balconies not included).

The model used SketchUp, OpenStudio, and EnergyPlus, with a flowchart illustrated in Figure 9. First, SketchUp was used to construct the building's geometric model, which was then imported into OpenStudio, where building settings, including the building envelope, indoor thermal disturbances, etc., were configured. Subsequently, the OpenStudio model was exported as an IDF file, which was imported into EnergyPlus to complete the simulation process.

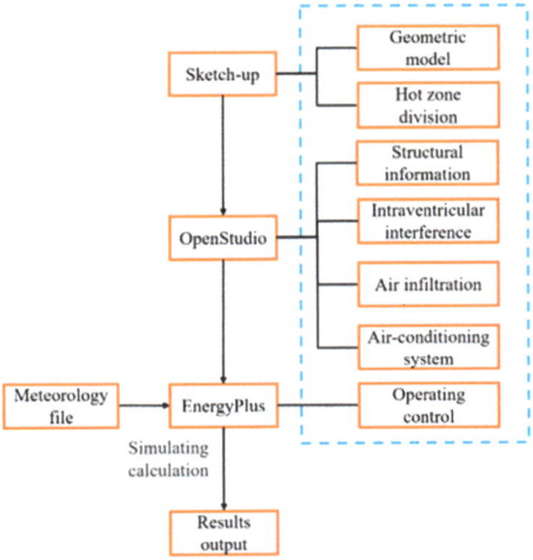

Figure 9. Modelling flowchart: blue dashed line represents settings and functions of each software.

Given that the sample building was a three-story structure in direct contact with the ground, the indoor temperatures were also affected by ground temperature. For our energy simulation in EnergyPlus, we incorporated temperature data that matched the actual conditions. To achieve this, we replaced the relevant portions of the typical meteorological year data from Huailai City with temperature data collected from outdoor temperature sensors placed around the sample building from December to April. The temperature data curve is shown in Figure 10. This replacement ensures a more accurate representation of local conditions in our temperature data.

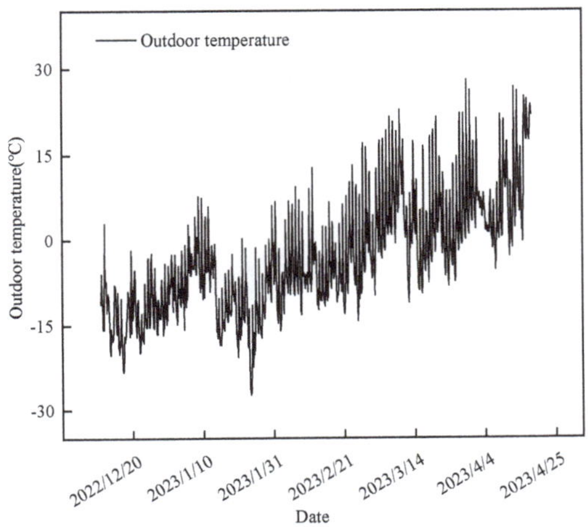

Figure 10. Outdoor temperature monitoring data.

Moreover, we set the average outdoor airflow velocity to match that around the sample building and included it in the EnergyPlus model. To further align the simulation with reality, we maintained consistency between the sample building and the model in terms of structural components, materials, thermal insulation coefficients, and heat transfer coefficients. The building remained closed and unoccupied for most of the monitoring period, with only three days occupied. The occupancy schedule for the three days and energy sources for different utilities are shown in Table 2. Since both lighting and domestic hot water were powered by electricity, it did not affect the gas consumption. Additionally, the building was occupied for only 3% of the monitoring period. Thus, the impact of occupancy and personnel, lighting, and electrical equipment on the building's energy consumption throughout the heating period was negligible. Therefore, thermal disturbance from personnel, lighting, and electrical appliances were not considered in this study. According to the Chinese standard JGJ/T 449-2018 [25], the space infiltration design flow rate, characterized by the indoor air change rate, was set as 0.5 times per hour, and the design specification outdoor air of the building was set as 0.008333 m^3/s per person. However, since this building was unoccupied during the monitoring period, there was no design specification outdoor air.

Table 2. Occupancy schedule and energy sources.

Time	House State	Energy Source		
		Lighting	Domestic Hot Water	Heating
2023/1/18–2023/1/20	Occupied	Electricity	Electricity	Gas
Other time	Vacant	-	-	Gas

During the modeling process, we segmented each room in the sample building to create distinct living zones. Our aim was to maximize the realism of the model and enable precise energy consumption simulations.

4. Results and Discussion

4.1. Average Indoor Temperature Monitoring Results

The building was monitored for most of the period from December 2022 to mid-April 2023, with monitoring on the first floor interrupted from early March for maintenance reasons. During the monitoring period, the building remained closed and unoccupied for the vast majority of the time, with only a few days of occupancy, and the building kept the energy systems on to maintain low temperature operation. The following analysis focuses on the thermal response of the building systems.

Figure 11 shows the weekly average indoor temperatures for each zone on each floor of the building. Due to the lack of cooling equipment, the indoor temperatures often exceeded the control temperatures. All areas of the building were kept at low temperatures from December to February. From March onwards, due to the significant increase in outdoor temperatures, the indoor temperatures also increased significantly, often exceeding the control temperatures of the heating equipment. The monitoring results show that the temperature on the south side is higher than that on the north side, while the temperature on the southwestern side is significantly higher than that on the southeastern side, and the temperature on the northeastern side has small difference with that on the northwestern side. Such a pattern is reflected in all three floors of the building. The average indoor temperature on the southwestern side area of the ground floor is significantly higher than the other areas, while the average indoor temperature on the southeastern side of the ground floor is similar to that of the north side from December to February, and slightly higher than that of the north side after March. This can be attributed to the fact that the south side is exposed to more solar radiation than the north side, and that the rooms on the southwestern side of the ground floor are smaller than those on the southeastern side, resulting in less heat loss. In addition, the ground floor has a specialized structural form to provide better thermal insulation. It is on the outside of the rooms on the north side, which may explain the smaller difference in average temperatures between the north and south sides of the ground floor compared with those of the first and second floors. The average indoor temperature on the southwestern side of the first floor is significantly higher than those of the other areas, while the average indoor temperature on the southeastern side is slightly higher than that on the north side, which is related to the orientation of the building, the layout of the rooms, and the location of the windows and doors. The rooms on the south side have large glass windows, which can receive more solar radiation, while the rooms on the southwestern side are much smaller than those on the southeastern side, and the heat loss is slower. The average indoor temperature on the southwestern side of the second floor is significantly higher than that on the north side, which is related to the fact that the south side receives more solar radiation. The temperature difference between the northwestern side and the northeastern side is not significant, which may be due to the location of the two areas and the common north-facing orientation.

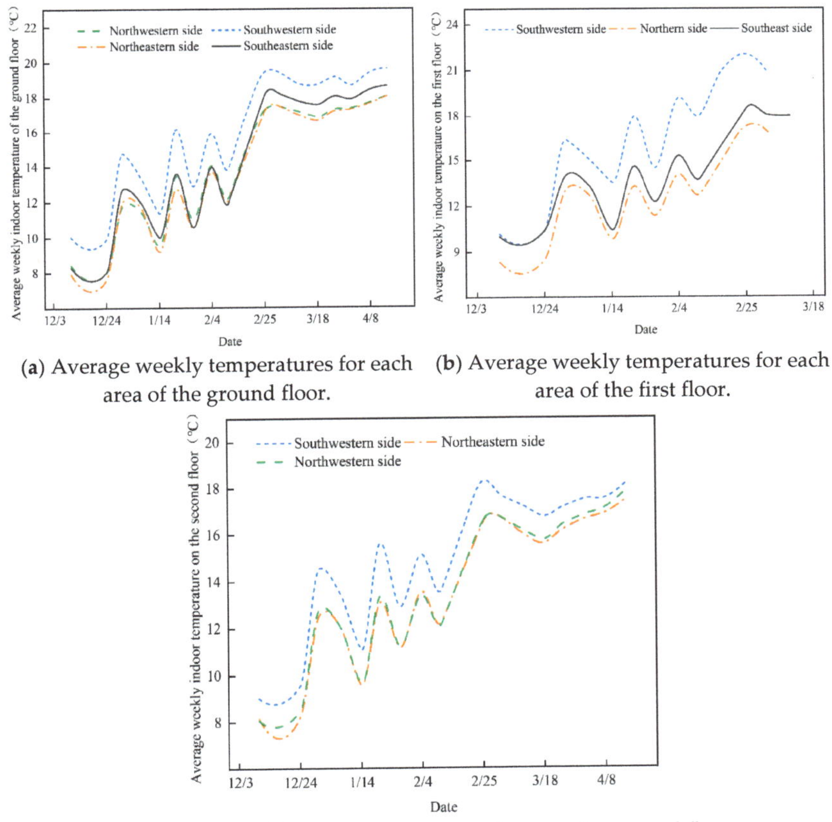

(**a**) Average weekly temperatures for each area of the ground floor.

(**b**) Average weekly temperatures for each area of the first floor.

(**c**) Average weekly temperatures for each area of the second floor.

Figure 11. Weekly average indoor temperatures for each zone on each floor of the building.

4.2. Building Heating Energy Consumption

During the monitoring period, we recorded outdoor and indoor temperatures and gas consumption. The actual energy consumption of building can be calculated based on the gas consumption. In the EnergyPlus model, the recorded outdoor and indoor temperatures were used as input values, based on which the building's energy consumption was calculated. In this case, it was considered that one cubic meter of natural gas produces

33,494.4 kJ of heat and the heating equipment used in the building, i.e., a gas wall oven, had a thermal efficiency of 85%.

The two results are shown in Table 3. After considering the thermal efficiency of the wall-hung gas boilers, the error between the simulated energy consumption and the actual energy consumption in each month is around 10%, which indicates that the 3D heat transfer model built with EnergyPlus can effectively simulate the energy consumption of the building. The monitoring began from December 10. Therefore, the actual outdoor temperature of the building in November and some days in December was missing. During the simulation, the outdoor temperature of the building was based on the weather data in the typical meteorological year file offered by EnergyPlus, resulting in lower simulation accuracy in November and December. In March, the error was at 17.68%, which was larger than the other months because the property only recorded the actual gas consumption of the building for one week.

Table 3. Monthly comparison of simulated and actual energy consumption.

	November	December	January and February	March
Simulate building heating energy consumption (GJ)	3.00	9.80	17.55	0.75
Simulate building gas consumption (m^3)	105.37	344.22	616.43	26.34
Actual building gas consumption (m^3)	124	396	657	32
Percentage (%)	84.98	86.92	94.37	82.32
NMBE (%)		−12.57		
CV(RMSE) (%)		12.18		
R^2 (%)		90		

To calibrate the accuracy of the model, three metrics were employed, i.e., NMBE, CV (RMSE) [26], and linear regression (R^2), based on monthly energy consumption. Since only one week's data were recorded in March, this month was excluded from the calculation. The results are shown in Table 3, where NMBE, CV (RMSE), and R^2 are −12.57%, 12.18%, and 90%, respectively. It can be seen that NMBE and CV (NMBE) are within the tolerance of ±20% specified by the International Performance Measurement and Verification Protocol (IPMVP) [26] and 15% specified by American Society of Heating, Refrigerating and Air-Conditioning Engineers (ASHRAE) Guideline 14-2014 [27], respectively. In addition, R^2 is much higher than the criteria of 75% [28]. Therefore, the 3D model developed above can be considered accurate and will be used in the following parametric analysis. A similar model was also calibrated in a previous work [29].

4.3. Parametric Analysis

4.3.1. Parameter Selection

About 50% of the building energy loss is through the envelope. Therefore, improving the thermal performance of the building envelope is an effective way to save energy in buildings [30–32]. The year-round dynamic energy simulation of the building model was carried out by EnergyPlus software, and the heat transfer coefficients of the envelope exterior walls and roofs as well as the solar heat gain coefficients of the exterior windows were taken as the key parameters [33–36]. The parameters of the original scheme of the building were used as benchmarks for adjusting each of the key parameters and analyzing the influence of each key parameter on the energy efficiency of the building, as shown in Table 4.

The heat transfer coefficient of the exterior wall of the original program of the building was 0.3114 $W/(m^2{\cdot}K)$. The increase in the heat transfer coefficient of the exterior wall was achieved by increasing the thickness of the exterior wall insulation material. Adding a layer of rock wool board for the exterior wall in the model, the thickness changed from 10 mm to 300 mm, the thermal resistance increased in the range of 0.2439~7.317 $m^2{\cdot}K/W$, and the corresponding heat transfer coefficient of the exterior wall changed in the range of 0.2894~0.0950 W/(m^2·K).

Table 4. The simulated parameters, variables, and variations of the variables.

	Insulating Layer Thickness (mm)	Thermal Resistance (m^2·K/W)	Heat Transfer Coefficient (W/(m^2·K))
Exterior wall	10~300	0.2439~7.317	0.2894~0.0950
Roof	10~300	0.2439~7.317	0.2277~0.0887

The roof heat transfer coefficient for the original building was 0.2523 W/(m^2·K). The increase in roof heat transfer coefficient was achieved by increasing the thickness of the roof insulation. By adding a layer of rock wool board for the roof in the model, the thickness changed from 10 mm to 300 mm, the thermal resistance increased in the range of 0.2439 to 7.317 m^2·K/W, and the corresponding roof heat transfer coefficient changed in the range of 0.2277 to 0.0887 W/(m^2·K).

The SHGC of the building's original program for the exterior windows was 0.4. The effect of the SHGC on the building's energy consumption is seasonal, with smaller SHGC yielding lower energy consumption in the summer months and larger SHGC yielding lower energy consumption in the winter months. Since the monitoring of the building as well as the modeling in this paper focuses on the winter heating energy consumption of the building, higher parameter values than the original scheme were chosen, with a parameter variation range of 0.42 to 0.6.

4.3.2. Analysis of Simulation Results

(1) Influence of Heat Transfer Coefficients of Exterior Walls on Building Heating Energy Consumption

Energy Plus software was used to carry out energy consumption simulation based on the above parameters, and only the heat transfer coefficient of the exterior wall was changed, while other parameters were kept unchanged to carry out energy consumption simulation for the whole year. Changing the heat transfer coefficient of the exterior wall was achieved by increasing the thermal resistance value of the exterior wall. The effect of increasing the thermal resistance of the exterior wall on the energy consumption of the building heating is shown in Figure 12, and the relationship between the heat transfer coefficient of the exterior wall, which corresponds to increasing the thermal resistance of the exterior wall, and the energy consumption of the building's heating is shown in Figure 13.

The annual heating energy consumption of the building decreases gradually as the thermal resistance of the exterior wall increases from 0.2439 m^2·K/W to 7.317 m^2·K/W, and the corresponding heat transfer coefficient of the exterior wall decreases from 0.2894 W/(m^2·K) to 0.0950 W/(m^2·K). The annual heating energy consumption of the building and the heat transfer coefficient of the exterior wall are basically linearly related, and the annual heating energy consumption decreases linearly with the reduction in heat transfer coefficient of the exterior wall. The energy saving rate is also linearly related to the heat transfer coefficient of the exterior wall, and the energy saving rate increases linearly with the decrease in heat transfer coefficient of the exterior wall. From the point of view of increasing the thermal resistance of the exterior wall, increasing the thermal resistance causes the annual heating energy consumption of the building to gradually decrease, but the decreasing trend gradually becomes slower. The energy saving rate increases with increasing thermal resistance, but the increasing trend also slows down significantly. For example, when a 130 mm thick rock wool board is added to the exterior wall of the building as an insulation layer, the energy saving rate increases by less than 0.5% compared to a 120 mm thick insulation layer. The simulation results show that after adding a certain thickness of insulation layer to the exterior wall of the building, increasing the thickness of the insulation layer does not result in a satisfactory energy saving effect.

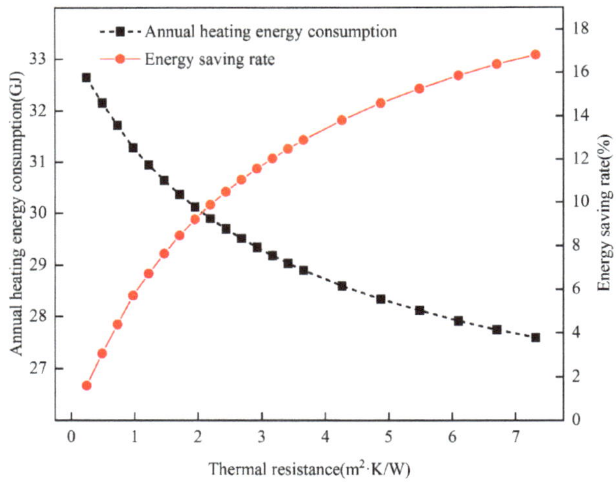

Figure 12. Influence of the thermal resistance of exterior walls on building heating energy consumption.

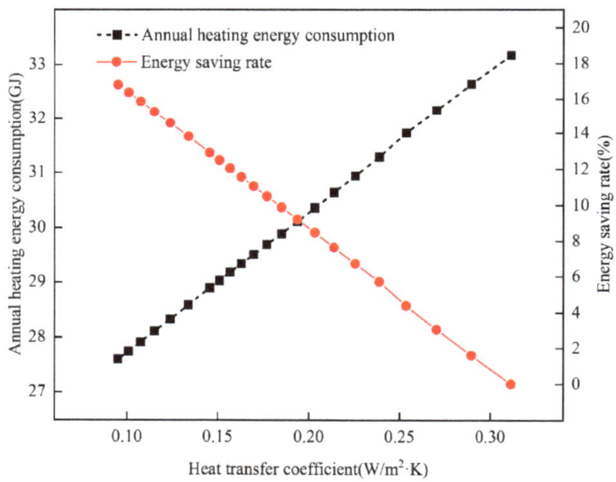

Figure 13. Influence of heat transfer coefficients of exterior walls on building heating energy consumption.

(2) Influence of Roof Heat Transfer Coefficient on Building Heating Energy Consumption

In this section, only the roof heat transfer coefficient was changed, and the values of the building's original scheme for the other parameters were adopted. Changing the heat transfer coefficient of the roof was achieved by increasing the thermal resistance value of the roof. The effect of increasing the thermal resistance of the roof on the building heating energy consumption is shown in Figure 14, and the relationship between the heat transfer coefficient of the roof—which corresponds to increasing the thermal resistance of the roof—and the building heating energy consumption is shown in Figure 15.

The annual heating energy consumption of the building decreases gradually as the value of the roof thermal resistance increases from 0.2439 $m^2 \cdot K/W$ to 7.317 $m^2 \cdot K/W$, and the corresponding heat transfer coefficient of the exterior wall decreases from 0.2277 $W/(m^2 \cdot K)$ to 0.0887 $W/(m^2 \cdot K)$. The annual heating energy consumption of the building is linearly related to the roof heat transfer coefficient, and the annual heating energy consumption decreases linearly with the decrease in roof heat transfer coefficient. The energy saving rate is also linearly related to the roof heat transfer coefficient, and the energy saving rate increases linearly with the decrease in roof heat transfer coefficient. From the point of view of increasing the thermal resistance of the roof, increasing the thermal resistance causes the annual heating energy consumption of the building to decrease gradually, but the decreasing trend gradually becomes slower. The energy saving rate increases with the increase in thermal resistance, but the increasing trend also gradually becomes slower. When 110 mm thick rock wool boards are added to the roof of the building as an insulation layer, the energy saving rate increases by less than 0.2% compared to a 100 mm thick insulation layer. The simulation results show that, after adding a 100 mm thick layer of insulation to the building roof, further increase in the thickness of the insulation produces little extra energy saving.

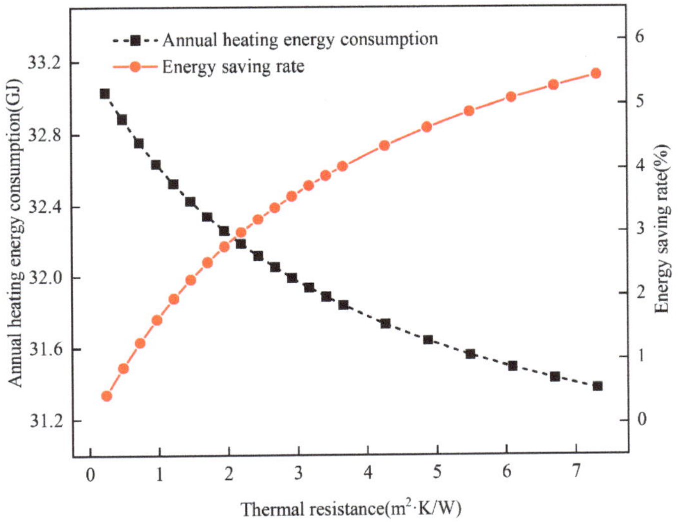

Figure 14. Influence of roof thermal resistance on building heating energy consumption.

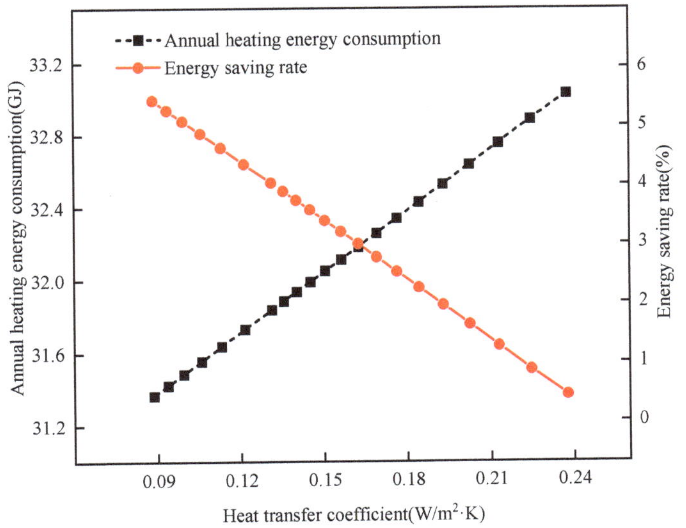

Figure 15. Influence of roof heat transfer coefficient on building heating energy consumption.

(3) Influence of Exterior Window Solar Heat Gain Coefficient on Building Heating Energy Consumption

In this section, only the solar heat gain coefficient of the exterior windows was changed, and the values of other parameters were adopted from the original scheme of the building to simulate the energy consumption for the whole year. The effect of increasing the solar heat gain coefficient of the exterior windows on the heating energy consumption of the building is shown in Figure 16.

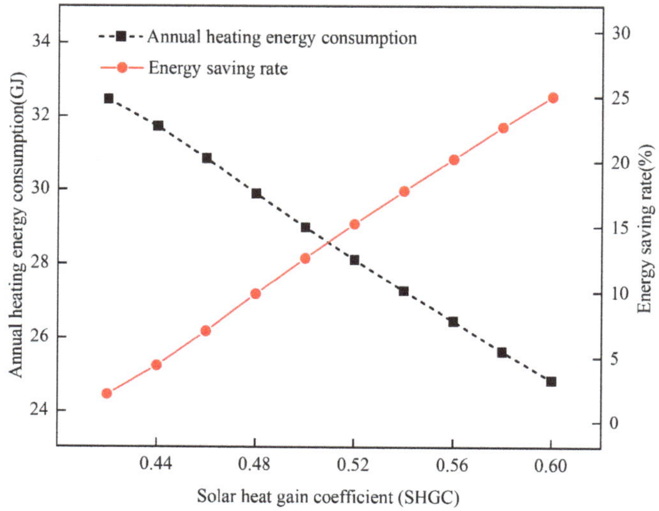

Figure 16. Influence of exterior window solar heat gain coefficients on building heating energy consumption.

As the solar heat gain coefficient of the exterior window increases from 0.42 to 0.6, the amount of indoor heat gain from solar radiation through the window gradually increases, and the annual heating energy consumption of the building decreases linearly. With the increase of each 0.02 in the solar heat gain coefficient of the exterior window, the annual heating energy consumption of the building decreases by 0.7143, 0.7407, 0.8726, 0.9451, 0.9107, 0.8727, 0.8395, 0.8150, 0.8217, and 0.7780 GJ, respectively, and the annual heating energy consumption of the building and the solar heat gain coefficient of the exterior window basically shows a linear relationship. The energy saving rate also increases linearly with the increase in solar heat gain coefficient of exterior windows. For every 0.02 increase in the solar heat gain coefficient of exterior windows, the energy saving rate increases by 2.23%, 2.63%, 2.85%, 2.75%, 2.63%, 2.53%, 2.46%, 2.48%, and 2.35%, respectively, and the energy saving rate is basically linearly related to the solar heat gain coefficient of exterior windows. The simulation results show that the use of building exterior windows with higher solar heat gain coefficients can effectively reduce building heating energy consumption.

5. Conclusions

This paper reports the energy monitoring and analysis of a typical residential house situated in the Haituo Valley of northern Beijing, a region with sub-zero temperatures in winter and mild temperatures in summer. Based on this study, the following conclusions can be drawn:

1. The monitoring results showed that the indoor temperature on the south side of the building was higher than that on the north side, while the temperature on the southwest side was significantly higher than that on the southeast side, and the temperature difference between the northwest and northeast sides was small. This is

related to the orientation of the building, the layout of the rooms, the position of the doors and windows, and the special structure of the ground floor.

2. A 3D model of the building was constructed using EnergyPlus, and the building structural information, room temperature control, and other parameters were consistent with the actual situation. Comparing simulation results with actual energy consumption data, the error between the two values is around 10%, proving the accuracy of the EnergyPlus model.
3. The results from the EnergyPlus model are more accurate when the outdoor temperature is based on the monitoring data, compared with those when the outdoor temperature is based on the typical meteorological year file.
4. The EnergyPlus model was further employed to analyze the impact of enhancing the thermal resistance of the exterior envelope on the building's energy efficiency. The results show that, after adding a certain thickness of thermal insulation to the building's wall and roof, the energy saving effect produced by further increasing the thickness of the thermal insulation layer is limited.
5. By analyzing the influence of exterior window solar heat gain coefficient on the building's energy efficiency, it can be concluded that the use of building exterior windows with a higher solar heat gain coefficient can effectively reduce building energy consumption.

This study illustrates the effectiveness of using EnergyPlus in assessing building energy performance and highlights the significance of window design in energy conservation efforts. These findings offer valuable insights for optimizing building designs and promoting energy-efficient practices in construction.

6. Limitations and Future Work

According to the energy monitoring and analysis of the residential house in China, the following suggestions and recommendations for future research are made:

1. Residential energy is a significant source of global growth in energy demand and carbon emissions. The location characteristics, housing characteristics, and unit size of a home all affect the energy use of residential buildings, which still needs to be explored.
2. Some parameters, such as external window heat transfer coefficient, window-to-wall ratio, and humidity, also impact building energy consumption, but are not included in this study, which can be considered in future research.
3. The specified air exchange value, which refers to the rate at which indoor air is replaced by outdoor air, is directly related to the thermal environment and energy consumption of a building. Appropriate air exchange rate helps maintain indoor air quality and provides a comfortable living and working environment. An excessively high air exchange rate can lead to significant energy loss through the ventilation system, thereby increasing the building's energy consumption. Therefore, it is worth studying the impact of different air exchange rates on the energy consumption of buildings in future research.

Author Contributions: Conceptualization, J.S., Z.C. and A.C.; Data curation, S.H. and Q.Z.; formal analysis, Y.W. and Q.Z.; funding acquisition, J.S. and A.C.; investigation, Y.W. and Q.Z.; methodology, Y.W. and Q.Z.; project administration, J.S. and A.C.; resources, J.S. and A.C.; software, Y.W., S.H. and Q.Z.; supervision, J.S., Z.C. and A.C.; validation, Y.W., S.H. and Q.Z.; writing—original draft, Y.W., S.H. and Q.Z.; writing—review & editing, J.S., Z.C. and A.C. All authors have read and agreed to the published version of the manuscript.

Funding: This work is supported by the National Natural Science Foundation of China (grant Nos. 52078032, 52078037).

Data Availability Statement: The data presented in this study are available on request to the corresponding author.

Conflicts of Interest: The authors declare no conflict of interest.

References

1. International Energy Agency. *Energy Efficiency: Buildings. The Global Exchange for Energy Efficiency Policies, Data and Analysis*; International Energy Agency: Paris, France, 2022.
2. Du, X.; Wen, L.; Wei, P.; Yang, M. A systematic approach for analyzing building energy conservation and emission reduction policies based on the principle of WSR. *Energy Build.* **2024**, *315*, 114328. [CrossRef]
3. Cai, W. *China Building Energy Consumption Annual Report 2020*; China Association of Building Energy Efficiency: Xiamen, China, 2020.
4. Vartholomaios, A. A parametric sensitivity analysis of the influence of urban form on domestic energy consumption for heating and cooling in a Mediterranean city. *Sustain. Cities Soc.* **2017**, *28*, 135–145. [CrossRef]
5. Al-Saggaf, A.; Nasir, H.; Taha, M. Quantitative approach for evaluating the building design features impact on cooling energy consumption in hot climates. *Energy Build.* **2020**, *211*, 109802. [CrossRef]
6. Wang, R. *Simulation and Prediction of Typical Building Energy Consumption in Beijing*; Xi'an University of Architecture and Technology: Xi'an, China, 2018.
7. Santana, B.O.; Torres-Quezada, J.; Coch, H.; Isalgue, A. Monitoring and Calculation Study in Mediterranean Residential Spaces: Thermal Performance Comparison for the Winter Season. *Buildings.* **2022**, *12*, 325. [CrossRef]
8. Charai, M.; Mezrhab, A.; Moga, L. A structural wall incorporating biosourced earth for summer thermal comfort improvement: Hygrothermal characterization and building simulation using calibrated PMV-PPD Model. *Build. Environ.* **2022**, *212*, 108842. [CrossRef]
9. Shan, Q.; Tong, K.; Zhang, X.; Li, Y. Field Test and Simulation Analysis of Thermal Performance of Bamboo Steel Composite Wall in Different Climate Regions. *Adv. Civ. Eng.* **2020**, *2020*, 1–10. [CrossRef]
10. Yu, D.; Tan, H.; Ruan, Y. A future bamboo-structure residential building prototype in China: Life cycle assessment of energy use and carbon emission. *Energy Build.* **2011**, *43*, 2638–2646. [CrossRef]
11. Seddik Hassan, A.M.; Abd El Aal, R.F.A.; Fahmi, A.A.E.; Ali, S.M.A.; Abdelhady, M.I. Courtyard geometry's effect on energy consumption of AlKharga city residential buildings, Egypt. *Sci. Rep.* **2024**, *14*, 11149. [CrossRef]
12. Cabeza, L.F.; Castell, A.; Medrano, M.; Martorell, I.; Pérez, G.; Fernández, I. Experimental study on the performance of insulation materials in Mediterranean construction. *Energy Build.* **2010**, *42*, 630–636. [CrossRef]
13. Tunçbilek, E.; Arıcı, M.; Krajčík, M.; Li, Y.; Jurčević, M.; Nižetić, S. Impact of nano-enhanced phase change material on thermal performance of building envelope and energy consumption. *Int. J. Energy Res.* **2022**, *46*, 19313–20900. [CrossRef]
14. Alyami, S.H.; Alqahtany, A.; Ashraf, N.; Osman, A.; Aldossary, N.A.; Almutlaqa, A.; Al-Maziad, F.; Alshammari, M.S.; Al-Gehlani, W.A.G. Impact of Location and Insulation Material on Energy Performance of Residential Buildings as per Saudi Building Code (SBC) 601/602 in Saudi Arabia. *Materials.* **2022**, *15*, 9079. [CrossRef] [PubMed]
15. Guo, L.; Liao, Y.; Cheng, Z.; Zheng, H.; Guo, L.; Long, E. Experimental study on dynamic effect of exterior insulation on indoor thermal environment and energy consumption. *Energy Build.* **2022**, *274*, 112299. [CrossRef]
16. Amani, N.; Kiaee, E. Developing a two-criteria framework to rank thermal insulation materials in nearly zero energy buildings using multi-objective optimization approach. *J. Clean. Prod.* **2020**, *276*, 122592. [CrossRef]
17. Shibuya, T.; Croxford, B. The effect of climate change on office building energy consumption in Japan. *Energy Build.* **2016**, *117*, 149–159. [CrossRef]
18. Li, D.H.W.; Lam, J.C.; Wong, S.L. Daylighting and its effects on peak load determination. *Energy.* **2005**, *30*, 1817–1831. [CrossRef]
19. Dong, H.R.; Liu, J.P. Analysis of Thermal Performance and Energy Consumption of Residential Envelope Structures in Zhangjiakou City. *Heat. Vent. Air Cond.* **2003**, *33*, 3.
20. Han, J.; Lu, X.; Gu, Y.; Liao, W.; Cai, Q.; Xue, H. Optimized data representation and understanding method for the intelligent design of shear wall structures. *Eng. Struct.* **2024**, *315*, 118500. [CrossRef]
21. National Bureau of Statistics of China. *China Statistical Yearbook*; China Statistics Press: Beijing, China, 2020.
22. Sowa, J.; Mijakowski, M. Humidity-Sensitive, Demand-Controlled Ventilation Applied to Multiunit Residential Building—Performance and Energy Consumption in Dfb Continental Climate. *Energies* **2020**, *13*, 6669. [CrossRef]
23. Lou, R.; Hallinan, K.P.; Huang, K.; Reissman, T. Smart Wifi Thermostat-Enabled Thermal Comfort Control in Residences. *Sustainability.* **2020**, *12*, 1919. [CrossRef]
24. Wang, Y.; Liu, K.; Liu, Y.; Wang, D.; Liu, J. The impact of temperature and relative humidity dependent thermal conductivity of insulation materials on heat transfer through the building envelope. *J. Build. Eng.* **2022**, *46*, 103700. [CrossRef]
25. *JGJ/T 449-2018*; Standard for green performance calculation of civil buildings. China building industry press: Beijing, China, 2018.
26. Efficiency Valuation Organization. *International Performance Measurement & Verification Protocol: Concepts and Options for Determining Energy and Water Savings*; Technical Report; Efficiency Valuation Organization: Washington, DC, USA, 2012.
27. American Society of Heating, Refrigerating, and Air-Conditioning Engineers. *ASHRAE Guideline 14-2014: Measurement of Energy, Demand, and Water Savings*; Technical Report; American Society of Heating; Refrigerating, and Air-Conditioning Engineers: Atlanta, GA, USA, 2014.
28. Pachano, J.E.; Bandera, C.F. Multi-step building energy model calibration process based on measured data. *Energy Build.* **2021**, *252*, 111380. [CrossRef]

29. Sun, J.; Zhang, Q.; Chen, A.; Wu, Z. Energy performance of buildings with multi-ribbed composite wall structural system. *J. Build. Eng.* **2024**, *91*, 109614. [CrossRef]
30. Manz, H.; Frank, T. Thermal simulation of buildings with double-skin façades. *Energy Build.* **2005**, *37*, 1114–1121. [CrossRef]
31. Omrany, H.; Ghaffarianhoseini, A.; Ghaffarianhoseini, A.; Raahemifar, K.; Tookey, J. Application of passive wall systems for improving the energy efficiency in buildings: A comprehensive review. Renew. *Sustain. Energy Rev.* **2016**, *62*, 1252–1269. [CrossRef]
32. Ginevičius, R.; Podvezko, V.; Raslanas, S. Evaluating the alternative solutions of wall insulation by multicriteria methods/Pastatų sienų šiltinimo variantų vertinimas taikant daugiakriterius metodus. *J. Civ. Eng. Manag.* **2008**, *14*, 217–226. [CrossRef]
33. Zhu, J.; Chew, D.A.; Lv, S.; Wu, W. Optimization method for building envelope design to minimize carbon emissions of building operational energy consumption using orthogonal experimental design (OED). *Habitat Int.* **2013**, *37*, 148–154. [CrossRef]
34. Maučec, D.; Premrov, M.; Leskovar, V.Ž. Use of sensitivity analysis for a determination of dominant design parameters affecting energy efficiency of timber buildings in different climates. *Energy Sustain. Dev.* **2021**, *63*, 86–102.
35. Liao, W. *Research on Energy Saving of Building Envelope Structure in Nanjing Area Based on DeST Simulation*; Anhui University of Science and Technology: Huainan, China, 2018.
36. Li, H. *Research on Energy Simulation and Energy Saving of an Office Building in Hot Summer and Cold Winter Area*; Harbin Institute of Technology: Harbin, China, 2020.

buildings

MDPI

Article

Operational Strategy of a DC Inverter Heat Pump System Considering PV Power Fluctuation and Demand-Side Load Characteristics

Yilin Li [1,2,*], Yang Lu [1], Jie Sun [3], Tianhang Wang [1], Shiji Zong [1], Tongyu Zhou [4] and Xin Wang [1]

1 School of Environment and Architecture, University of Shanghai for Science and Technology, Shanghai 200093, China; 222271953@st.usst.edu.cn (S.Z.); wangxinshiyun@126.com (X.W.)
2 Key Laboratory of Ecology and Energy Saving Study of Dense Habitat, Ministry of Education, Shanghai 200092, China
3 BSE Energy Technology (Suzhou) Co., Ltd., Suzhou 215000, China; irene_sun@126.com
4 Department of Architecture and Built Environment, University of Nottingham Ningbo China, Ningbo 315100, China; tongyu.zhou@nottingham.edu.cn
* Correspondence: yilin.li@usst.edu.cn

Abstract: With the increase in application of solar PV systems, it is of great significance to develop and investigate direct current (DC)-powered equipment in buildings with flexible operational strategies. A promising piece of building equipment integrated in PV-powered buildings, DC inverter heat pump systems often operate with strategies either focused on the power supply side or on the building demand side. In this regard, the aim of this study was to investigate the operational strategy of a DC inverter heat pump system for application in an office building with a PV power system. Firstly, the PV power fluctuation and demand-side load characteristics were analyzed. Then, a series of heat transfer and heat pump system models were developed. A reference building model was developed for simulating the performance of the system. A control logic of the DC inverter heat pump was proposed with a certain level of flexibility and capability considering both the characteristics of the PV power generation and the demand-side heating load. MATLAB/Simulink 2021 software was used for simulation. The simulation results show that the DC inverter heat pump is able to regulate its own power according to the change signal of the bus voltage such that the DC distribution network can achieve power balance and thus provide enough energy for a room. This study can provide a reference for developing flexible operational strategies for DC inverter heat pump systems. The proposed strategy can also help to improve the systems' performance when they are applied in buildings with distributed PV systems.

Keywords: operational strategy; DC inverter heat pump; PV power; demand side

Citation: Li, Y.; Lu, Y.; Sun, J.; Wang, T.; Zong, S.; Zhou, T.; Wang, X. Operational Strategy of a DC Inverter Heat Pump System Considering PV Power Fluctuation and Demand-Side Load Characteristics. *Buildings* **2024**, *14*, 1139. https://doi.org/10.3390/buildings14041139

Academic Editor: Zhenjun Ma

Received: 22 March 2024
Revised: 7 April 2024
Accepted: 12 April 2024
Published: 18 April 2024

1. Introduction

Global climate change and increasing demand for energy have led to a dramatic transformation from conventional energy to renewable energy systems [1–3]. Over the last 20 years, solar photovoltaics (PVs) has witnessed rapid growth as the most widely applied form of renewable energy source (RES) [4,5]. Since the building sector consumes nearly 40% of the total energy consumption and generates over 30% [6] of the carbon dioxide emissions around the world, RESs integrated with buildings have gained prominence for their capacity to mitigate building energy consumption and reduce carbon dioxide emissions. Among various RES applications, building integrated photovoltaic (BIPV) systems have become more prevalent in new building and retrofit projects as their energy conversion rate has increased and the cost of the technology dramatically reduced [7–10]. BIPV refers to photovoltaic modules and systems that can replace conventional building components [11], which is an efficient way of generating renewable energy on-site while meeting architectural requirements [12–14]. However, there are still some problems with BIPV systems from both the power supply side and demand side, which restricts their application.

The main problem associated with the power supply quality is the intermittence and stochastic character of PV systems, which may result in significant fluctuations in the generated power [15]. Except for long-period PV output variations due to seasonal impacts, unpleasant weather and cloud transience induce short-term variability and considerable PV output fluctuations in power supply systems. Previous researchers found that passing clouds resulted in fluctuations in 45–90% of the rated PV capacity per minute in a PV plant in Portugal [16]. The electrical energy provided by the power system responds instantaneously, transmits quickly, has no time lag or loss, and has little inertia [17]. Therefore, it is crucial to balance power supply and consumer demand and to ensure reliable power delivery to systems and equipment in buildings. Fluctuating PV power requires load-side equipment in buildings to meet flexibility demands on multiple time scales. On the other hand, solar photovoltaic systems generate direct-current (DC) distributed power supply, which triggers a significant trend towards DC power equipment and DC loads in buildings, such as LED lamps, IT equipment, and motor loads, prompting development in inverter drives [18]. Although the role of the demand side is increasingly important in PV-powered buildings, previous research mostly focused on the power grid side, such as DC microgrid architecture [19], passive solar technology [20], multi-objective optimization of dynamic voltage [21], numerical analysis [22], segmentation of photovoltaic panels [23], and comparative analysis of control strategies and current status of development [24,25]. Although research has been conducted on control strategies of DC light flicker [26] to respond to PV power output fluctuations, there is comparably little research on DC-powered heating, ventilation and air-conditioning (HVAC) systems.

HVAC systems, one of the largest energy consumers in buildings, are considered suitable equipment to respond to the fluctuations in PV output. From the perspective of demand-side load, thermal energy generated by the heat supply system is transmitted slowly and with high inertia, with some delay and heat loss [27]. As the amount of heat supply from HVAC systems changes, the room temperature can still remain stable and within a comfortable level for a long period due to the thermal inertia of the building. Among the various pieces of HVAC equipment, heat pumps have been regarded as efficient mechanical devices that produce heating energy from renewable sources such as solar energy, ambient air, geothermal energy or waste heat [28,29]. Wang et al. [30] developed a new solar–air-source heat pump building energy supply system with energy storage that maintains efficient operation during extreme weather in winter and summer. Roccatello et al. [31] analyzed the primary energy consumption of a heat pump with various control strategies, and found that the most efficient control strategies were maximizing the utilization of heat pumps. Xia L et al. [32] presented a model-based adaptive control strategy for a ground source heat pump with integrated solar photovoltaic collectors, which can improve the energy efficiency of the integrated system. Wei et al. [33] developed a matching control strategy to improve the system performance of a CO_2 heat pump, and found that the overall COP of the system could be increased by using a control method: simultaneous optimal electronic expansion valve opening and regulation frequency. Lyu et al. [34] built a model of an energy storage heating system using TRNSYS and investigated the influence of the size of the air source heat pump on the energy-saving potential of the energy storage heating system. From previous research, it can be seen that the operational strategy of the heat pump system has a large impact on indoor comfort as well as on the building's energy consumption. There is a lack of research focused on the control strategies of heat pumps considering the power supply-side characteristics, such as PV power fluctuations.

In recent years, DC inverter heat pumps have attracted increasing attention from researchers. Operational strategies of the DC inverter heat pump system considering PV power output fluctuations and demand-side characteristics have been addressed in multiple studies, with inconclusive results. Wang et al. [35] proposed an injection-assisted air source heat pump (ASHP) to improve heating performance in cold areas under various temperature conditions. They conducted an experimental study and proved that the system performed well over a wide range of temperature conditions. Zhang et al. [36] proposed a PV/T solar-assisted heat pump water heating system driven by a DC inverter compressor and evaluated its characteristics and operational performance under different operating conditions in Lanzhou. They found that the low speed of the compressor favors the continuity of the refrigerant circulation and improves the performance of the system. Despite the fact that DC inverter heat pumps have become a research focus, there are few studies on their operational strategies in terms of the characteristics of PV power characteristic and demand-side responses.

To address the above problems, this paper presents an operational strategy for a DC inverter heat pump that considers both PV power fluctuations and demand-side characteristics. The characteristics of PV power fluctuation and demand-side load for a case study building were firstly analyzed. A theoretical model of the system was then developed, including a heat pump model, an ice storage unit model, and a room model. The DC heat pump system was simulated by using MATLAB/Simulink software under both cooling and heating conditions. The operational strategy of the system takes into account the DC bus voltage of PV output and room temperature variations. This study, to some extent, can provide a reference for developing suitable operational strategies for DC inverter heat pump systems. The contribution of this study advances the field of DC-powered HVAC systems by providing a new control method to enhance the performance of heat pump systems, with potential applications in both residential and commercial settings.

2. Analysis of Power and Load Characteristics

The power generation of a PV array is affected by many factors, such as solar radiation, permanent shading, surface temperature of the panel, and the panel's tilt angle. The energy consumption of buildings varies depending on the weather conditions, building type, indoor occupants, and operation of equipment. Therefore, the development of an operational strategy for an HVAC system should take into account the characteristics of PV output and energy consumption in buildings. In order to analyze the power and load characteristics of a BIPV system, a typical PV-integrated building was selected as a case study. The building has three floors with a total air-conditioned area of 7000 m^2.

The system architecture is demonstrated in Figure 1. On the power side, a 461.54 kW solar PV system is installed on the roof of the building with a tilt angle of 15° facing the south. It can directly supply DC power and AC power to equipment and systems in the building. The utility grid is integrated in the system as a supplement. On the load side, the HVAC for the offices and workshops is provided by a set of fan coil units and a fresh air system fed by magnetic bearing centrifugal chillers and air-to-water heat pumps. The chillers and fan coil units consume electricity directly from the DC bus, while the water pumps and fresh air systems use AC electricity that is converted from DC power. The room temperature setpoint for the heating season is 18 °C. These systems and equipment are scheduled to operate from 6:00 A.M. to 20:00 P.M. The goal of the system is to take full advantage of the PV-generated DC power using the HVAC system while maintaining stable and reliable operation.

Figure 2 presents the photovoltaic generation and load data obtained in July 2023for each piece of building equipment. It can be seen that the energy consumption on weekdays is obviously higher than at the weekends, and the PV power generation is mainly influenced by meteorological factors. The energy consumption of different systems and equipment (HVAC, lighting, and other equipment) are comparably steady, while the PV power generation profile fluctuates due to the unpleasant weather. The daily average energy consumption for HVAC and lighting are 1102.85 kWh and 583.22 kWh, respectively, during the investigated week. The daily average electricity generated by the PV system is 2123.43 kWh, which is adequate for the demand side when the weather is fine (expect for Tuesday). The mismatch values between PV generation and load-side electricity consumption are shown in Figure 3. On weekdays when the weather is fine, there is almost no gap between the supply and demand. The mismatch between the PV generation and user electricity consumption is on average 1.25%, and the shortage can be made up by the utility grid. During weekends when the solar radiation is sufficient, the PV system can generate more electricity than the demand, which can be stored by energy storage systems. It should be noticed that on Tuesdays, the difference between supply and demand can be as high as −188.90% due to the bad weather.

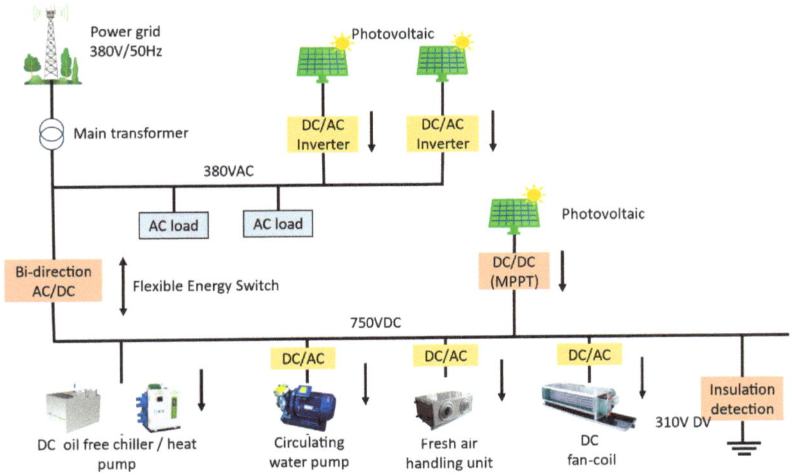

Figure 1. Schematic overview of the system architecture for the case study.

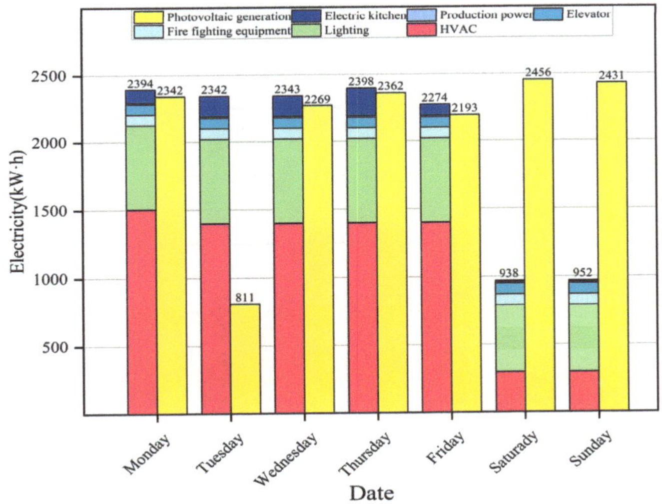

Figure 2. The weekly electricity statistics of the case study building.

Figure 4 shows the power supply profiles of the building on a weekday during the cooling season. It can be seen that there is an obvious rise in electricity provided by the grid at 6:00 A.M. Due to the weak solar radiation, the PV system cannot generate adequate electricity, and thus the grid needs to make up for it and supply enough power for the crowds flooding into the building during morning rush hours. The leading role of the grid lasts about one hour until nearly 7:00 A.M., when the PV system can provide enough electricity. During 7:00 A.M. to 16:00 P.M., the electricity consumed by the building is mostly provided by the PV system. The trend in the generated electricity profile is similar to the variation in daily solar radiation, and the peak value occurs at 11:30 A.M. During this PV-dominant period, there are some fluctuations in PV-generated electricity despite the electricity supply to the building remaining steady. The PV- and grid-generated electricity profiles are almost symmetrical, and together meet the building's demand-side electricity with a stable value of 180 kW. The PV power generation is higher than the grid supply power by about 8.23%, which indicates that the PV power is the main electricity source of the building. After 16:00 P.M., the grid power supply gradually takes the lead due to the decrease in solar radiation. The systems and equipment in the building are turned off at 20:00 P.M. and consume little energy afterwards.

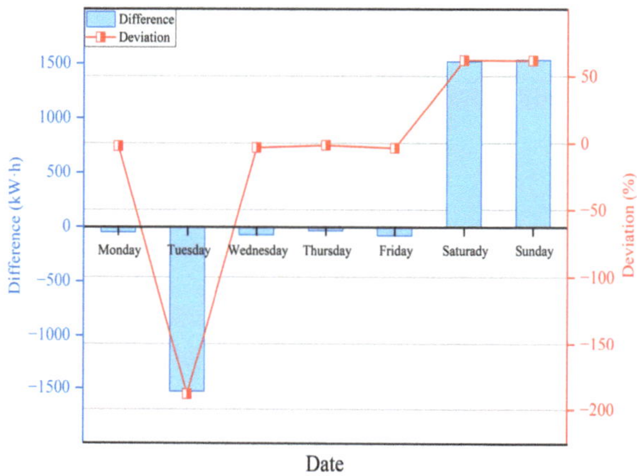

Figure 3. The weekly mismatch values of the supply and demand electricity.

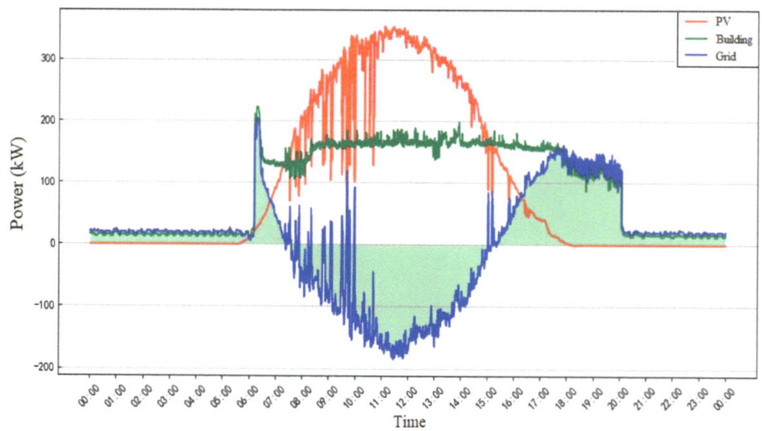

Figure 4. Profiles of power supply on a typical day.

Figure 5 shows the daily sub-metering PV electricity generation and building electricity consumption profiles. In general, the PV power generation profile exhibits a quadratic trend, but experiences significant fluctuations from 7:00 A.M. to 11:00 A.M. due to the varied weather conditions. The maximum difference between the maximum and minimum PV power values reaches 260 kW, which poses a significant challenge to the safety and stability of the equipment in the building. In terms of electricity consumption, apart from the HVAC system, the trends in electricity consumption of other equipment and systems remain relatively constant. The mismatch between the supply and demand sides is obvious, including the four stages, as indicated below.

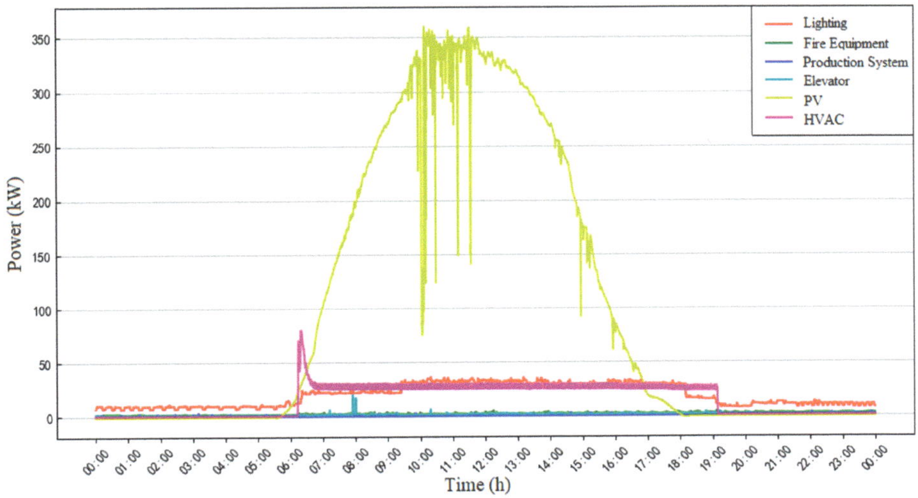

Figure 5. Profiles of building consumed electricity on a typical day.

During 6:00–6:30 A.M., most of the equipment in the building is turned on, and the HVAC system starts to work at its maximum capacity. In the meanwhile, the building's cooling load increases significantly due to the influx of occupants. However, the photovoltaic system cannot generate adequate electricity since the solar radiation is not strong enough in the early morning. This mismatch between the extensive energy demand and low PV power generation needs to be optimized. During 10:00 A.M.–14:00 P.M., the amount of PV power generation peaks. The interference rate of PV power generation can reach 71.43%, which may result in a PV power abandonment problem. Between 16:00 P.M. and 20:00 P.M., the solar radiation gradually reduces, resulting in a significant difference between the electricity demand of the HVAC system and the actual PV power generation. Therefore, the power grid needs to supply the electricity for the demand side, which may lead to fluctuation problems. After 20:00 P.M., most of the equipment in the building is turned off. Although there are minimal thermal comfort requirements, the building still requires a certain amount of electricity to maintain its basic functionality for lighting, fire safety, and elevator equipment.

The above analysis of PV power generation and demand-side energy consumption provides references for setting up the input parameters.

3. System Design and Modeling

A reference building situated in Shanghai serves as the building model, and its standard floor layout is depicted in Figure 6. The building has three floors above the ground and one floor underground for a total area of 2026 m^2, accommodating functionality of office spaces and conference rooms. In order to mitigate the mismatch problems between supply and demand, the system demonstrated in Figure 7 was proposed. Rooftop PV arrays and an energy storage water tank are utilized in the building. The PV system, batteries, and chargers are connected to the DC bus. A DC inverter heat pump is used with the compressor, which is able to adjust its frequency and power according to the power supply voltage. The operation time for the HVAC system is 8:00 A.M.–18:00 P.M. The building model was developed in SketchUp and its envelope structure was then imported into EnergyPlus (8.9.0). Simulations were conducted to assess the heating load in winter.

The input weather data adopted the typical meteorological year data of Shanghai. The indoor and outdoor environmental parameters were set up as shown in Table 1.

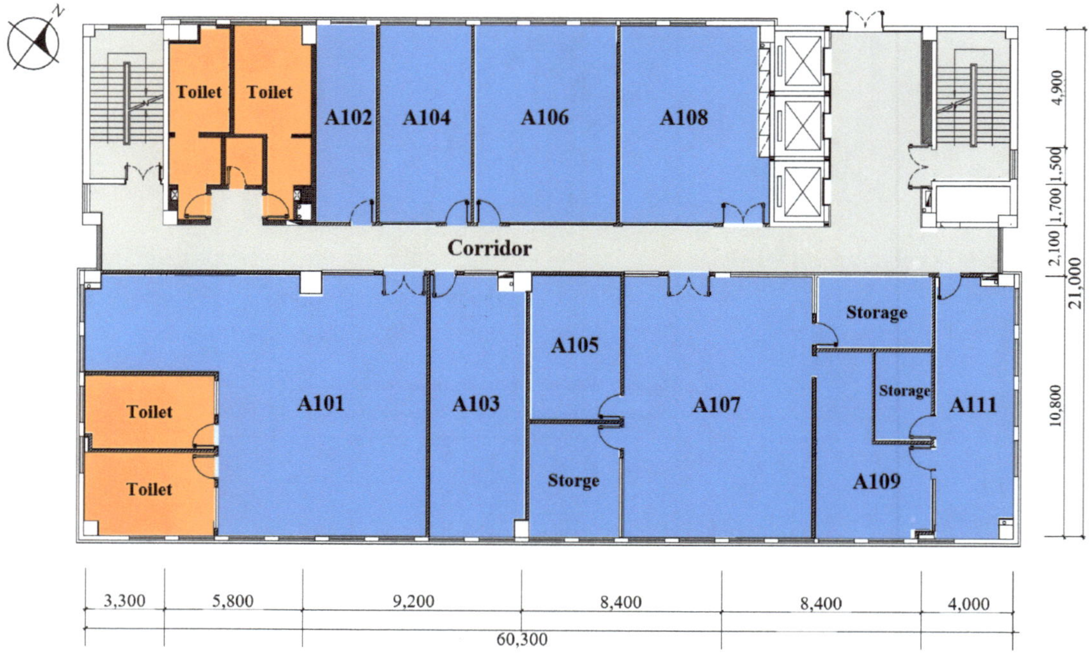

Figure 6. The standard floor layout of the reference building.

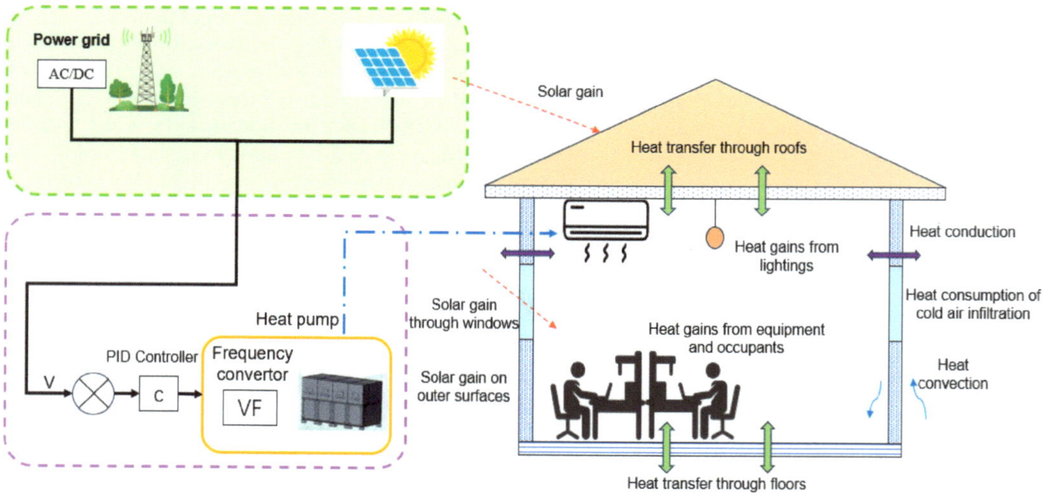

Figure 7. Schematic diagram of the proposed system.

Table 1. Building envelopes and building information.

Building Component	Value
External wall	W/(m^2·K)
Interior wall	0.617 W/(m^2·K)
Roof	0.408 W/(m^2·K)
Floor	5.24 W/(m^2·K)
Window WWR	0.30 (South and north facades) 0.35 (East and west facades)
Occupancy gain	0.1 person/m^2
Air circulation	30 m^3/person
Lighting intensity	9 W/m^2
Equipment load	15 W/m^2

3.1. Heat Transfer Model of the Building

As shown in Figure 8, the internal heat transfer process in the building is an unsteady process involving conduction, convection, and radiation. The ambient temperature and indoor air temperature are expressed as the lumped temperature of a node. The following assumptions were made before the heat transfer model was established.

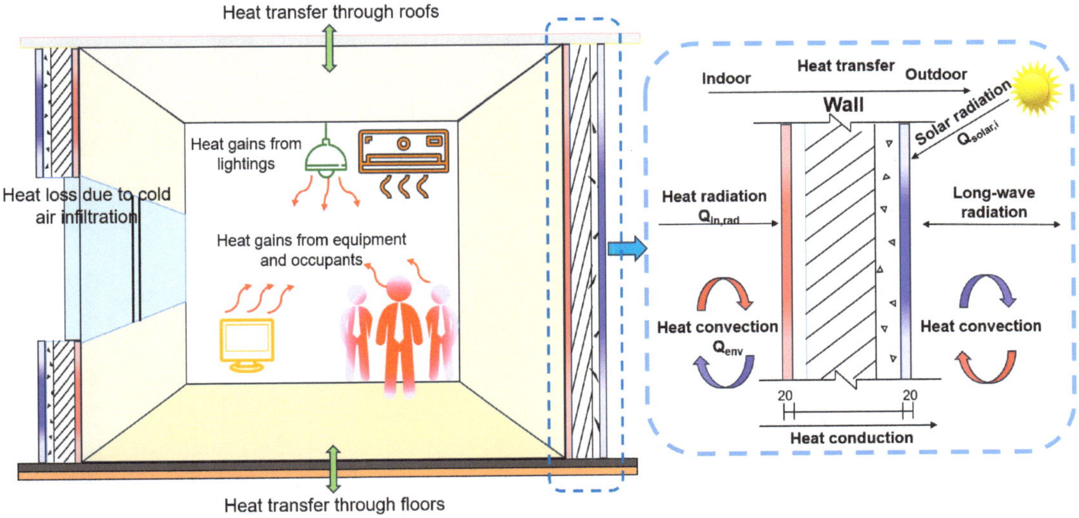

Figure 8. Heat transfer analysis of the room.

- The heat transfer is in a one-dimensional direction.
- The indoor air is incompressible and homogeneous, and the influence of the indoor temperature on the air density is neglected.
- The air infiltration in the building is neglected.
- The external heat absorbed by the building can be absorbed uniformly by the building envelope without a reflection effect.
- The thermal storage and release from furniture and walls are neglected.

According to the law of conservation of energy, the heat transfer model of an air-conditioned room can be expressed as Equation (1) [37]:

$$\frac{dQ}{dt} = (T_{heater} - T_{room}) \cdot M_{dot} \cdot c \tag{1}$$

where $\frac{dQ}{dt}$ represents the heat flow from the heater into the room; c is the constant specific heat capacity of air, J/(kg·k); M_{dot} is the hot air from heater, kg/h; T_{heater} is the heater air supply temperature, °C; and T_{room} is the real-time room temperature, °C:

$$\left(\frac{dQ}{dt}\right)_{losses} = \frac{T_{room} - T_{out}}{R_{eq}} \tag{2}$$

$$\frac{dT_{room}}{dt} = \frac{1}{M_{air} \cdot c} \cdot \left(\frac{dQ_{heater}}{dt} - \frac{dQ_{losses}}{dt}\right) \tag{3}$$

where M_{air} is the room air quality, kg, and R_{eq} is the room equivalent thermal resistance, K/W.

3.2. DC Heat Pump Modeling

The DC inverter heat pump consists of a compressor, condenser, expansion valve, and evaporator. The heat transfer process of each component of the heat pump is one-dimensional, steady state, and insulated. As demonstrated in Figure 9, the DC power supply is directly connected to the DC inverter heat pump. The power of the compressor is regulated by voltage fluctuations within a range of 300 V to 800 V.

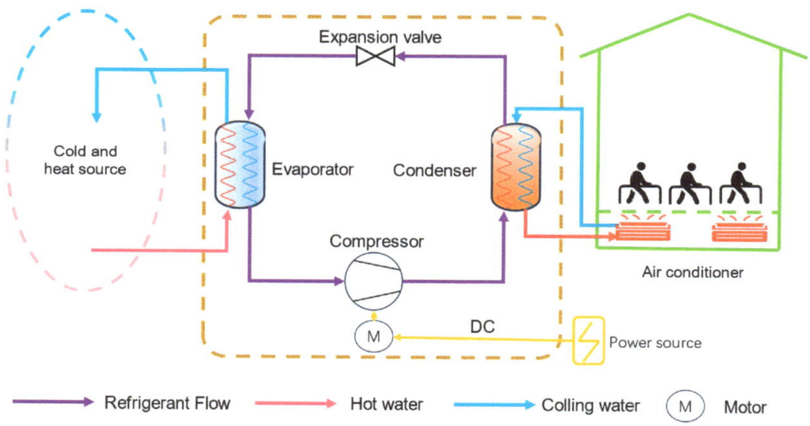

Figure 9. Scheme of the DC inverter heat pump system.

The key to modeling the compressor is to establish the relationship between refrigerant flow rate and frequency of the compressor, which can be described by a fitting curve provided by the manufacturer, derived as shown in Equations (4)–(6) [38]. As expressed in Equations (7)–(10), the water-side heat exchange Q_w of the condenser is equal to the heat exchange Q_{rc} of the refrigerant side. The expansion valve regulates and reduces the pressure of the high-temperature refrigerant after it is condensed in the condenser. The mass flow rate of refrigerant through the expansion valve can be expressed by Equation (11) [39]:

$$W_{com} = m_{com}(h_{dis} - h_{suc}) \tag{4}$$

$$h_{dis} = \frac{h_{dis,is} - h_{suc}}{\eta_{is}} + h_{suc} \tag{5}$$

$$\dot{m}_{com} = 13.59 f_{com} + 53.52 \tag{6}$$

where m_{com} is the refrigerant flow, kg/s; W_{com} is the compressor heating capacity, kW; h_{dis} is the specific enthalpy of the refrigerant at the compressor outlet; h_{suc} is the specific enthalpy of the refrigerant at the suction port of the compressor, kJ/kg; $h_{dis,is}$ is the ideal specific enthalpy of the refrigerant at the compressor discharge port under isentropic compression conditions, kJ/kg; η_{is} is the entropic efficiency of the compressor; and f_{com} is the compressor frequency, Hz.

$$Q_w = c_w m_w (t_{cw,in} - t_{cw,out}) \tag{7}$$

where Q_w represents the water-side heat exchange, kW; c_w is the specific heat capacity of water at constant pressure, J/(kg·°C); m_w is the water mass flow, kg/h; $t_{cw,in}$ is the cooling water inlet temperature, °C; and $t_{cw,out}$ is the cooling water outlet temperature, °C.

$$Q_{rc} = K_c A_c \Delta tc = m_r (h_{con,in} - h_{con,out}) \tag{8}$$

where Q_{rc} represents the heat exchange heat on the refrigerant side, kW; K_c is the convection heat transfer coefficient on the condenser side, W/(m^2·°C); A_c is the total heat exchange area of the condenser, m^2; $h_{con,in}$ is the specific enthalpy of the refrigerant at the condenser inlet, kJ/kg; $h_{con,out}$ is the specific enthalpy of the refrigerant at the condenser outlet, kJ/kg; and Δt_c is the heat exchange temperature on the condenser side, °C.

The heat exchange temperature on the condenser side is calculated by Equation (9):

$$\Delta t_c = t_c - \frac{t_{cw,in} + t_{cw,out}}{2} \tag{9}$$

where t_c is the condensation temperature, °C; $t_{cw,in}$ is the cooling water inlet temperature, °C; and $t_{cw,out}$ is the cooling water outlet temperature, °C.

The heat transfer balance on the condenser side is:

$$Q_w = Q_{rc} \tag{10}$$

$$m_{ex} = A_{ex} C_{ex} \sqrt{2\rho_{ex,in} (P_{ex,in} - P_{ex,out})} \tag{11}$$

where C_{ex} is the flow coefficient of the expansion valve provided by the manufacturer; $\rho_{ex,in}$ is the refrigerant density at the valve inlet, kg/m^3; $P_{ex,in}$ and $P_{ex,out}$ are the expansion valve inlet and outlet pressure, respectively, kJ; and A_{ex} is the effective passage flow area of the expansion valve, m^2.

In this study, four DC inverter heat pumps were used for heating. The setup of these is demonstrated in Table 2.

Table 2. Heat pump parameters.

Parameter	Value
Evaporating temperature	10 °C
Condensing temperature	46.1 °C
Superheat	1.1 °C
Speed range	20–120 rps
Rated speed	90 rps
Refrigerant mass flow rate	1021 kg/h
Compressor power	10.3 kW
Compressor normal operating voltage	586–636 V
Rated heat capacity	196.0 kW (47.2 × 4)
Type of refrigerant	R410A

3.3. Control Strategy for Heat Pump

The traditional control methods of heat pumps include sequential control of multiple heat pumps, control of return water temperature, and control of water flow rate. In this study, the control strategy of the DC inverter heat pumps depends on both DC bus voltage and the indoor temperature setup, as shown in Figure 10. When the DC bus voltage exceeds the setpoint, the compressor operates at a higher frequency and stores the excess energy in the envelope. When the voltage falls below the setpoint, the compressor operates at a lower frequency and discharges the insufficient energy from the envelope to meet the user demand. The heating/cooling load of the building is calculated by using the indoor temperature, which determines the theoretical power value of the heat pump. When the theoretical power value P_{need} is lower than the actual power value P_{actual}, the heat pump should be operated at the theoretical power value. If the theoretical power value P_{need} is higher than the actual power value P_{actual}, the power of the heat pump should be decreased until the actual power value is lower than the theoretical power value.

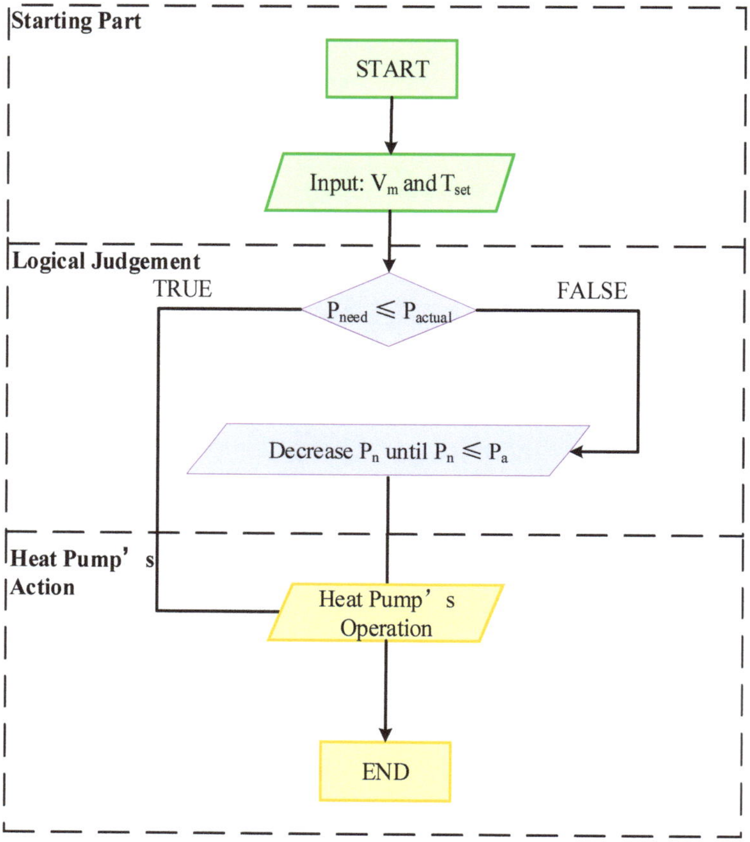

Figure 10. Control logic of heat pump power based on DC bus voltage and indoor temperature setup.

The sequence control method for the heat pumps is presented in Figure 11. The heating load is calculated by using monitored return water temperature and the mass flow rate of the water. The sequence of heat pumps is determined by the actual heating load and the optimal load rate for each heat pump. When the actual heating load is lower than 0.15 times of the rated heating load, the heat pump system should be turned off. If the heating load is higher than 0.15 times of the rated heating load, one heat pump will be turned on. When a single heat pump is not capable of providing adequate heat for the building, the second heat pump should be turned on. The process will be continued until enough heat can be provided to the building.

Based on the developed models, MATLAB/Simulink was utilized for simulation, and the developed model is presented in Figure 12.

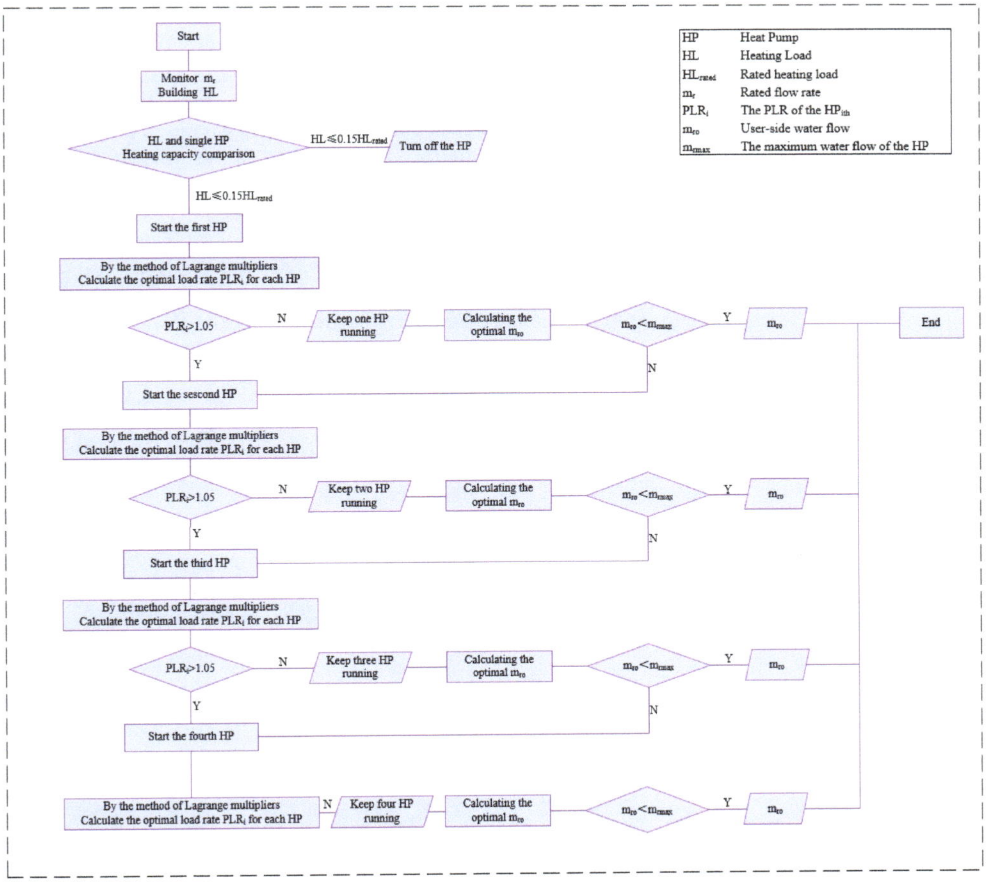

Figure 11. Control flow diagram of multiple heat pumps.

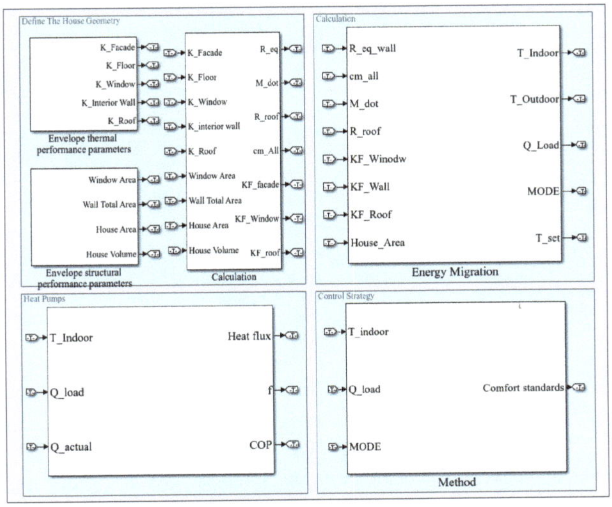

Figure 12. MATLAB/Simulink model.

3.4. Setup of Supply and Demand Sides

The supply-side input parameter is mainly DC bus voltage, and the demand-side input parameters include ambient and indoor environmental conditions.

The DC bus voltage in the range of 80–107% of the rated voltage can be regarded as the working bus voltage for the heat pump system. The heat pump can achieve an optimal performance when the bus voltage is within 90–105% of the rated voltage. When the bus voltage is within 80–90% or 105–107% of the rated voltage, the heat pump should operate at reduced frequency. For safety, the heat pump should be shut down when the bus voltage is lower than 80% or higher than 105% of the rated voltage [40]. Figure 13 shows the profile of building supply power on a typical day. The DC bus voltage during the time period of 13:00–14:00 was selected as the worst operating condition to set the DC bus voltage corresponding to the normal operating voltage range (596–636 V) of the equipment. The variation in input DC bus voltage is presented in Figure 14.

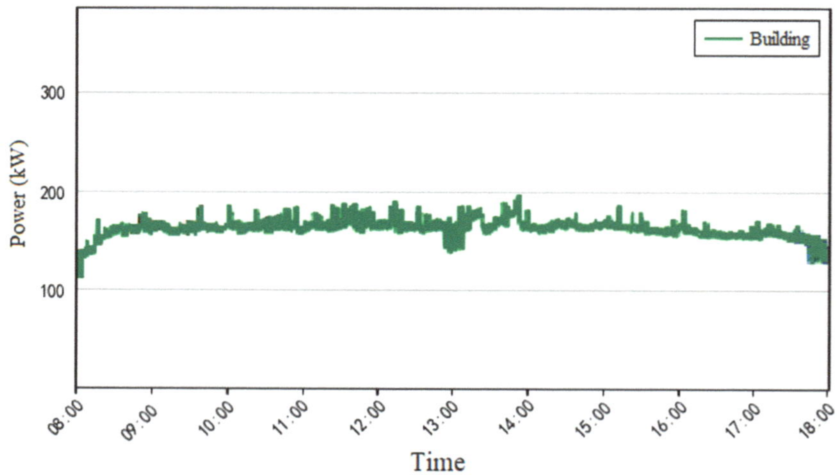

Figure 13. Profile of supply power for the building.

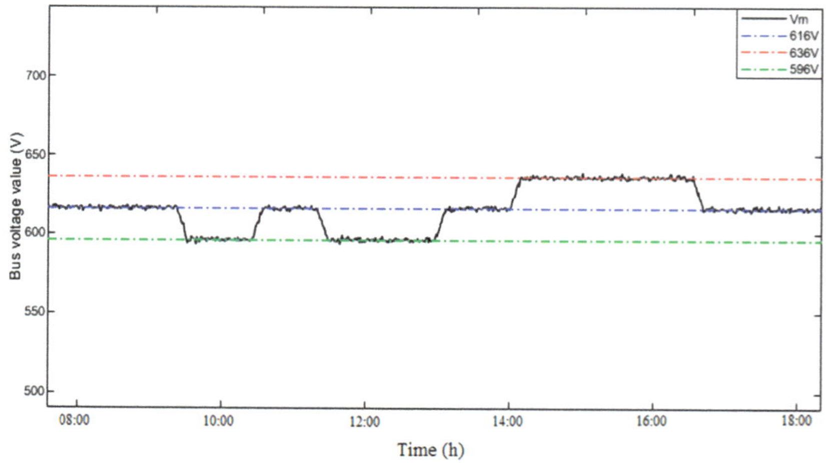

Figure 14. Variation in the input DC bus voltage.

The setup for an ambient environment utilized typical year weather data in EnergyPlus software. Figure 15 demonstrates the hourly ambient temperature, solar radiation intensity, and the simulated heating load of the reference building on a typical winter day. The daily average ambient temperature was −3.99 °C, while the average solar radiation during daytime was 471 W/m^2. The heating load of the day peaked at 6:00 A.M., with a peak value of 123.95 kW due to the low solar radiation and the large temperature difference between indoor and ambient temperature.

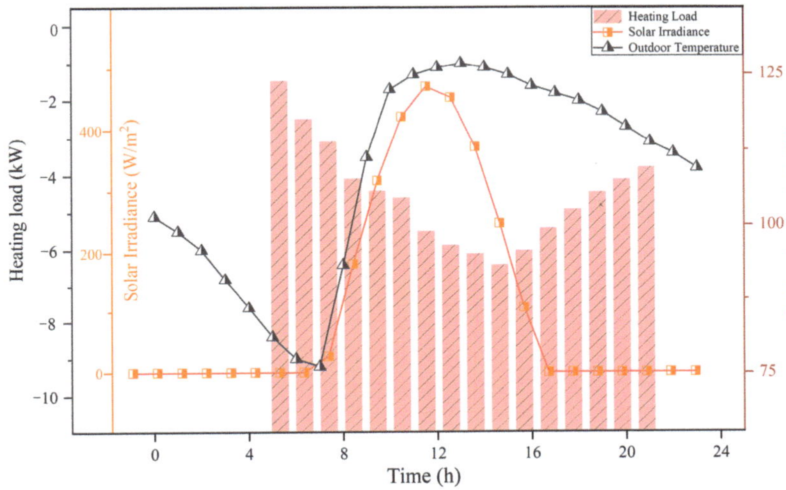

Figure 15. Setup for ambient environment and the simulated heating load of the building.

The setup values for the supply and demand sides of the simulation model are presented in Table 3. Three working conditions were simulated: transfinite mode, normal mode, and energy-saving mode. When the DC bus voltage (V_m) is within the range of 596–606 V, the supply power is insufficient, and thus the indoor temperature can be operated at a lower value (18 °C) to save energy. When the DC bus voltage is within the range of 606–626 V, the energy supply is normal, and thus the indoor temperature can be operated at a medium level (21 °C). When the DC bus voltage is in the range of 626–636 V, the power supply is abundant, and the indoor temperature can be operated at a higher value (24 °C) and the excess heat can be stored in the building envelope. The range of indoor temperatures (18–24 °C) is based on the comfortable thermal value in GB50736-2016. The comfortable levels are identified as Class I (22 °C ≤ t < 24 °C) and Class II (18 ≤ t < 22 °C).

Table 3. Heat pump flexible adjustment characteristic parameters.

Condition	Simulation Time (h)	Room Control Temperature (°C)	Comfort Level	DC Bus Voltage Range (V)	Heat Pump Mode
1		18	II	$596 < V_m < 606$	Low
2	10	21	I	$606 \leq V_m \leq 626$	Normal
3		24	I	$626 < V_m < 636$	High

4. Results and Discussion

The daily variations in DC bus voltage, the corresponding power of the heat pumps, and outdoor and indoor temperature are demonstrated in Figure 16. Three typical change processes are analyzed: process A (normal to low), process B (low to normal), and process C (normal to high), and the detailed change points are depicted in Table 4.

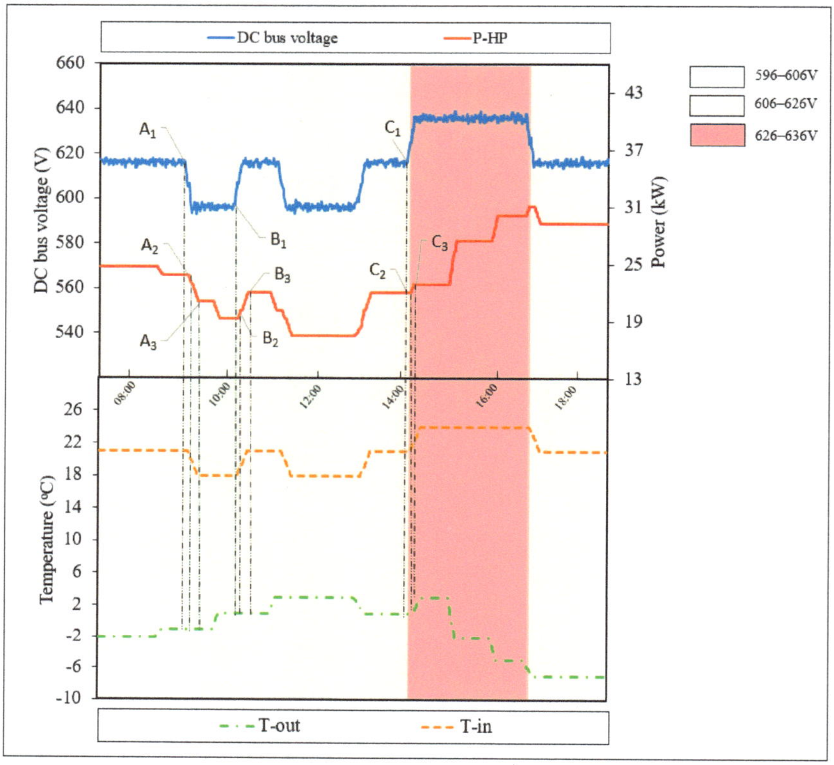

Figure 16. Profiles of variations in DC bus voltage, power of heat pump, and ambient and indoor temperature.

Table 4. DC heat pump flexible adjustment characteristic parameters.

Typical Changing Process	DC Bus Voltage Change Point t_1	Response Startup Point t_2	Response Completion Point t_3	DR1 $\Delta = t_2 - t_1$ (min)	DR2 $\Delta = t_3 - t_2$ (min)
A	A_1: 09:25 A.M.	A_2: 09:31 A.M.	A_3: 09:46	6	15
B	B_1: 10:36 A.M.	B_2: 10:42 A.M.	B_3: 10:56	6	14
C	C_1: 14:02 P.M.	C_2: 14:08 A.M.	C_3: 14:13	6	5

As shown in Figure 16, when the DC bus voltage drops from 616 V to about 595 V at change point A_1, the outdoor temperature increases gradually. The joint effect of reduced DC bus voltage and outdoor temperature results in a lowered set temperature of 18 °C. Thus, the power of the heat pump is reduced from 23.97 kW (at A_2 9:31 A.M.) to 21.07 kW (at A_3 9:46 A.M.), taking 21 min in total.

At point B_1, the DC bus voltage increases from 595 V to 616 V. During this period of time, the outdoor temperature remains relatively constant. To fully utilize the building's heat storage effect and prevent sudden increases in electricity consumption, the indoor temperature is raised from 18 °C to 21 °C. The power of the heat pumps is increased from 19.26 kW to 21.97 kW (the increase in DC bus voltage and the constant HL resulting in HP compressors that can be operated in normal mode). It took 6 min for the heat pumps to respond to the temperature change (at B_1 10:36 A.M.) at the response startup point B_2 (10:42 A.M.), and the response completion point was at 10:56 A.M. (B_3).

At point C_1, the DC bus voltage has increased to nearly the upper limit of the HPs' operating range, which is 636 V. To absorb the excess photovoltaic resources, the indoor temperature has been set to from 21 °C to 24 °C. Therefore, the power of the HPs is increased from 21.97 kW to 22.88 kW (the increase in DC bus voltage and the HL resulted in HP compressors that can be operated at higher power). The initial response time was at 14:02 (C_1), and the HPs' actual change occurred at 14:08 (C_2). The final actual response was at 14:13 (C_3), which took 6 and 5 min, respectively.

To sum up, the average startup time for the heat pumps' response to the DC bus voltage for the three processes is 6 min. The average response time of the first two response times is 14.5 min, and the deviation of the third response time of 5 min is 65.52%. This is because during the third response time, the outdoor irradiance is relatively sufficient, which shortens the time for the indoor temperature to reach the set value.

Figure 17a shows the simulation results of the HP standby time based on comfort criteria. The HP employs a startup and standby control strategy, adjusting its operating status based on ambient temperature changes to meet the user thermal comfort requirements. By initiating the HP before 8:00 and utilizing the building's heat storage characteristics, the excess heat is stored within the building in advance. Once indoor temperature rises, the HP is turned off, relying on the residual heat in the building to meet the users' heating needs, which greatly reduces the unit operating costs. Figure 17b shows the standby time deviation results of Class I/II comfort criteria, indicating a downward trend with deviation values ranging from 20% to 30%. Expanding the setting range of room temperature can significantly affect the allowable standby time of HPs, which is highly sensitive to changes in indoor heat load. The smaller the heat load demand, the longer the standby time can last.

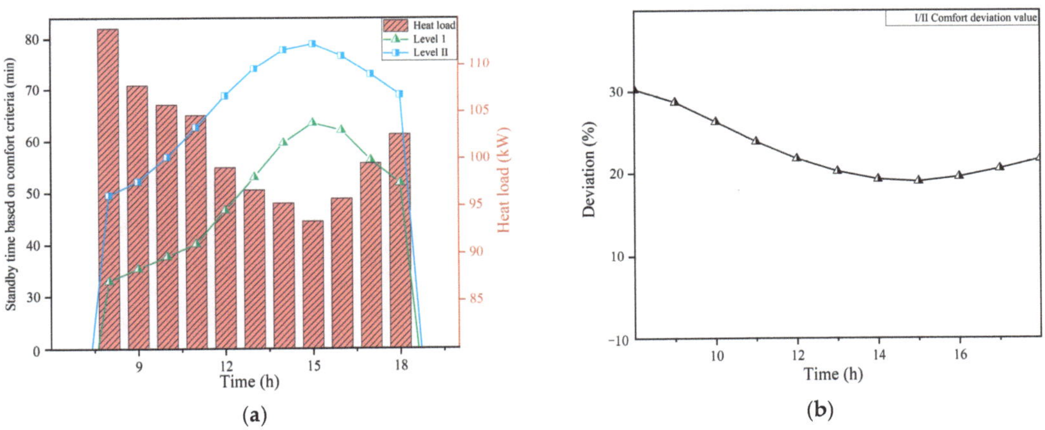

Figure 17. Heat pump standby time based on comfort criteria. (**a**) Heat pump standby time; (**b**) standby time deviations between level I and level II.

5. Conclusions

This paper investigates the operational strategy of DC inverter heat pumps in a building integrated with a PV system. Both the PV system and utility grid were utilized for power supply. A flexible control method for DC inverter heat pumps considering varied DC bus voltage and room temperature proposed. Models of a reference building with the DC inverter heat pump were developed and simulated in MATLAB/Simulink software. The results show that the proposed strategy can achieve stable operation of a DC inverter heat pump while maintaining thermal comfort in the indoor environment under the fluctuating DC bus voltage. The specific conclusions are as follows.

(1) The heat pump can operate normally at 596–636 V. When the DC bus voltage is low (596–616 V), the building's electricity is supplied by the grid, and the HP can operate at lower operating power. When the DC bus voltage is high (626–636 V), the PV system is the main power source for the building, the HP should increase its operating power,

and the excess heat can be stored in the building envelope to mitigate the fluctuations in DC bus voltage. When the DC bus voltage fluctuates normally (606–626 V), the HP regulates the indoor temperature normally.

(2) According to the data analysis of the case study, significant fluctuations in power supply may occur in the PV integrated building, which may pose negative effects on the operation of the building equipment and systems. Therefore, the heat pumps employ a startup and standby control strategy. By expanding the setting range of the room temperature, it can significantly affect the allowable standby time of heat pumps. For instance, the duration of standby time of heat pumps can extend to about 20–30% by reducing the room temperature from Class I to Class II comfort criteria.

In future, other environmental factors, such as ambient temperature and seasonal solar irradiance variations, can be further considered to evaluate the stability and adaptability of the DC inverter heat pump in practical applications.

Author Contributions: Conceptualization, Y.L. (Yilin Li); methodology, Y.L. (Yilin Li) and Y.L. (Yang Lu); software, T.Z. and Y.L. (Yang Lu); validation, Y.L. (Yang Lu) and S.Z.; formal analysis, Y.L. (Yang Lu); investigation, Y.L. (Yang Lu) and T.W.; resources, J.S.; data curation, Y.L. (Yang Lu) and S.Z.; writing—original draft preparation, Y.L. (Yilin Li) and T.W.; writing—review and editing, Y.L. (Yilin Li), S.Z. and T.Z.; visualization, S.Z.; supervision, Y.L. (Yilin Li); project administration, X.W. and J.S.; funding acquisition, Y.L. (Yilin Li) and X.W. All authors have read and agreed to the published version of the manuscript.

Funding: This research was funded by National Natural Science Foundation of China grants 52008245 and 52378103 and a Key Laboratory of Ecology and Energy Saving Study of Dense Habitat, Ministry of Education grant (20230102).

Data Availability Statement: Data available on request from the authors.

Conflicts of Interest: Jie Sun was employed by BSE Energy Technology (Suzhou) Co., Ltd. The remaining authors declare that the research was conducted in the absence of any commercial or financial relationships that could be construed as a potential conflict of interest.

Nomenclature

PVs	Photovoltaics
RESs	Renewable energy sources
BIPV	Building integrated photovoltaic
DC	Direct current
HVAC	Heating ventilation and air-conditioning
ASHP	Assisted air source heat pump
PV/T	Solar photovoltaic/thermal
HL	Heating load
HP	Heat pump
Greek symbols	
ρ	Density, kg/m^3
Φ	Opening degree, %
η	Efficiency, %
Subscripts	
Com	Compressor
Dis	Discharge
w	Water
Cw	Cooling water
Cond	Condenser
rc	Refrigerant side
c	Condensation
ex	Expansion valve
m	DC bus
ref	Reference

References

1. Lo, K.Y.; Yeoh, J.H.; Hsieh, I.Y.L. Towards Nearly Zero-Energy Buildings: Smart Energy Management of Vehicle-to-Building (V2B) Strategy and Renewable Energy Sources. *Sustain. Cities Soc.* **2023**, *99*, 104941. [CrossRef]
2. Chen, W.H.; You, F. Sustainable building climate control with renewable energy sources using nonlinear model predictive control. *Renew. Sustain. Energy Rev.* **2022**, *168*, 112830. [CrossRef]
3. Pramanik, A.S.; Sepasi, S.; Nguyen, T.L.; Roose, L. An ensemble-based approach for short-term load forecasting for buildings with high proportion of renewable energy sources. *Energy Build.* **2024**, *308*, 113996. [CrossRef]
4. IRENA. *Trading into a Bright Energy Future (The Case for Open, High-Quality Solar Photovoltaic Markets)*; World Trade Organization: Geneva, Switzerland, 2021.
5. Du, H.; Han, Q.; de Vries, B.; Sun, J. Community solar PV adoption in residential apartment buildings: A case study on influencing factors and incentive measures in Wuhan. *Appl. Energy* **2024**, *354*, 122163. [CrossRef]
6. Zhu, J.; Dong, H.; Zheng, W.; Li, S.; Huang, Y.; Xi, L. Review and prospect of data-driven techniques for load forecasting in integrated energy systems. *Appl. Energy* **2022**, *321*, 119269. [CrossRef]
7. Diab, I.; Damianakis, N.; Chandra-Mouli, G.R.; Bauer, P. A shared PV system for transportation and residential loads to reduce curtailment and the need for storage systems. *Appl. Energy* **2024**, *353*, 122131. [CrossRef]
8. Chen, Q.; Kuang, Z.; Liu, X.; Zhang, T. Optimal sizing and techno-economic analysis of the hybrid PV-battery-cooling storage system for commercial buildings in China. *Appl. Energy* **2024**, *355*, 122231. [CrossRef]
9. Zhang, S.; He, W.; Fan, Y.; Wang, K.; Hu, Z.; Chu, W.; Yu, H. Field experimental investigation on electricity and thermal performances of a large scale photovoltaic solar-thermal direct expansion heat pump system. *Energy Convers. Manag.* **2022**, *267*, 115941. [CrossRef]
10. Ma, J.; Fung, A.S.; Brands, M.; Moyeed, O.M.A.; Mhanna, A.; Juan, N. Effects of photovoltaic/thermal (PV/T) control strategies on the performance of liquid-based PV/T assisted heat pump for space heating. *Renew. Energy* **2021**, *172*, 753–764. [CrossRef]
11. Ding, L.; Zhu, Y.; Zheng, L.; Dai, Q.; Zhang, Z. What is the path of photovoltaic building (BIPV or BAPV) promotion? The perspective of evolutionary games. *Appl. Energy* **2023**, *340*, 121033. [CrossRef]
12. Aguacil, S.; Lufkin, S.; Rey, E. Active surfaces selection method for building-integrated photovoltaics (BIPV) in renovation projects based on self-consumption and self-sufficiency. *Energy Build.* **2019**, *193*, 15–28. [CrossRef]
13. Sankar, M.M.; Chatterjee, K. A posteriori multiobjective approach for techno-economic allocation of PV and BES units in a distribution system hosting PHEVs. *Appl. Energy* **2023**, *351*, 121851. [CrossRef]
14. Trong, T.N.; Son, H.V.X.; Do Dinh, H.; Takano, H.; Duc, T.N. Short-term PV power forecast using hybrid deep learning model and Variational Mode Decomposition. *Energy Rep.* **2023**, *9*, 712–717. [CrossRef]
15. Notton, G.; Nivet, M.L.; Voyant, C.; Paoli, C.; Darras, C.; Motte, F.; Fouilloy, A. Intermittent and stochastic character of renewable energy sources: Consequences, cost of intermittence and benefit of forecasting. *Renew. Sustain. Energy Rev.* **2018**, *87*, 96–105. [CrossRef]
16. Brinkel, N.B.G.; Gerritsma, M.K.; AlSkaif, T.A.; Lampropoulos, I.; van Voorden, A.M.; Fidder, H.A.; van Sark, W.G.J.H.M. Impact of rapid PV fluctuations on power quality in the low-voltage grid and mitigation strategies using electric vehicles. *Int. J. Electr. Power Energy Syst.* **2020**, *118*, 105741. [CrossRef]
17. Hong, J.; Liang, F.; Yang, H. Research progress, trends and prospects of big data technology for new energy power and energy storage system. *Energy Rev.* **2023**, *2*, 100036. [CrossRef]
18. Asea Brown Boveri, Z. Enviline™DCGear DC Switchgear for DC Traction Power Distribution Network. 2022. Available online: https://new.abb.com/medium-voltage/zh/switchgear/railway-switchgear/dc-traction-power-supply/dc-switchgear-and-vld/envilinetm-dcgear-GB (accessed on 1 March 2024).
19. Han, S.; Li, X.; Zhao, W.; Wang, L.; Liang, A.; Zeng, S. Simulation Study of the Control Strategy of a DC Inverter Heat Pump Using a DC Distribution Network. *Energy Eng.* **2023**, *120*, 1421–1444. [CrossRef]
20. Kockel, C.; Nolting, L.; Goldbeck, R.; Wulf, C.; De Doncker, R.W.; Praktiknjo, A. A scalable life cycle assessment of alternating and direct current microgrids in office buildings. *Appl. Energy* **2022**, *305*, 117878. [CrossRef]
21. Candanedo, L.M.; Athienitis, A.K.; Kalogirou, S.A. Modeling and simulation of passive and active solar thermal systems. *Compr. Renew. Energy* **2012**, *3*, 357–417.
22. Azaioud, H.; Farnam, A.; Knockaert, J.; Vandevelde, L.; Desmet, J. Efficiency optimisation and converterless PV integration by applying a dynamic voltage on an LVDC backbone. *Appl. Energy* **2024**, *356*, 122416. [CrossRef]
23. Cavazzuti, M.; Bottarelli, M. Performance analysis of a multi-source renewable energy system for temperature control in buildings of varied thermal transmittance and climate zone. *Renew. Sustain. Energy Rev.* **2023**, *187*, 113725. [CrossRef]
24. Tan, H.; Guo, Z.; Zhang, H.; Chen, Q.; Lin, Z.; Chen, Y.; Yan, J. Enhancing PV panel segmentation in remote sensing images with constraint refinement modules. *Appl. Energy* **2023**, *350*, 121757. [CrossRef]
25. Sigounis, A.M.; Vallianos, C.; Athienitis, A. Model predictive control of air-based building integrated PV/T systems for optimal HVAC integration. *Renew. Energy* **2023**, *212*, 655–668. [CrossRef]
26. Modu, B.; Abdullah, M.P.; Sanusi, M.A.; Hamza, M.F. DC-Based microgrid: Topologies, control schemes, and implementations. *Alex. Eng. J.* **2023**, *70*, 61–92. [CrossRef]
27. Zhou, G.; Bai, M.; Zhao, X.; Li, J.; Li, Q.; Liu, J.; Yu, D. Study on the distribution characteristics and uncertainty of multiple energy load patterns for building group to enhance demand side management. *Energy Build.* **2022**, *263*, 112038. [CrossRef]

28. Efkarpidis, N.A.; Vomva, S.A.; Christoforidis, G.C.; Papagiannis, G.K. Optimal day-to-day scheduling of multiple energy assets in residential buildings equipped with variable-speed heat pumps. *Appl. Energy* **2022**, *312*, 118702. [CrossRef]
29. Walden, J.V.; Wellig, B.; Stathopoulos, P. Heat pump integration in non-continuous industrial processes by Dynamic Pinch Analysis Targeting. *Appl. Energy* **2023**, *352*, 121933. [CrossRef]
30. Wang, Y.; Quan, Z.; Zhao, Y.; Wang, L.; Liu, Z. Performance and optimization of a novel solar-air source heat pump building energy supply system with energy storage. *Appl. Energy* **2022**, *324*, 119706. [CrossRef]
31. Roccatello, E.; Prada, A.; Baggio, P.; Baratieri, M. Analysis of the Influence of Control Strategy and Heating Loads on the Performance of Hybrid Heat Pump Systems for Residential Buildings. *Energies* **2022**, *15*, 732. [CrossRef]
32. Xia, L.; Ma, Z.; Kokogiannakis, G.; Wang, S.; Gong, X. A model-based optimal control strategy for ground source heat pump systems with integrated solar photovoltaic thermal collectors. *Appl. Energy* **2018**, *228*, 1399–1412. [CrossRef]
33. Wei, D.; Cui, Q.; Qi, P.; Gao, E.; Zhang, X. Experimental study on operating characteristics and control strategy of CO_2 heat pump in circulating heating mode. *Appl. Therm. Eng.* **2024**, *236*, 121719. [CrossRef]
34. Lyu, W.; Wang, Z.; Li, X.; Deng, G.; Xu, Z.; Li, H.; Yang, Y.; Zhan, B.; Zhao, M. Influence of the water tank size and air source heat pump size on the energy saving potential of the energy storage heating system. *J. Energy Storage* **2022**, *55*, 105542. [CrossRef]
35. Wang, J.; Qv, D.; Ni, L.; Yao, Y. Experimental study on an injection-assisted air source heat pump with a novel two-stage variable-speed scroll compressor. *Appl. Therm. Eng.* **2020**, *176*, 115415. [CrossRef]
36. Zhang, D.; Xinyue, S.; Pengfei, L.; Bo, H.; Gang, H.; Chunyang, L.; Baorui, X.; Tao, Z.; Zhoujian, A. Experiment study on startup characteristics and operation performance of PV/T solar assisted heat pump waterheater system driven by direct current variable frequency compressor. *Sol. Energy* **2023**, *263*, 111771. [CrossRef]
37. Ma, T.; Li, J.; Cao, X.; Ma, Y.; Cai, Y.; Lu, Z. Research on optimal dispatch strategy of electric heating load groups considering user behavior difference and distribution network power flow. *J. Electr. Power Sci. Technol.* **2023**, *38*, 77–87.
38. Bin, Z.; Qing, M.; Daoke, T.; Ling, F. Parameter optimization and off-design operation strategy of multistage scroll heat pumps in heating station. *Heat. Vent. Air Cond.* **2017**, *47*, 125–129.
39. Shang, Y.; Wu, A.; Fang, X.; You, Y. Dynamic Simulation of Electronic Expansion Valve Controlled Refrigeration System under Different Heat Transfer Conditions. *Int. J. Refrig.* **2016**, *72*, 41–52. [CrossRef]
40. Yu, L.; Yi, D.; Yu, Z.; Yuan, L. Interpretation of the DC Power Distribution Design Standard for Civil Buildings. *Build. Electr.* **2022**, *41*, 25–32.

buildings

Article

Evaluation of Phase Change Materials for Pre-Cooling of Supply Air into Air Conditioning Systems in Extremely Hot Climates

Usman Masood [1], Mahmoud Haggag [2,*], Ahmed Hassan [2] and Mohammad Laghari [1]

1 Electrical and Communication Engineering Department, United Arab Emirates University, Al Ain P.O. Box 15551, United Arab Emirates; 700040271@uaeu.ac.ae (U.M.); mslaghari@uaeu.ac.ae (M.L.)
2 Architectural Engineering Department, United Arab Emirates University, Al Ain P.O. Box 15551, United Arab Emirates; ahmed.hassan@uaeu.ac.ae
* Correspondence: mhaggag@uaeu.ac.ae; Tel.: +971-50-5638461

Abstract: This research investigates the use of phase change materials (PCMs) in thermal energy storage (TES) unit-based cooling systems to increase the efficiency of air conditioners (ACs) by reducing the air inlet temperature. This study aims to evaluate different configurations of PCM enclosures, and different PCMs (paraffin and salt hydrate), by changing the speed of inlet air to achieve heat reduction of inlet air. The study includes experimental and simulation investigations. Every configuration simulates the hot-season atmospheric conditions of the UAE. A duct containing enclosures of paraffin RT-31 and salt hydrate (calcium chloride hexahydrate) was used for the simulation study using ANSYS/Fluent. A conjugate heat transfer model employing an enthalpy-based formulation is developed to predict the optimized PCM number of series and optimum airflow rate. Four designs of the AC duct were modelled and evaluated that contained one to four series of PCM containers subjected to different levels of supplied air velocities ranging from 1 m/s–4 m/s. The simulation study revealed that employing four series (Design 4) of PCM enclosures at a low air velocity of 1 m/s enhanced the pre-cooling performance and reduced the outlet air temperature to 33 °C, yielding a temperature drop up to 13 °C. The performance of salt hydrate (calcium chloride hexahydrate) was observed to be better than paraffin (RT-31) in terms of the cooling effect. Characterization of paraffin wax (RT-31) and salt hydrate was performed to establish the thermophysical properties. The experimental setup based on a duct with integrated PCM enclosures was studied. The experiment was repeated for three days as the repeatability test incorporating RT-31 as the PCM and a 3 °C maximum temperature drop was observed. The drop in the outlet air temperature of the duct system quantifies the cooling effect. Net heat reduction was around 16%.

Keywords: phase change materials (PCMs); thermal energy storage (TES); heat transfer augmentation; cooling load; buildings; latent heat thermal energy storage (LHTES)

Citation: Masood, U.; Haggag, M.; Hassan, A.; Laghari, M. Evaluation of Phase Change Materials for Pre-Cooling of Supply Air into Air Conditioning Systems in Extremely Hot Climates. *Buildings* **2024**, *14*, 95. https://doi.org/10.3390/buildings14010095

Academic Editors: Elena Lucchi and Antonio Caggiano

Received: 7 November 2023
Revised: 20 December 2023
Accepted: 28 December 2023
Published: 29 December 2023

1. Introduction

A massive CO_2 emission of 37 gigatons was reported by the IEA for the year 2019. A similar emission was reported in the year 2020. These emissions highlight the need to find innovative solutions to reduce CO_2 emissions and mitigate climate change. Gagliano et al. [1] projected that the total cooled floor area in European countries would reach 2 billion square meters in 2020, up from 1000 million square meters in 2012. This would require more than 100 terawatt-hours (TWh) of energy per year for building cooling alone. The building sector in Europe is a major energy consumer (40% of total energy) and emitter of greenhouse gases. The demand for cooling in buildings is increasing due to rising expectations for indoor thermal comfort levels due to the rise in global temperature. Accounting for an average of 40% of energy use, air conditioners are the single largest consumer of energy in buildings. As a result, reducing global energy consumption may be accomplished with even a small increase in AC performance.

A recent critical review of the strategies needed to achieve zero-energy buildings found that both passive and active strategies are important [2]. The review recommended the use of energy storage, among other solutions. TES can be considered a key technology for promoting the more efficient use of renewable energy sources, which are one of the solutions to reducing cooling load [3]. TES systems can be classified into three types: sensible, latent, and thermochemical. Latent thermal energy storage (LTES) systems, which use PCMs, have several advantages over other types of TES systems (e.g., sensible heat storage). PCMs possess a considerably higher latent heat of fusion than the majority of materials that possess a sensible heat capacity [4]. As a result, LTES systems can store a greater amount of energy in a smaller volume. PCMs can store and transfer heat, making them a promising solution for reducing the gap between energy demand and supply.

PCMs are novel latent heat-based thermal energy storage materials that absorb and release significantly large amounts of thermal energy compared with sensible heat-based energy storage materials that do not undergo a phase change while absorbing thermal energy. The PCMs predominantly employed are solid–liquid based on the phase change mechanism. The solid–liquid PCMs are broadly categorized into paraffin, salt hydrates, eutectic mixtures, and chemical compound allotropes from a materials point of view. Most paraffin and salt hydrates belong to the solid–liquid PCM category. In the current application, paraffin-based and salt hydrate-based solid–liquid phase change materials are deployed owing to their superior thermo-physical properties and abundant cost-effective availability [5].

TES applications using PCMs can be classified into passive and active cooling technologies. Passive cooling technologies rely on the natural heat transfer mechanisms of conduction, convection, and radiation to store and release heat [6]. Active cooling technologies use pumps, fans, or other devices to move heat around, which can make them more efficient but also more complex and expensive. PCMs are used as the thermal mass in passive cooling, which helps to reduce the heat load on air conditioning (AC) systems. In general, PCMs are incorporated within a building's ceilings, walls, or floors, or they are incorporated in air handlers or heat exchanger units [7–9]. The need for air conditioning can be reduced by using PCMs, which absorb heat during the day and release it at night, keeping the building cooler during the day. The incorporation of PCMs in passive cooling and heating systems has been demonstrated to enhance thermal comfort and lower energy demands in buildings, as evidenced by several studies. The night ventilation efficiency of a building was studied numerically by Stritih et al. [10] utilizing a PCM storage chamber located on the roof. Power usage decreased by 14% to 87%, while temperatures dropped by 3 °C. Liu et al. [11] conducted a numerical evaluation to determine the optimal geometries and phase change material (PCM) types for enhancing the thermal efficiency of an air-to-PCM heat exchanger in a ventilation system.

When PCMs are integrated into a thermal management system, the temperature of a room or piece of equipment can be actively controlled by absorbing and releasing heat. The transfer of heat between the PCM and air or fluid for cooling typically involves a heat exchanger or similar device. PCMs are used in traditional HVAC systems, such as air conditioners and heat exchangers, because of their ability to store thermal energy [12–14]. The study conducted by Akbari et al. [15] aimed to explore the thermodynamic and economical optimization of an AC unit using several PCMs. Their findings indicate that a higher air temperature difference across the PCM leads to an increased coefficient of performance (COP) and exergy efficiency. Said et al. [16] found that a PCM-based TES improved the performance of an AC unit in both computational and numerical studies. The PCM plates were connected to the AC unit's condenser and reduced the temperature of the surrounding air together. Nagano et al. [17] investigated the use of a PCM in a floor AC system to improve the building's thermal energy performance. Their calculations showed that nighttime energy storage could provide cooling for 89% of the building.

Hoseini et al. [18] investigated the effects of two hybrid TES systems (ice and PCM) on an air conditioning (AC) system. They found that the PCM and ice TES systems reduced power consumption by 7.58% and 4.59%, respectively, compared with the basic system. Vakilaltojjar and Saman [19] developed three models to assess the impacts of the PCM flat slab thickness and air gaps between slabs on the performance of LTES systems for integration with air conditioning systems. Their results showed that the smallest air gap resulted in the best performance. Mosaffa et al. [20] conducted a detailed exergy analysis of a cooling system that uses LTES and vapor refrigeration. They showed that the irreversibility in the LTES unit is a significant contributor to the total exergy destruction rate. Mosaffa and Farshi [21] conducted energy and cost-benefit analyses of a similar system using SP25, RT27, and S27 PCM types. The results showed that SP25, RT27, and S27 achieved the highest coefficient of performance (COP), exergy efficiency, and lowest overall cost rate, respectively.

A simulation-based analysis was conducted to investigate the primary energy consumption of a new HVAC system for various thermal plant configurations. The system uses a single hydronic circuit for both heating and cooling. Four configurations were considered: district heating and cooling networks, reversible air-to-water heat pumps, and free cooling devices. The results showed that energy consumption could be reduced by up to 67% per year by adding free cooling devices, such as a PCM-based heat exchanger [22]. Hai et al. [23] performed a simulation-based study of a tube structure TES consisting of $CaCl_2{\cdot}6H_2O$ to improve the efficiency of AC and found that incorporating PCM cylinders into an AC unit could increase the cooling capacity by up to 34% during the warmest months. Farah et al. [24] investigated the impact of adding an LTES system to an air source heat pump (ASHP) on cooling efficiency. They found that installing an LTES system increases the ASHP's cooling capacity and reduces its electricity consumption at the same time. Nie et al. experimentally and economically evaluated the performance of an integrated system consisting of an LTES and a standard air conditioner, where the conditioned air passes through the LTES before being released into the room. They found that the integrated system reduced electrical costs by 17.8% [25].

Based on a comprehensive analysis of the existing literature and the authors' thorough investigation, the existing body of literature mostly focuses on the development of free cooling systems using PCMs. Comparatively fewer studies focus on improving the efficiency of air conditioners (ACs) using PCMs, and few have investigated the effects of placing a TES unit at different locations of the refrigeration cycle's elements, such as the condenser. Furthermore, research on the melting and solidification of phase change materials (PCMs) in hot and desert climates is limited, and the importance of selecting the right PCM for cooling applications in such climates, considering the temperature variations, has not been fully addressed. Therefore, this study aims to design and evaluate a novel air pre-cooling system using PCMs integrated into the AC duct system, specifically to provide cooler air to the evaporator. The criteria for selecting PCMs, which are essential for reliable and efficient thermal energy storage, exclude many options suitable for higher ranges of temperature. Paraffin and salt hydrates are recommended as the preferred choices due to advantageous properties, including a higher TES capacity and a suitable melting range. During the daytime, the TES unit is exposed to the outside hot air, which drops the air temperature and then routes it to the evaporator. This design is used to evaluate the effects of air inlet speed, the number of PCM columns placed in the AC duct, air outlet temperature, PCM melting and solidification rates, and power savings. Additionally, the performance of the proposed configuration is compared using different PCMs, such as RT31 and $CaCl_2{\cdot}6H_2O$. A comprehensive mathematical model of the physical system is developed and then solved with ANSYS software (Version 20, Computer software company, Cecil Township, PA, USA, 2015). To enhance the credibility of the numerical solution, furthermore, an experimental repetition test is conducted using a PCM block-type container design.

2. Experimental Analysis

2.1. Phase Change Material (PCM) Selection and Characterization

The proposed air pre-cooling units reduce energy consumption and improve thermal comfort using PCMs to stabilize indoor temperature fluctuations. To achieve the desired objectives, the selection of PCM and design of a thermal energy storage system considers the ambient temperature, humidity levels, air speed velocity, and conditioning load of specified residential buildings carefully. These factors must be balanced to maintain consistent indoor air temperature values and long-term thermal comfort, while also reducing energy use. United Arab Emirates (UAE) has one of the hottest and most humid climates in the world. According to weather data from a climate consultant [16], Al Ain's summer daytime temperatures range from 35 °C to 50 °C. Summer temperatures are highest from May to August, peaking at 49 °C in June with an average of 45 °C. Summer nighttime temperatures average 34 °C in June. Winter temperatures range from 14 °C to 27 °C. A weather analysis provides insights that PCMs intended to be used for high-temperature ranges do not exhibit effectiveness for thermal energy storage purposes in Al Ain. Consequently, we opted for a PCM characterized by a melting point within the range of 27–33 °C. This choice adheres to the necessity of upholding a temperature marginally higher than the prevailing ambient air temperature during the morning period.

To determine the thermal properties of PCM (RT 31), manufactured by Rubitherm in Germany, we used a technique called differential scanning calorimetry (DSC). For this purpose, we utilized the DSC-Q200 instrument. We measured the latent heat of fusion, liquidus and solidus temperatures, and the temperature at which the material shows appreciable solidification. We then compared these measurements to the characteristics of the material provided. The DSC curve revealed that RT 31 is a paraffin wax with a solidus temperature of 34.2 °C, a melting point of 32.5 °C, and a latent heat of fusion of 158 KJ/kg [26]. The article investigates a novel concept of placing the PCM containers inside the air supply duct of the AC unit to supply pre-cooled air to the evaporator. The TES design is easy to integrate into AC ducts for both air-based ACs and chilled water-based ACs. The objective of this article is to study the cooling effect produced by the PCM while it is melting due to outdoor solar radiation and the ambient temperature.

2.2. Experimental System

An experimental setup was designed to study the performance of a PCM-based air precooling system. The heat load and system efficiency were calculated using a typical one-story structure of the size 20 × 20 × 3 m^3 in Al Ain (UAE). It is important to understand the crucial characteristics and systems within a building. The building has a total floor area of 300 square meters, making it a moderately sized structure. It features windows that cover 10% of the gross wall area. These windows are constructed with 6 mm single, green-tinted glazing, potentially providing benefits such as reduced glare and controlled solar heat gain. Notably, the external walls and roof of the building have a solar absorbance of 0.50. The building is designed for an occupancy density of 6 people, which may have implications for space planning and HVAC requirements. Additionally, the lighting power density in the building is 4.5 watts per square meter (W/m^2), suggesting the electrical power used for lighting. Lastly, the equipment power density is 7 W/m^2, providing information about the electrical power consumption of equipment and appliances within the building. These details are vital for understanding the building's design, energy efficiency, and potential environmental impact. The explained characteristics are listed in Table 1.

The equation below was used to calculate heat losses over three days by contrasting the average interior temperatures obtained with and without the use of PCM-based systems.

$$Q_{inner} = h_c \times A \times (T_s - T_i) \tag{1}$$

The inner surface temperature T_s, the ambient air temperature T_i, and the convective heat transfer coefficient h_c are related by the following Equation (2). Assuming free cooling and applying the wind speed v_w for UAE [16], we obtain that the h_c value at the inner surface facing inside is 6.5 W/m^2 °C.

$$h_c = 3.3\, v_w + 6.5 \quad (2)$$

Table 1. Type of system and building characteristics.

Characteristics	Description
Area of the window	10% of the gross wall area—uniformly distributed
Area of the floor	300 m^2
Solar absorbance value	0.50 for the external walls and roof
Window	6 mm single green-tinted glazing
Occupancy density	6 people
Equip. power density	7 W/m^2
Lighting power density	4.5 W/m^2

2.3. PCM-Based TES Unit

For designing the PCM-based TES, a test chamber was built that replicates a standard building's air conditioning duct system. The chamber measures 45 cm × 90 cm × 20 cm. During testing, the AC duct system was exposed to varying air velocities supplied by an air pump. Metallic PCM containers (5 cm × 30 cm × 20 cm) were fabricated using a 4 mm thick sheet of aluminum alloy (1050A), and every PCM container was filled with 2.7 kg of PCM (RT-31). The solid PCM was placed within four PCM enclosures that were positioned vertically inside the duct with a 5 cm space between them. Figure 1 shows a schematic diagram of the system. Multiple calibrated T-type copper–constantan thermocouples were placed at different positions across the air duct system and inside the PCM. The sensors were connected to a data logger (National Instruments-NI, Compact-Rio) to store and retrieve the data. The PCM was filled as a liquid in the containers and subsequently cooled until it was completely solidified. The solidified PCM left a 7 cm free space on top of the container, which was intended to accommodate volume expansion during PCM melting. The efficiency of the system depends on the reduction in temperature of the inlet air. Therefore, the effects of two parameters are being studied: airspeed and the number of PCM slabs in the air duct of a split air conditioning system with an efficiency ratio (EER) of 8. Figure 1 shows a schematic diagram of the experimental setup and Figure 2 shows the experimental setup.

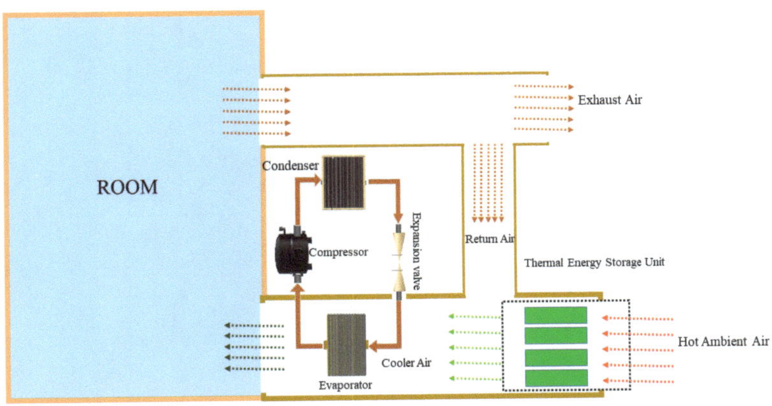

Figure 1. Schematic diagram of employing TES in an air conditioning duct.

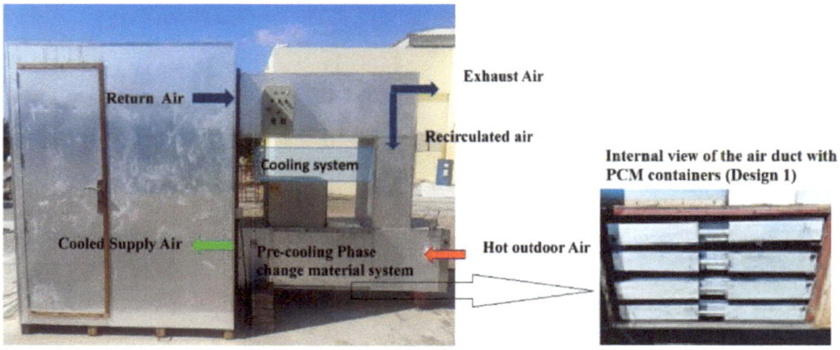

Figure 2. Experimental setup of employing a PCM into an AC duct.

3. Numerical Modeling

Four 2D mathematical, finite-volume, heat transfer models of four designs containing 1, 2, 3, and 4 series of PCM enclosures embedded in an air duct are developed in ANSYS Fluent, including air and PCM as heat exchange media, as illustrated in Figure 3. The system design involves a comparative study between two phase change materials (PCMs): paraffin (RT-31) and salt-hydrate ($CaCl_2{\cdot}6H_2O$). The behavior of the two PCMs under different inlet air velocities, ranging from 1 m/s to 4 m/s, were studied using ambient air conditions of a hot climate. The air and PCM both are characterized as unsteady and incompressible fluids since no pressure-induced volume changes are expected in the computational domain with melt flow. Figure 3a–d show the models of AC ducts with 1, 2, 3, and 4 series of PCMs, respectively. The PCM volume of each enclosure was kept the same as shown in Figure 3a–d.

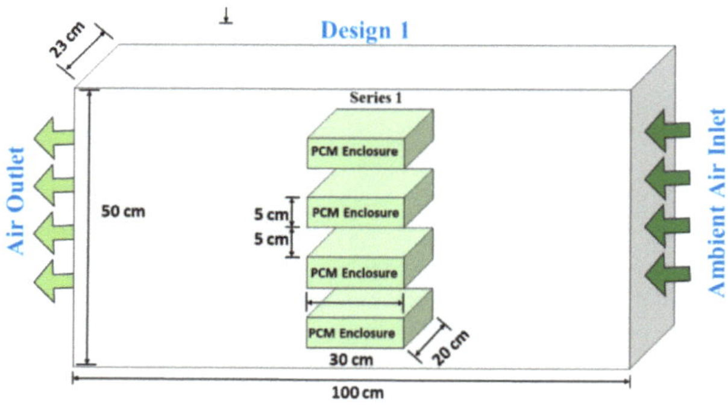

(a)

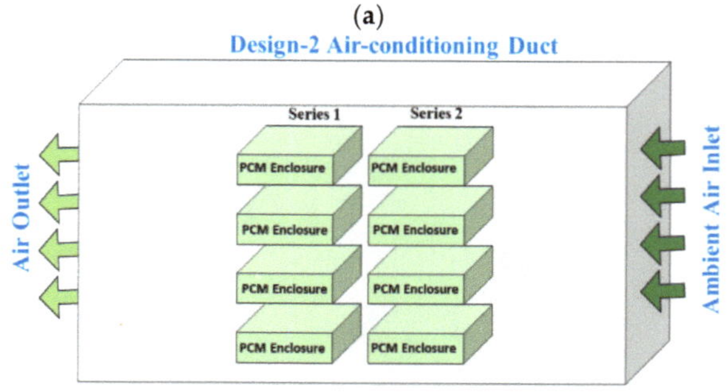

(b)

Figure 3. *Cont.*

Design-3 Air-Conditioning Duct

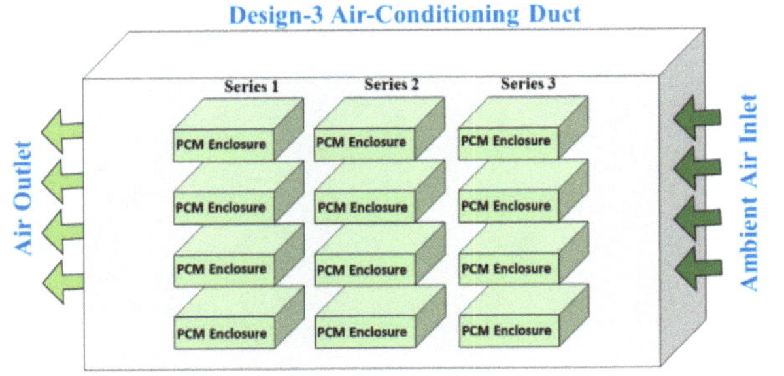

(c)

Design-4 Air-conditioning Duct

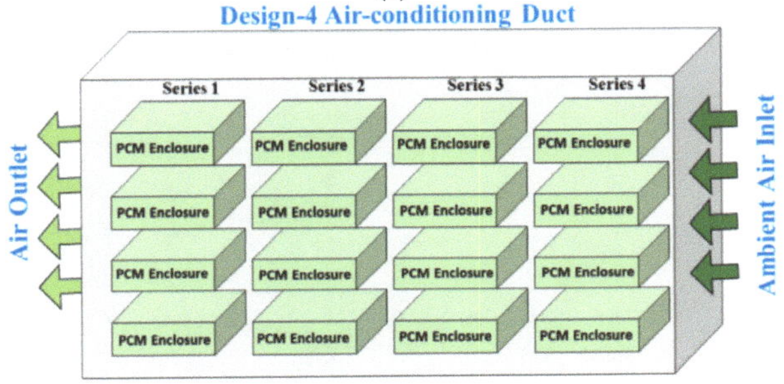

(d)

Figure 3. (**a**) Model of an AC duct with 1 series of the PCM (Design 1); (**b**) model of an AC duct with 2 series of the PCM (Design 2); (**c**) model of an AC duct with 3 series of the PCM (Design 3); and (**d**) model of an AC duct with 4 series of the PCM (Design 4).

Differential Equation (3) governs the transient heat transfer in two dimensions [27]. Heat flux (q), representing incident irradiance, is applied as a boundary condition at the PV surface.

$$\rho c \frac{\partial T}{\partial t} - \left[\frac{\partial}{\partial x_i} \left(k_{ij} \frac{\partial T}{\partial x_j} \right) \right] = 0 \tag{3}$$

where ρ is density, c is heat capacity, k is thermal conductivity, T is temperature, t is time and x_i, and x_j are unit vectors. Equation (3) accounts for conduction in the domain ignoring boundary heat losses. The heat losses (convection and radiation) are imposed at the boundary as given in Equations (4) and (5):

$$Z_c = h_c A (T - T_{amb}) \tag{4}$$

$$Z_r = \sigma \varepsilon A \left(T^4 - T_\infty^4 \right) \tag{5}$$

Z_c and Z_r are convective and radiative heat losses, h_c is the convective heat loss coefficient, A is the PV area, ε is PV emissivity, σ is the Stefan–Boltzman constant, T_{amb} is ambient temperature, and T_∞ is the sky temperature. The unified governing heat transfer equation imposing heat losses yields Equation (6).

$$\rho c \frac{\partial T}{\partial t} - \left[\frac{\partial}{\partial x_i} \left(k_{ij} \frac{\partial T}{\partial x_j} \right) \right] + Z_c + Z_r = 0 \tag{6}$$

The weak formulation is generated by multiplying the test function δT and integrating it over the domain, resulting in Equation (7):

$$\underbrace{\int_{\Omega} \delta T \cdot \rho c \frac{\partial T}{\partial t} \partial\Omega}_{1} - \underbrace{\int_{\Omega} \delta T \cdot \left[\frac{\partial}{\partial X_i} \left(k_{ij} \frac{\partial T}{\partial X_j} \right) \right] \partial\Omega}_{2} + \underbrace{\int_{\Gamma_\sigma} \delta T \cdot (z_c + z_r) \partial A = 0}_{3} \tag{7}$$

The weak formulation is simplified by applying the Green–Gauss and divergence theorem in part 2 of Equation (8), resulting in Equation (10):

$$\underbrace{\int_{\Omega} \delta T \cdot \rho c \frac{\partial T}{\partial t} \partial\Omega}_{1} + \underbrace{\int_{\Omega} \left[k_{11} \frac{\partial \delta T}{\partial X_1} \left(\frac{\partial T}{\partial X_1} \right) + k_{22} \frac{\partial \delta T}{\partial X_2} \left(\frac{\partial T}{\partial X_2} \right) \right]}_{2} + \underbrace{\int_{\Gamma} \delta T \cdot (z_c + z_r) \partial A = 0}_{3} \tag{8}$$

The energy balance is retained in the weak formulation for 2D transient differential heat diffusion given in Equation (9).

$$M\dot{T} + KT - \hat{q} + (H + R)T = 0 \tag{9}$$

where $\dot{T}$ is the time derivative of temperature, $\hat{q}$ is the irradiance or boundary flux matrix, and M, K, H and R are the mass, conductivity, convection, and radiation matrices, respectively. The temporal discretization of the domain is carried out using the Crank–Nicholson method [28,29]. The heat storage during a phase change is simulated by applying the effective heat capacity method [30] given in Equation (10):

$$c_{p,e} = c_0 + \frac{L}{T_s - T_l} \text{ if } T_s \leqslant T \leqslant T_l \text{ or else } c_{p,e} = c_0 \tag{10}$$

Initial and Boundary Conditions

At t = 0, both the liquid and solid phases of the PCM were assumed to be motionless. The boundary was considered with a no-slip assumption for the wall boundaries of the PCM. The outer wall of the AC duct was taken as adiabatic. The inlet air was considered at atmospheric pressure. The initial condition of the temperature was considered at 25 °C. Table 2 shows the thermal and physical properties of the used materials in the study.

Table 2. Properties of the material used in the simulation and experimental study.

Property	RT-31	$CaCl_2{\cdot}6H_2O$	Aluminum	Air
Melting temperature range (°C)	27–33	27.7–32.5	660	-
Latent heat of fusion (kJ kg^{-1})	158	187	-	-
Specific heat (kJ kg^{-1} K^{-1})	2	1.4 (solid), 2.2 (liquid)	8.7	1.007
Density (kg m^{-3})	900	1710 (solid), 1530 (liquid)	2710	1.14
Thermal conductivity (W m^{-1} K^{-1})	0.2	1.09 (solid), 0.53 (liquid)	237	0.028
Velocity (m s^{-1})		-	-	1, 2, 3, 4 m/s

The finite volume method using ANSYS/Fluent was employed to solve the equations of continuity, energy, and momentum, while melting/solidification and energy models were used to simulate heat transfer and phase transition. Smaller effects, such as the expansion or contraction of the PCM after a phase shift, natural convection inside the molten PCM, and the floating or sinking of solid dendrites within the molten PCM due to buoyancy, are not considered in the model. External ambient conditions are added to the input heat flow boundary.

In the ANSYS/Fluent numerical simulations, a meticulous approach was taken to ensure both accuracy and efficiency, extending also to the meshing strategy. The computational domain was discretized into a finely structured mesh, with a total of 25 divisions in each spatial direction. This mesh refinement was chosen to strike a balance between capturing the intricacies of the geometry and minimizing computational costs. The convergence criteria were set at a maximum residual of 0.001 for continuity and velocities in the x, y, and z directions, guaranteeing a robust convergence of the solution. For thermal considerations, the energy residual was maintained at an exceptionally low threshold of 0.000001. The simulations advanced in time with a discrete time step of 10 s. These carefully chosen numerical parameters collectively ensure the reliability of the results, providing a solid foundation for the analysis by striking a judicious balance between mesh resolution and computational efficiency in the ANSYS/Fluent simulations. HP Z840 workstation Core i-7 (20 cores) (sourced from HP and made in China) was used for simulations. Each case used 4 cores and 2 days for the completion of the simulation.

4. Results and Discussion

4.1. Experimental Validation

Figure 4 shows the experiment conducted on Design 1 at an air speed of 4 m/s for RT-31, where inlet air temperature (from 7 a.m. to 7 a.m.) subjected to the AC duct system started at 27 °C and rose as the solar radiation increased, reaching the maximum temperature of 45 °C at noon time (1–3) p.m. in 8 h. The inclusion of PCM as the latent heat storage system in the AC duct system cools the outlet air temperature and reduces the peak air temperature, achieving a temperature drop of 4 °C as the maximum and 1 °C as the average. The experimental findings effectively corroborate the simulation results, which are particularly evident 11 h into the study when the ambient air temperature begins to drop. Notably, the outlet air temperature surpasses that of the ambient air, initiating a phase where the PCM starts to transfer its stored thermal energy to the surrounding air. This process continues and becomes particularly pronounced after 12 h. This insightful observation sheds light on a critical juncture—the point at which the PCM, having absorbed heat during the warmer period, now serves as a heat source, warming the ambient air. This knowledge offers a valuable perspective on determining an opportune cutoff point in the system operation. Strategically, this signifies the juncture where directing only ambient air through the evaporator, while concurrently allowing the PCM to cool for the next cycle, becomes advantageous. This innovative approach has the potential to enhance the efficiency of the air conditioning system, particularly during peak hours, by leveraging the PCM's thermal properties to optimize the cooling process. The synchronization of the system's operation with ambient temperature fluctuations presents an intriguing avenue for improving overall energy efficiency and performance.

It is noted that the PCM will cool the air and it will result in a lower temperature of the pre-cooled air compared with the ambient air as long as the PCM is melting and its temperature is below the ambient temperature. Once the PCM is completely melted by 13:00, a reverse trend is seen as the pre-cooled air that gets out of the PCM is warmer than the ambient air, which is in complete compliance with the heat transfer mechanism of the flow of heat according to the temperature gradient.

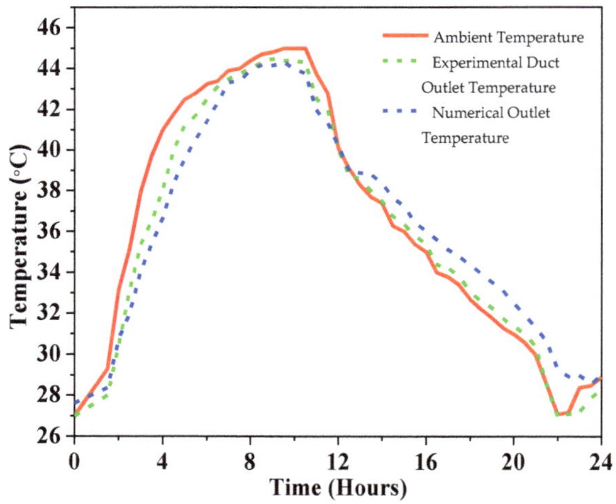

Figure 4. Experimental and numerical results of the outlet temperature of the duct containing RT-31 in Design 1 at 4 m/s under ambient temperature.

4.2. Numerical Results

The present work provides a study assessing the performance of a thermal energy storage system comprised of four different designs placed within an air conditioning (AC) duct. This evaluation involved monitoring key parameters, such as the outlet temperature, PCM average temperature, and melting fraction at varying inlet air speeds (1 m/s, 2 m/s, 3 m/s, and 4 m/s), with the inlet air temperature set to the average highest daytime temperature of 45 °C. The realistic transient summertime temperature was incorporated as input to the model and, accordingly, the results are predicted for one day. The one-day simulation results are compared with the experiment and accordingly are presented to validate the model. The validated model is used to predict off-normal conditions and is presented accordingly. The idea here was to generalize the results that can apply to a wide range of input conditions, not a specific weather condition or climate.

4.2.1. Outlet Temperature

Figure 5a–d show the 24 h assessment of air outlet temperature using RT-31 as the PCM. Figure 5a shows the effects of 1 m/s air on Designs 1, 2, 3, and 4. Using PCM enclosures in the duct system, the outlet air temperature is reduced below the input temperature, with a temperature drop of about 7 °C for Design 1, 10 °C for Design 2, 12 °C for Design 3, and 12.2 °C for Design 4. The drop in air temperature is attributed to the heat transfer to the cooled PCM, which absorbs the passing air heat and exhibits a solid–liquid phase transition. The PCM kept absorbing thermal energy at a nearly constant temperature until it completely melted and started to heat up sensibly, thus reducing the heat transfer from hot outside air to the PCM. The heated PCM eventually reached a steady state with the ambient temperature in approximately 12.5 h for Design 1, 15 h for Design 2, 16.5 h for Design 3, and 18 h for Design 4. It can be noted that the PCM reached 40 °C in 12 min, 54 min, 57 min, and 1.5 h in the case of Design 1, Design 2, Design 3, and Design 4, respectively. Since the inlet air temperature remained around 45 °C at the peak daytime for 2 to 3 h, hence the proposed designs can provide a considerable drop in temperature and a resulting reduction in the cooling load, leading to an improved COP of AC, reducing the installed capacity, and saving energy. Figure 5b shows the impact of increasing the air speed to 2 m/s at a constant inlet air temperature of 45 °C subjected to Designs 1, 2, 3, and 4. The peak air temperature dropped to a maximum of 4 °C for Design 1, 7 °C for Design 2, and 8 °C for Design 3 and Design 4. In Figure 5c, all four designs are subjected to 3 m/s, thereby achieving the peak inlet air temperature drop of 6.5 °C at the most with Design 4 and 3 °C at the least with

Design 1. In Figure 5d, all four designs are exposed to an air inlet velocity of 4 m/s, with the highest temperature drops thereby showing a 5 °C drop in peak air temperature at most with Design 4 and 2.5 °C at least with Design 1. As the effective duration of heat transfer decreases with increases in velocity, the temperature drop observed at the outlet air reduces.

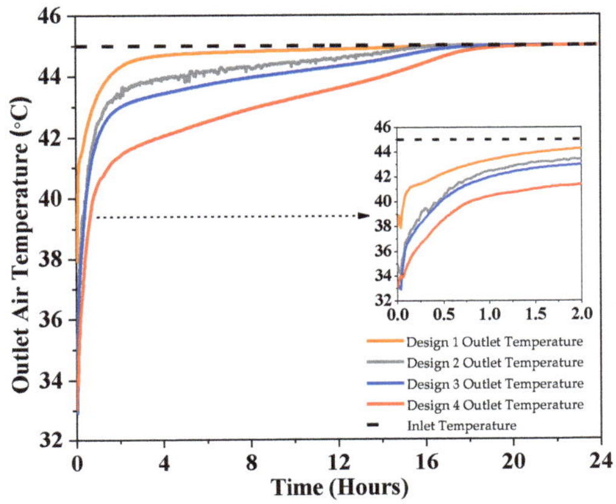

(a)

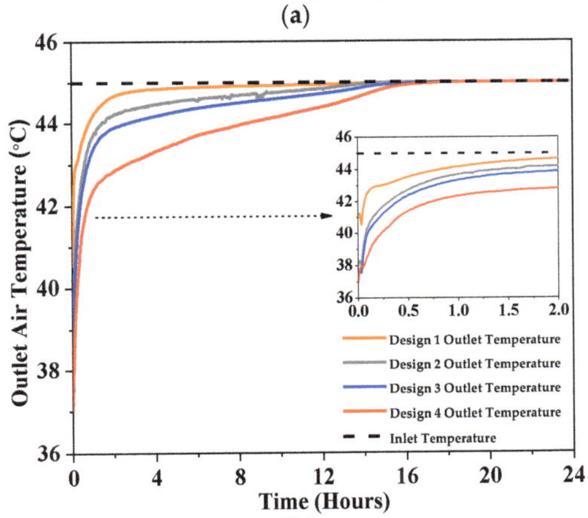

(b)

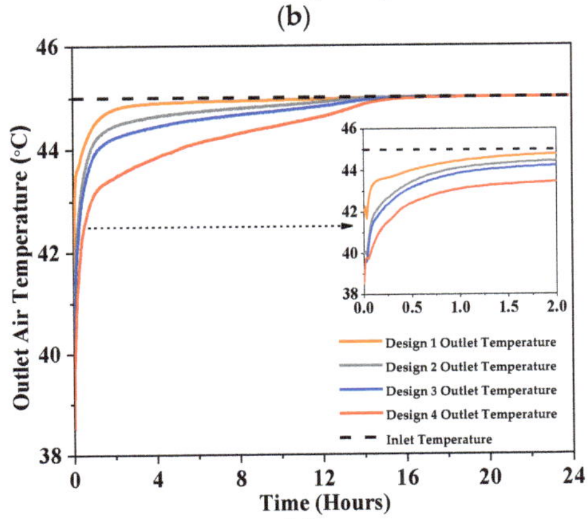

(c)

Figure 5. *Cont.*

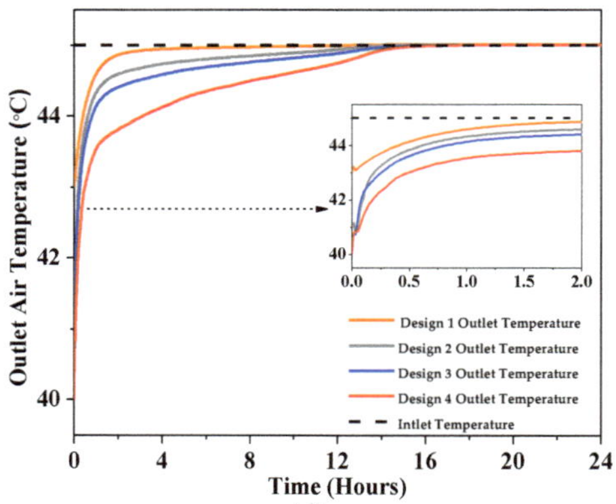

(**d**)

Figure 5. (**a**) Outlet air temperatures of Designs 1, 2, 3, and 4 with RT-31 at an air inlet velocity of 1 m/s. (**b**) Outlet air temperatures of Designs 1, 2, 3, and 4 with RT-31 at an air inlet velocity of 2 m/s. (**c**) Outlet air temperatures of Designs 1, 2, 3, and 4 with RT-31 at an air inlet velocity of 3 m/s. (**d**) Outlet air temperatures of Designs 1, 2, 3, and 4 with RT-31 at an air inlet velocity of 4 m/s.

The numerical scheme is tested by replacing the paraffin-based PCM (RT-31) with a salt hydrate PCM ($CaCl_2 \cdot 6H_2O$) to evaluate the impact of the PCM on the performance of the proposed air pre-cooling system.

Results are presented in Figure 6a–d. Figure 6a shows that Design 1 observed the lowest drop of 7 °C in the peak air temperature while Design 4 observed the highest drop of 12 °C in the peak air temperature. Design 1 maintained an outlet air temperature below 40 °C for 12 min, while Design 4 maintained an air temperature below 40 °C for 1.4 h. Figure 6b–d, show lesser air temperature drops when the airspeed is increased to 2 m/s, 3 m/s, and 4 m/s, respectively.

The transient outlet air temperature in all cases is lower than the average steady-state ambient air acting as inlet air. The difference is huge in the beginning as the PCM significantly cools the air for the first 1–2 h due to initial PCM cooling by a sensible mechanism, while the difference starts rapidly dropping with time, indicating sensible heating of the PCM. At the advent of the PMC melting point, the PCM starts melting and the temperature of the outlet air stabilizes for 4–5 h due to latent heat absorption by the PCM at near isothermal conditions representative of a steady outlet air temperature. After 7 h in most cases, the PCM completes its phase change and starts absorbing sensible heat, which again increases the gradient of the temperature rise of the outlet air.

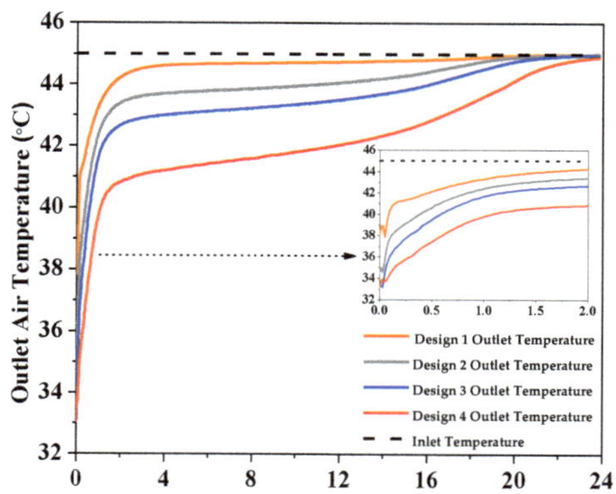

(**a**)

Figure 6. *Cont.*

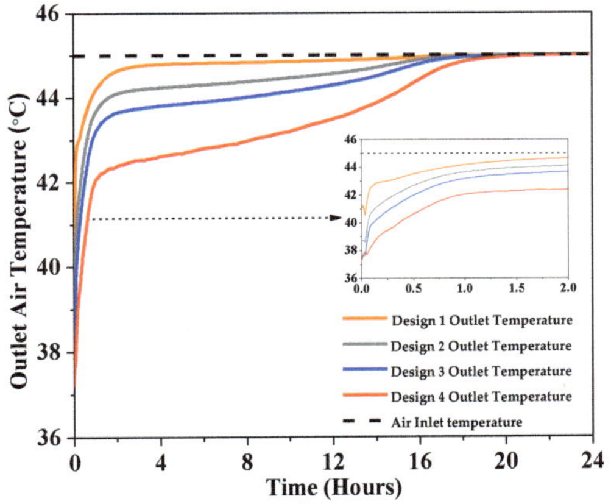

(b)

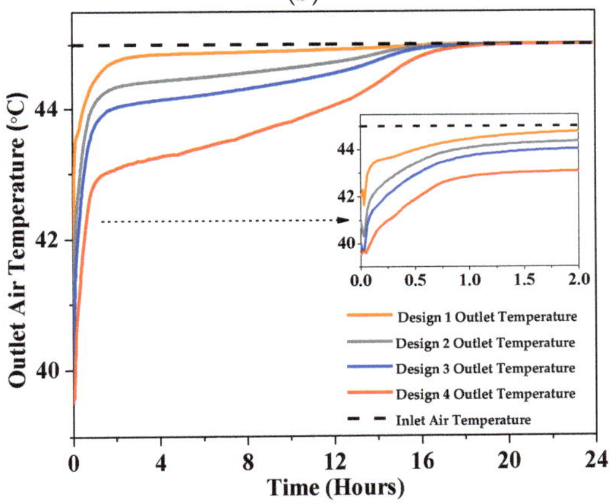

(c)

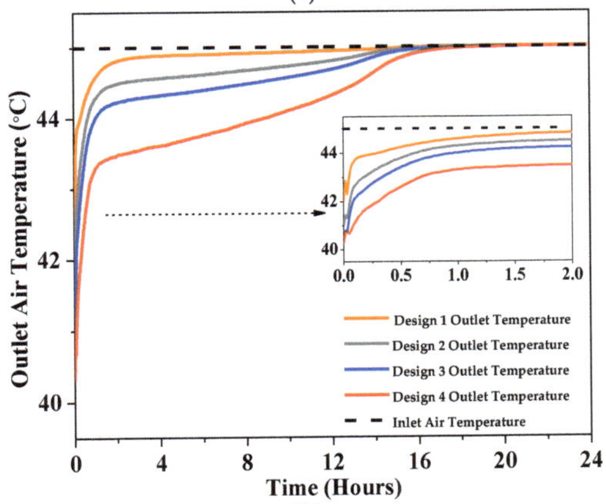

(d)

Figure 6. (**a**) Outlet air temperatures of Designs 1, 2, 3, and 4 with $CaCl_2{\cdot}6H_2O$ at an air inlet velocity of 1 m/s. (**b**) Outlet air temperatures of Designs 1, 2, 3, and 4 with $CaCl_2{\cdot}6H_2O$ at an air inlet velocity of 2 m/s. (**c**) Outlet air temperatures of Designs 1, 2, 3, and 4 with $CaCl_2{\cdot}6H_2O$ at an air inlet velocity of 3 m/s. (**d**) Outlet air temperatures of Designs 1, 2, 3, and 4 with $CaCl_2{\cdot}6H_2O$ at an air inlet velocity of 4 m/s.

The temperature rise happened initially with a higher gradient due to sensible heating of the PCM in the solid state. At the phase change, the gradient of the temperature rise is reduced and kept lower until the PCM has fully melted, after 12 h in most cases. After the PCM had melted completely, the gradient started to pick up again due to sensible heating of the PCM in liquid form and finally stabilized due to thermal equilibrium with the outdoor air temperature.

4.2.2. Average PCM Temperature

Figures 7a–d and 8a–d show the average PCM temperatures for Designs 1, 2, 3, and 4 of RT-31 and $CaCl_2{\cdot}6H_2O$ at different velocities (1 m/s, 2 m/s, 3 m/s, and 4 m/s). In Figure 7a, it can be observed that sensible heating occurs in the PCM temperature range of 20 °C–27 °C for a very short time followed by the latent heat absorption, as reflected by a lower temperature gradient from PCM temperature of 33 °C to 45 °C. The delayed temperature gradient helps maintain the PCM temperature substantially below the ambient temperature for several hours and helps transfer heat from inlet air to the PCM, thereby producing cooled air at the outlet. A similar trend is observed with a lesser temperature drop when increasing the air velocity to 2 m/s, 3 m/s, and 4 m/s, as shown in Figure 7b–d. As the airspeed increases, the PCM takes less time to reach the reference temperature due to a greater enhanced convective heat transfer coefficient between the air and the PCM.

In the case of $CaCl_2{\cdot}6H_2O$ shown in Figure 8a–d, it can be observed that sensible heat storage prevails in the PCM from 20 °C to 27.7 °C, while the latent heat storage dominates from 27.7 °C to 32.5 °C. After complete PCM melting, sensible heat storage takes over from 32.5 °C to 45 °C. As the air velocity increases, the PCM temperature gradient becomes larger and the PCM tends to reach the reference temperature in a shorter time duration as compared to the low velocity.

As shown in Table 2, $CaCl_2{\cdot}6H_2O$ possesses a higher latent heat of fusion, density, heat capacity, and thermal conductivity than RT-31; hence, the cooling effect of $CaCl_2{\cdot}6H_2O$ is higher than the RT-31, as it produces a higher net cooling effect on the incoming air that is reflected by the lower temperature gradient in Figure 8a–d.

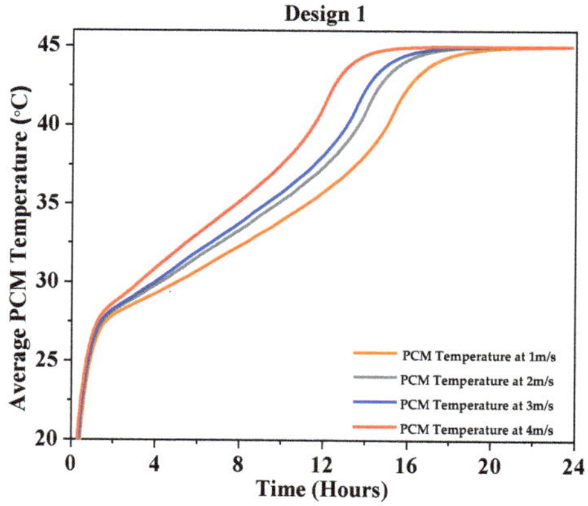

(a)

Figure 7. *Cont.*

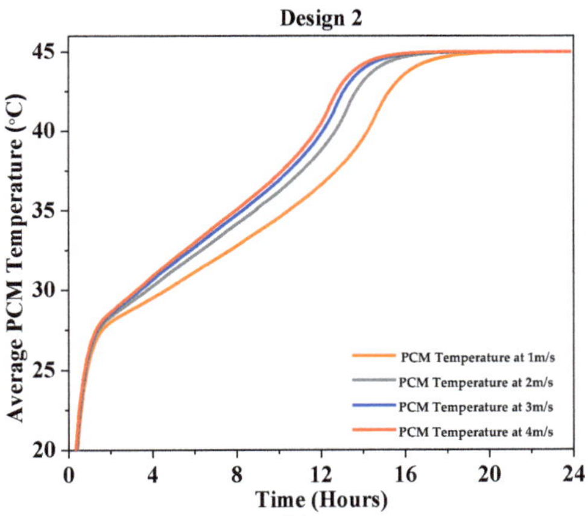

(b)

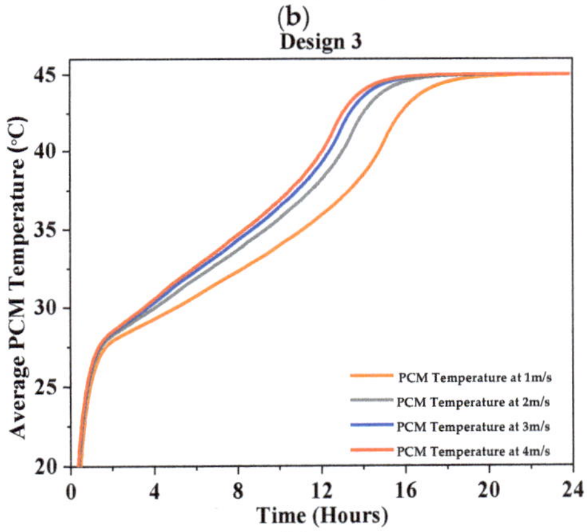

(c)

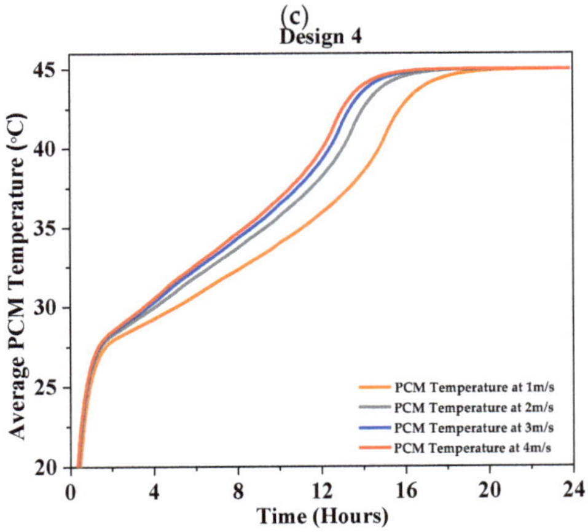

(d)

Figure 7. (**a**) Average temperature of RT-31 (Design 1) at air inlet velocities of 1 m/s, 2 m/s, 3 m/s, and 4 m/s. (**b**) Average temperature of RT-31 (Design 2) at air inlet velocities of 1 m/s, 2 m/s, 3 m/s, and 4 m/s. (**c**) Average temperature of RT-31 (Design 3) at air inlet velocities of 1 m/s, 2 m/s, 3 m/s, and 4 m/s. (**d**) Average temperature of RT-31 (Design 4) at air inlet velocities of 1 m/s, 2 m/s, 3 m/s, and 4 m/s.

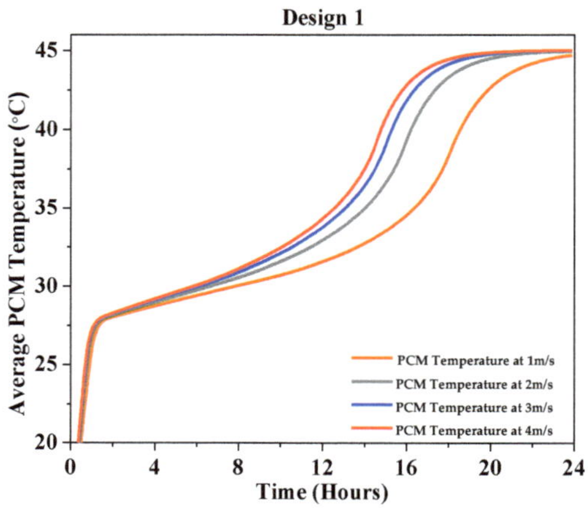

(a)

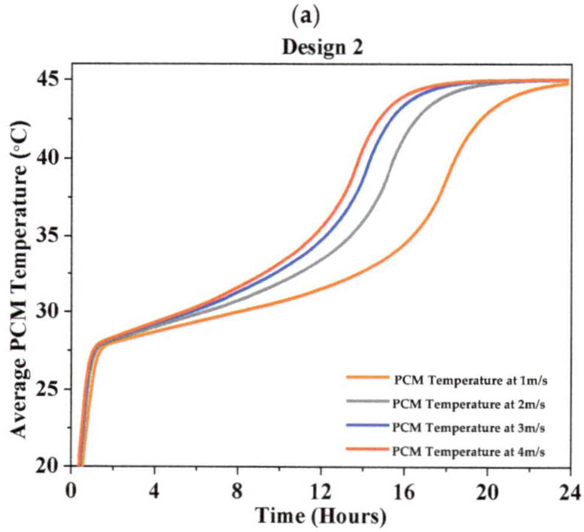

(b)

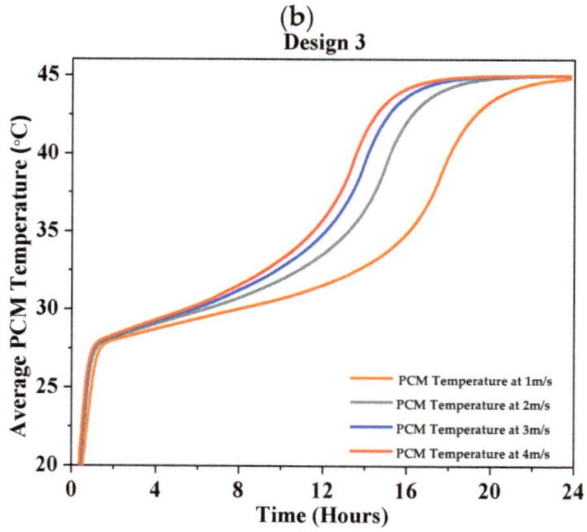

(c)

Figure 8. *Cont.*

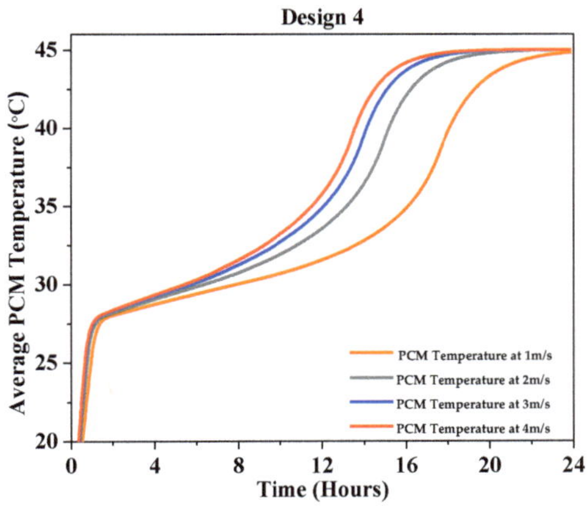

(**d**)

Figure 8. (**a**) Average temperature of $CaCl_2{\cdot}6H_2O$ (Design 1) at air inlet velocities of 1 m/s, 2 m/s, 3 m/s, and 4 m/s. (**b**) Average temperature of $CaCl_2{\cdot}6H_2O$ (Design 2) at air inlet velocities of 1 m/s, 2 m/s, 3 m/s, and 4 m/s. (**c**) Average temperature of $CaCl_2{\cdot}6H_2O$ (Design 3) at air inlet velocities of 1 m/s, 2 m/s, 3 m/s, and 4 m/s. (**d**) Average temperature of $CaCl_2{\cdot}6H_2O$ (Design 4) at air inlet velocities of 1 m/s, 2 m/s, 3 m/s, and 4 m/s.

4.2.3. Melting Fraction

During the daytime, the hot outside air routes through the duct containing the PCM to be pre-cooled before entering the evaporator. The effects of PCM containment designs, air speed, and outside air temperature on the PCM melting rate are presented in Figure 9a–d for RT-31 and Figure 9 for $CaCl_2{\cdot}6H_2O$ and show the time-based average of the PCM melting fraction during the melting process. The air with an inlet temperature of 45 °C is subjected to Designs 1, 2, 3, and 4 with variable inlet velocities of 1 m/s, 2 m/s, 3 m/s, and 4 m/s. As the air velocity increases, the volume of air flowing over the PCM containers also increases, and hence the heat transfer is increased. Therefore, at an air velocity of 1 m/s, the PCM takes more time to completely melt compared with the air velocity of 4 m/s. Figure 10a–d show the melting fraction of $CaCl_2{\cdot}6H_2O$. The heat transfer is greater compared with RT-31; hence, the temperature drop is greater. Hence, it takes more time to melt than RT-31.

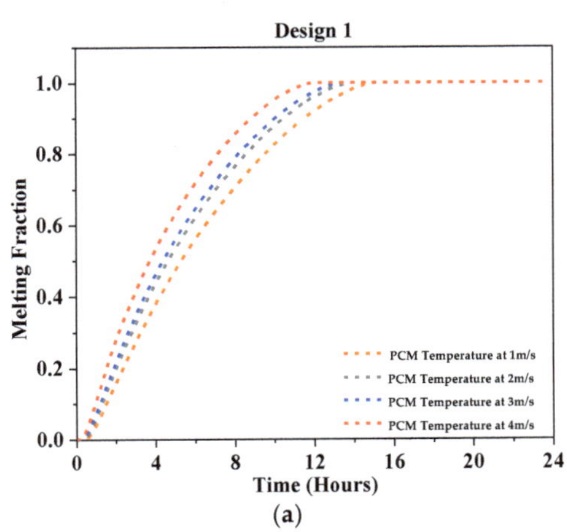

(**a**)

Figure 9. *Cont.*

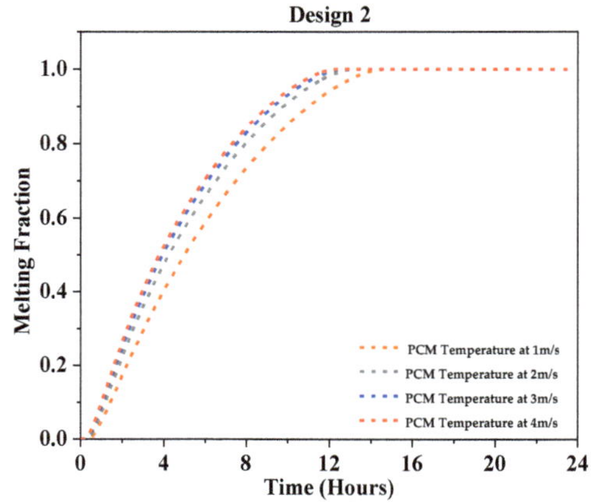

(b)

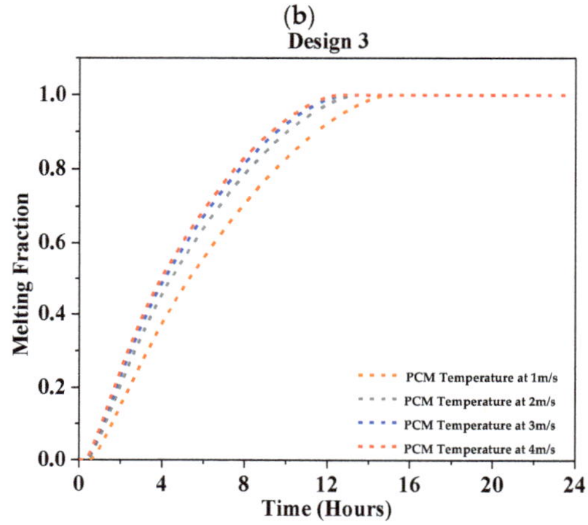

(c)

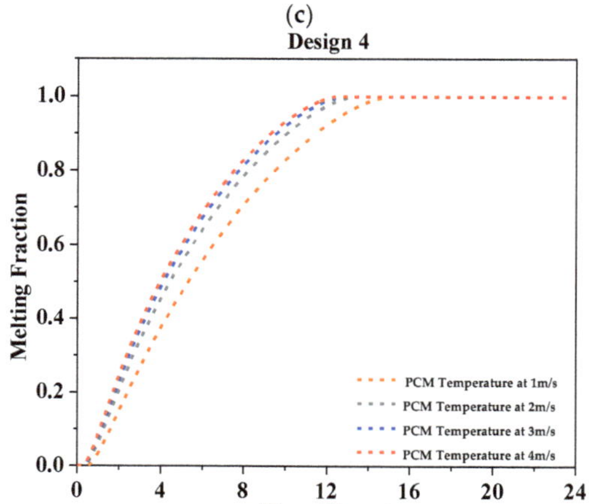

(d)

Figure 9. (**a**) Average melting fraction of RT-31 (Design 1) at air inlet velocities of 1 m/s, 2 m/s, 3 m/s, and 4 m/s. (**b**) Average melting fraction of RT-31 (Design 2) at air inlet velocities of 1 m/s, 2 m/s, 3 m/s, and 4 m/s. (**c**) Average melting fraction of RT-31 (Design 3) at air inlet velocities of 1 m/s, 2 m/s, 3 m/s, and 4 m/s. (**d**) Average melting fraction of RT-31 (Design 4) at air inlet velocities of 1 m/s, 2 m/s, 3 m/s, and 4 m/s.

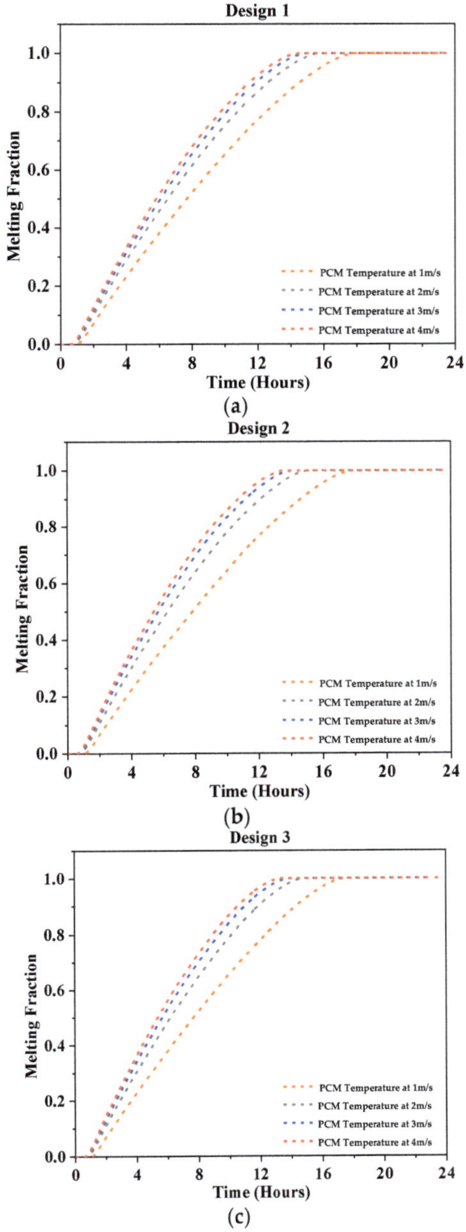

Figure 10. *Cont.*

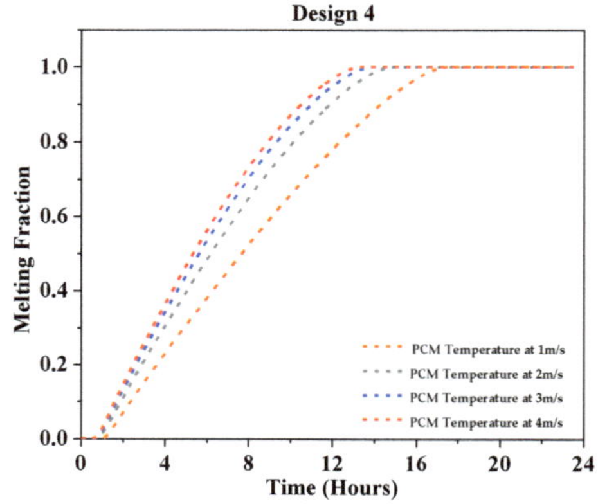

(**d**)

Figure 10. (**a**) Average melting fraction of $CaCl_2{\cdot}6H_2O$ (Design 1) at air inlet velocities of 1 m/s, 2 m/s, 3 m/s, and 4 m/s. (**b**) Average melting fraction of $CaCl_2{\cdot}6H_2O$ (Design 2) at air inlet velocities of 1 m/s, 2 m/s, 3 m/s, and 4 m/s. (**c**) Average melting fraction of $CaCl_2{\cdot}6H_2O$ (Design 4) at air inlet velocities of 1 m/s, 2 m/s, 3 m/s, and 4 m/s. (**d**) Average melting fraction of $CaCl_2{\cdot}6H_2O$ (Design 4) at air inlet velocities of 1 m/s, 2 m/s, 3 m/s, and 4 m/s.

4.2.4. Velocity Streamlines and Melting Fraction Contours

Figure 11a–d capture a detailed analysis of velocity streamlines at a velocity of 4 m/s for Designs 1, 2, 3, and 4, showcasing a duct that incorporates PCM containers. The visualization offers a nuanced depiction of fluid dynamics within the system, providing valuable insights into the interaction between the flowing fluid and the PCM containers. The color-coded streamlines vividly represent the distribution of velocities, aiding in the identification of regions with varying flow intensities. Notably, the impact of the PCM containers on the velocity field is discernible, underlining its influence on heat transfer processes. Through the observed streamlines, distinct flow patterns, such as recirculation zones and vortex formations, become apparent, contributing to a comprehensive understanding of the system's behavior. Additionally, it is noted that the vortex formation perpendicular to the flow can affect the fan power requirements, which eventually may increase energy consumption to keep the flow for indoor comfort. The extent of the impact will be explored in the future to maintain indoor comfort.

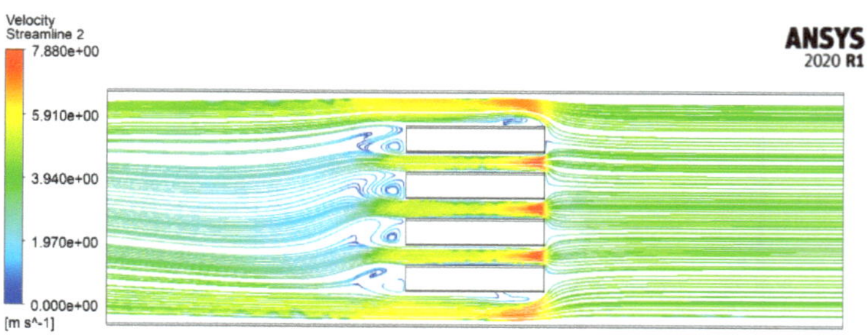

(**a**)

Figure 11. *Cont.*

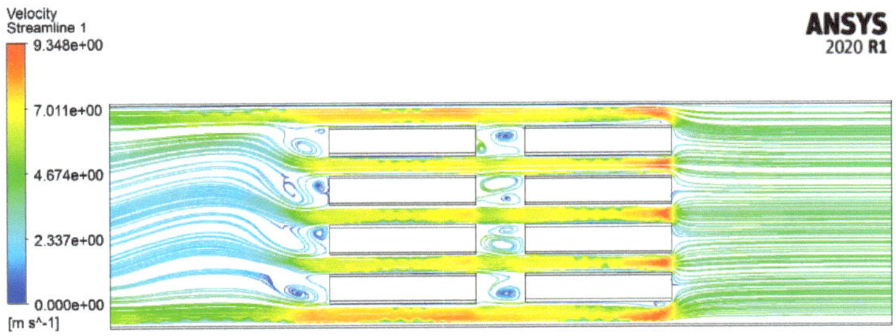

(b)

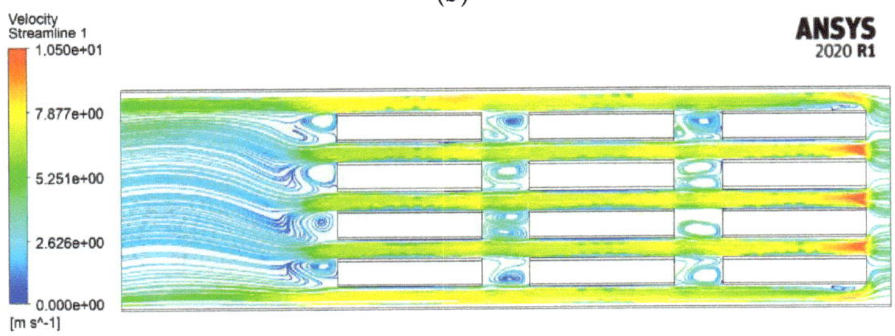

(c)

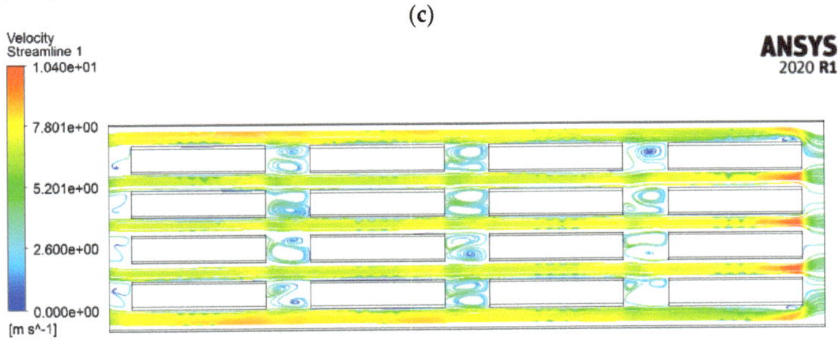

(d)

Figure 11. (**a**) Velocity streamlines at 4 m/s for Design 1. (**b**) Velocity streamlines at 4 m/s for Design 2. (**c**) Velocity streamlines at 4 m/s for Design 3. (**d**) Velocity streamlines at 4 m/s for Design 4.

Figure 12a–l shows the melting fraction contours for Designs 1, 2, 3, and 4 (RT-31) at a fluid velocity of 4 m/s after 2 h, 4 h, and 8 h durations and provides a detailed and insightful analysis of the dynamic phase change phenomena within the system. The contour plot not only captures the temporal evolution of the melting process but also offers a spatial representation of melting fraction variations across the designated region of interest.

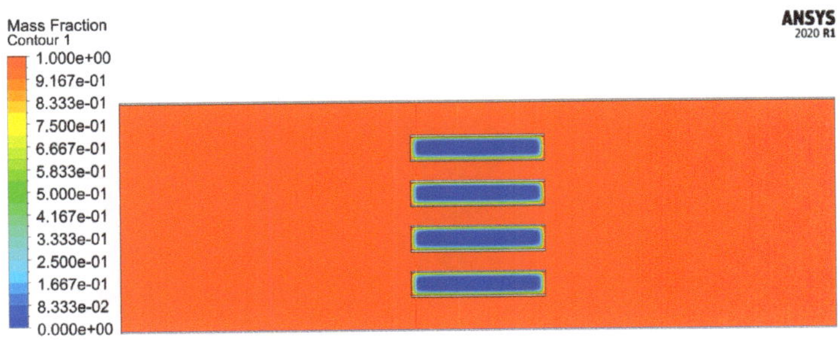

(a)

Figure 12. *Cont.*

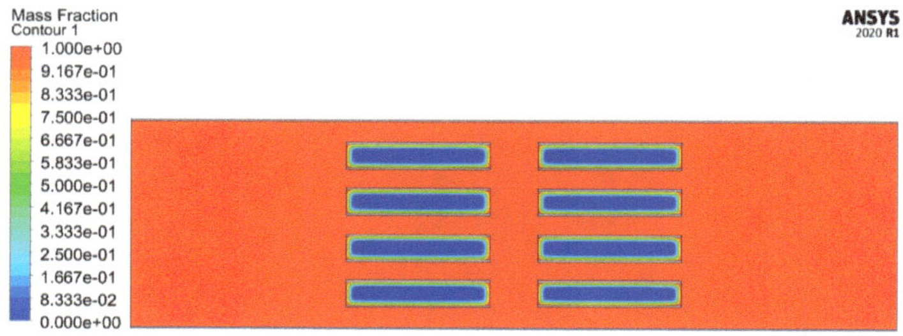

(b)

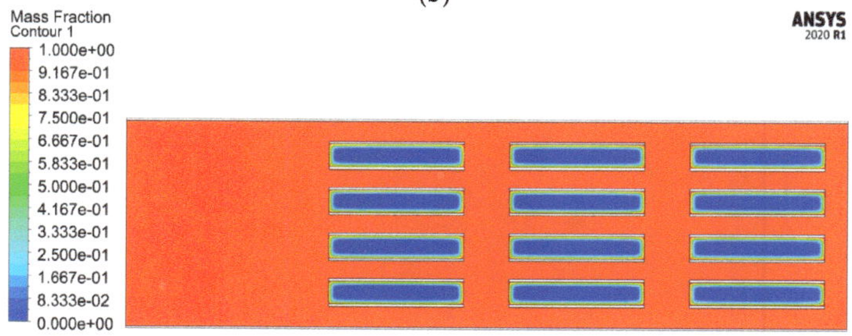

(c)

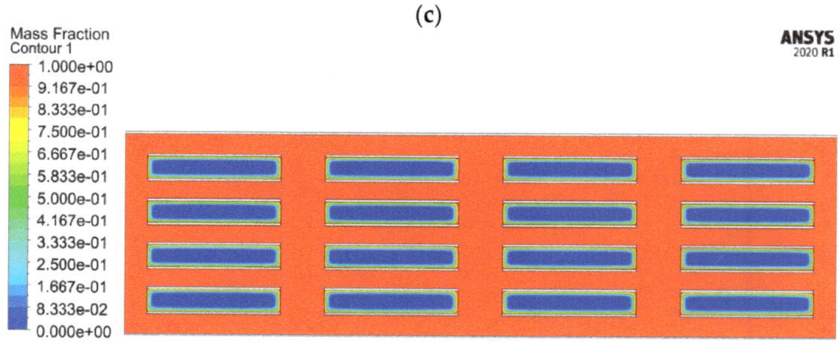

(d)

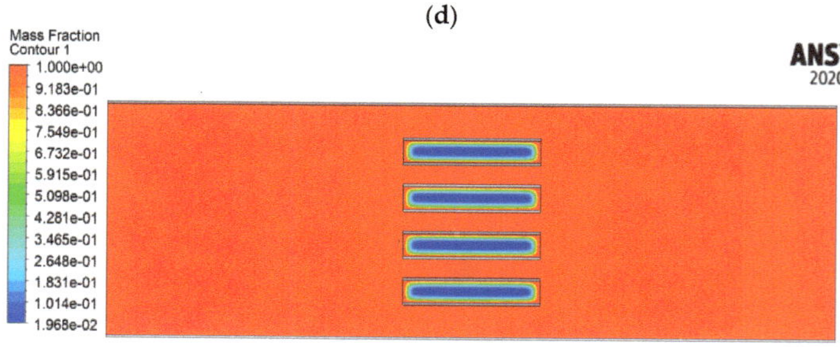

(e)

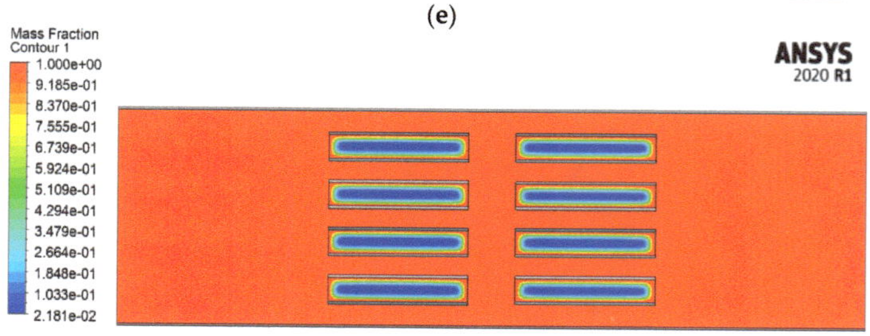

(f)

Figure 12. *Cont.*

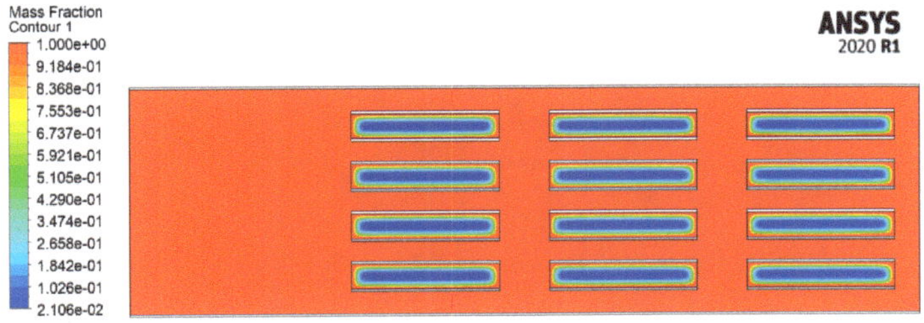

(**g**)

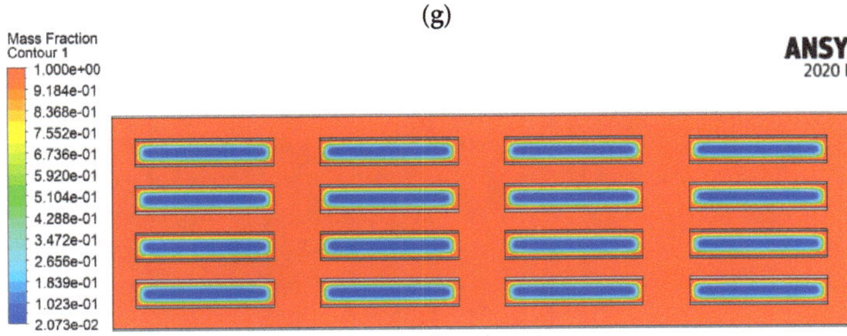

(**h**)

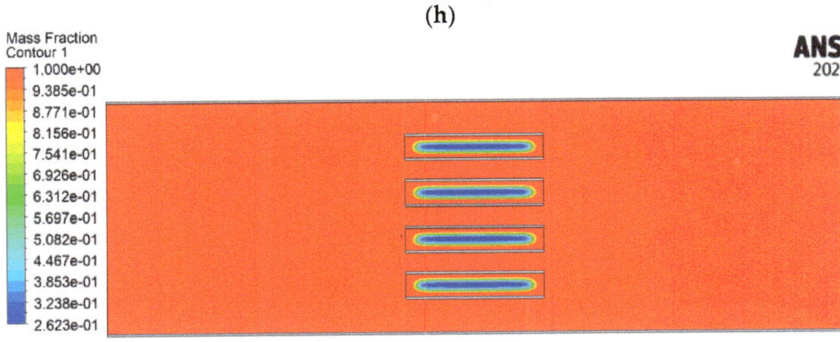

(**i**)

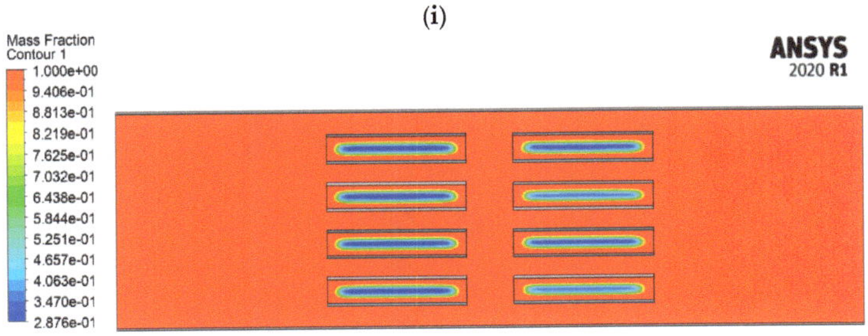

(**j**)

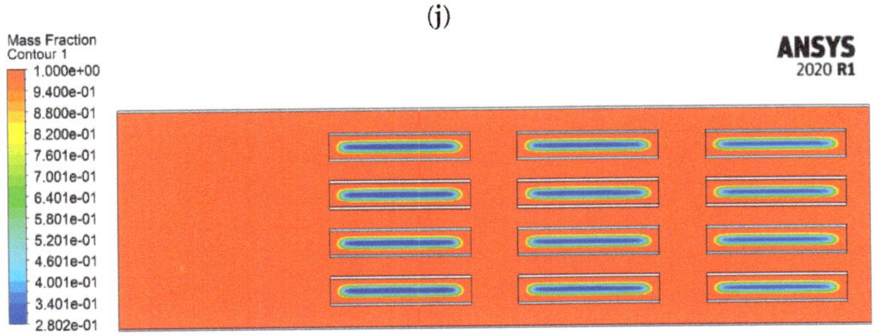

(**k**)

Figure 12. *Cont.*

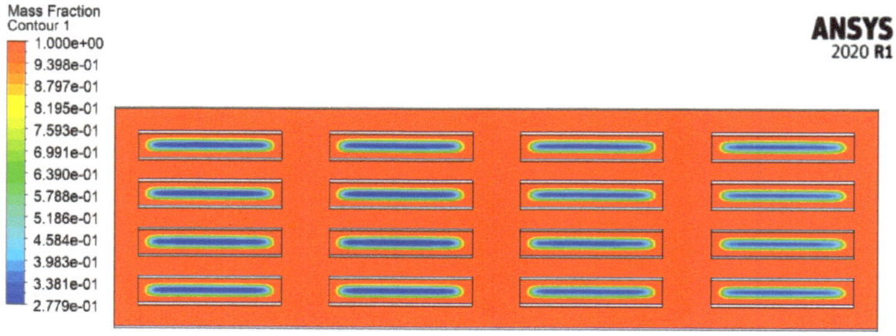

(l)

Figure 12. (**a**) Contours of the melting fraction for Design 1 (RT-31) at 4 m/s after 2 h. (**b**) Contours of the melting fraction for Design 2 (RT-31) at 4 m/s after 2 h. (**c**) Contours of the melting fraction for Design 3 (RT-31) at 4 m/s after 2 h. (**d**) Contours of the melting fraction for Design 4 (RT-31) at 4 m/s after 2 h. (**e**) Contours of the melting fraction for Design 1 (RT-31) at 4 m/s after 4 h. (**f**) Contours of the melting fraction for Design 2 (RT-31) at 4 m/s after 4 h. (**g**) Contours of the melting fraction for Design 3 (RT-31) at 4 m/s after 4 h. (**h**) Contours of the melting fraction for Design 4 (RT-31) at 4 m/s after 4 h. (**i**) Contours of the melting fraction for Design 1 (RT-31) at 4 m/s after 8 h. (**j**) Contours of the melting fraction for Design 2 (RT-31) at 4 m/s after 8 h. (**k**) Contours of the melting fraction for Design 3 (RT-31) at 4 m/s after 8 h. (**l**) Contours of the melting fraction for Design 4 (RT-31) at 4 m/s after 8 h.

Figure 13a,b present a comprehensive overview of Design 1's performance with two distinct phase change materials: RT-31 and $CaCl_2{\cdot}6H_2O$. The plotted data includes the outlet temperature, average PCM temperature, and melting fraction, providing critical insights into the system's behavior during the phase transition. Notably, a compelling observation emerges from the graphical representation—a correlation between the maximum melting fraction and the ending of the phase transition region, coinciding with a minimum in the outlet temperature drop.

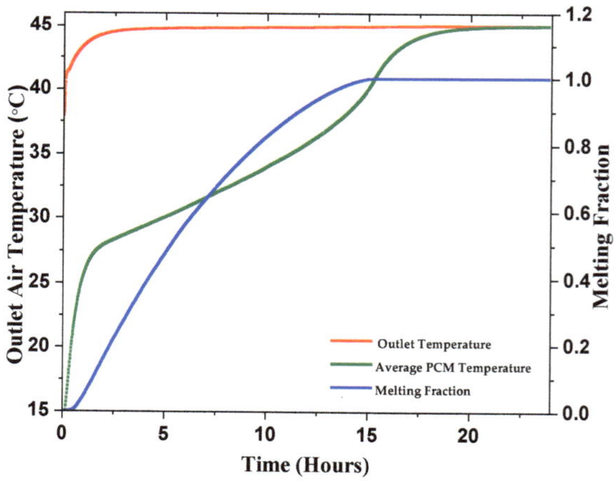

(a)

Figure 13. *Cont.*

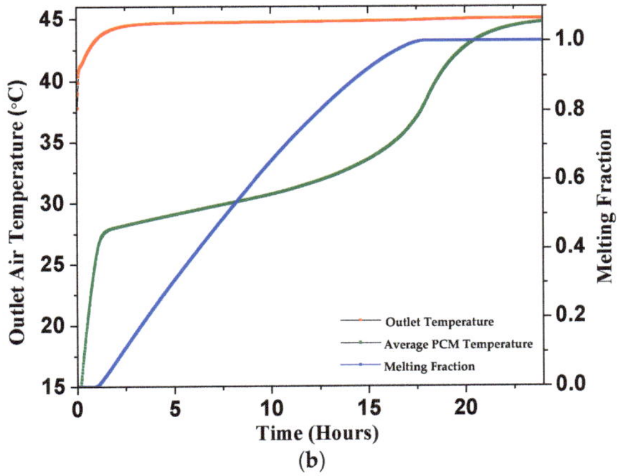

(**b**)

Figure 13. (**a**) Outlet air temperature, melting fraction, and PCM average temperature for Design 1 using RT-31 at 1 m/s. (**b**) Outlet air temperature, melting fraction, and PCM average temperature for Design 1 using $CaCl_2{\cdot}6H_2O$ at 1 m/s.

The increased amount of PCM in subsequent cases generally kept the PCM temperature lower for several hours, which represents the impact of an increased mass leading to increased energy absorption that impacts the outlet air temperature and is almost in line with the increased amount in the order of 2, 3, and 4 in the 2nd, 3rd, and 4th cases. The additional compound impact of an increased stay time and reduced velocity in the case of Design 4 might have contributed as a compound impact and produced a more visible temperature drop.

Figure 14 shows the Reynolds number of the duct with the variation in air inlet velocities. The Reynolds numbers associated with varying inlet velocities (1 m/s, 2 m/s, 3 m/s, and 4 m/s) play a crucial role in characterizing the flow regime within the system. As the inlet velocity increases, the Reynolds number, a dimensionless parameter representing the ratio of inertial forces to viscous forces, demonstrates a proportional rise. The Reynolds numbers for the specified velocities are 2255, 4510, 6765, and 9020, respectively. The observed trend indicates a transition from laminar to turbulent flow as the Reynolds number increases. At lower velocities, such as 1 m/s, the flow is more likely to exhibit laminar characteristics due to the dominance of viscous forces. As the inlet velocity escalates, the Reynolds number surpasses a critical threshold, marking the onset of turbulent flow. In the context of fluid dynamics, this transition is associated with increased mixing, enhanced heat transfer, and altered flow patterns within the system.

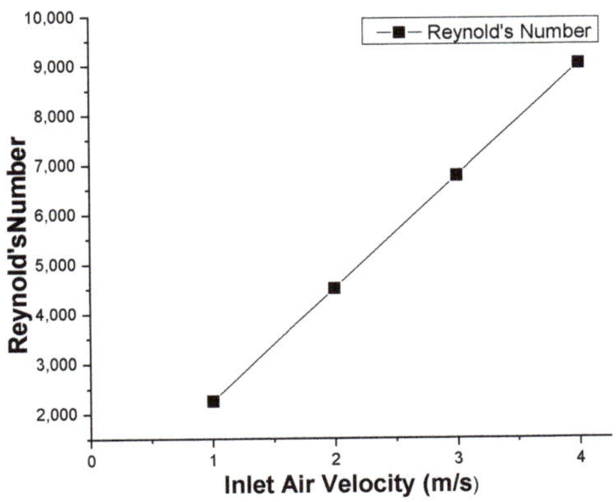

Figure 14. Reynolds number variations with the variation of inlet air velocities.

5. Conclusions

In conclusion, this study presents a comprehensive examination of a PCM-based air pre-cooling system in the context of the hot climate of the UAE. The experiment showed that adding a PCM to the air conditioning duct system pre-cools the air flowing through the duct system through latent heat absorbed by the melting PCM.

The performance of the AC duct system is investigated through numerical simulations when the unit evaporator is coupled with the TES system. The PCM, which is solidified during the night, serves to cool the inlet air entering the evaporator during the day. The study was conducted for two PCMs, RT-31 and $CaCl_2.6H_2O$, with four duct designs subjected to different air velocities of 1 m/s, 2 m/s, 3 m/s, and 4 m/s, with a focus on the outlet temperature, average PCM temperature, and melting fraction. Experimentally, the study was conducted on an RT-31-based PCM with a single series of the PCM (Design 1) at 4 m/s. Numerical results showed that a maximum temperature drop of 13 °C was achieved using Design 4 at 1 m/s with both PCMs. However, $CaCl_2{\cdot}H_2O$ showed more of a cooling effect compared to RT-31. It kept a low outlet temperature for more time compared to RT-31.

- At 1 m/s, a significant outlet air temperature drop was reported, about 7 °C for Design 1, 10 °C for Design 2, 12 °C for Design 3, and 12.2 °C for Design 4 using RT-31.
- At 2 m/s, the peak air temperature dropped by a maximum of 4 °C for Design 1, 7 °C for Design 2, and 8 °C for Design 3 and Design 4, as reported using RT-31.
- At 3 m/s, the peak outlet temperature dropped by 6.5 °C at most with Design 4 and 3 °C at least with Design 1 as reported using RT-31.
- At 4 m/s, the highest temperature drop showed a 5 °C drop in peak air temperature at most with Design 4 and 2.5 °C at least with Design 1 was reported using RT-31.
- Similar highest temperature drop trends were reported using $CaCl_2{\cdot}6H_2O$ compared to paraffin, which was able to sustain the temperature drop for more time.

The results indicate that employing PCM-based thermal energy storage within the air duct system effectively reduces the air temperature, resulting in potential energy savings and improved air conditioning system efficiency, especially during peak hours. The simulation results were validated experimentally, showcasing a considerable agreement between the simulated and experimental air temperature drop. Overall, this research underscores the promise of PCM-based air precooling systems as an effective means of enhancing cooling system performance and reducing energy efficiency in hot climates.

Author Contributions: Conceptualization, M.H. and A.H.; methodology, M.H., A.H., U.M. and M.L.; software, U.M. and A.H.; validation, U.M., A.H. and M.H.; formal analysis, U.M.; investigation, A.H., M.H. and U.M.; resources, M.H.; writing—original draft preparation, U.M.; writing—review and editing, A.H., M.H. and M.L.; visualization, A.H. and M.H.; supervision, M.H. and A.H.; project administration, M.H.; funding acquisition, M.H. All authors have read and agreed to the published version of the manuscript.

Funding: This article is related to a research project funded by the UAE University (12N091), entitled "Air Pre-Cooling through Phase Change Material Integrated into Chiller Water Cooling System to Reduce Peak Air-Conditioning Demand: Case Study of UAEU".

Data Availability Statement: The data presented in this study are available on request from the corresponding author. The data are not publicly available due to privacy.

Acknowledgments: The authors would like to express their appreciation to the Research Office at UAEU and the College of Engineering for their support.

Conflicts of Interest: The authors declare no conflict of interest.

References

1. Gagliano, A.; Patania, F.; Nocera, F.; Ferlito, A.; Galesi, A. Thermal performance of ventilated roofs during the summer period. *Energy Build.* **2012**, *49*, 611–618. [CrossRef]
2. Cabeza, L.F.; Chàfer, M. Technological options and strategies towards zero energy buildings contributing to climate change mitigation: A systematic review. *Energy Build.* **2020**, *219*, 110009. [CrossRef]
3. Masood, U.; Haggag, M.; Hassan, A.; Laghari, M. A Review of Phase Change Materials as a Heat Storage Medium for Cooling Applications in the Built Environment. *Buildings* **2023**, *13*, 1595. [CrossRef]
4. Hasan, A.; Al-Sallal, K.A.; Alnoman, H.; Rashid, Y.; Abdelbaqi, S. Effect of phase change materials (PCMs) integrated into a concrete block on heat gain prevention in a hot climate. *Sustainability* **2016**, *8*, 1009. [CrossRef]
5. Velmurugan, K.; Kumarasamy, S.; Wongwuttanasatian, T.; Seithtanabutara, V. Review of PCM types and suggestions for an applicable cascaded PCM for passive PV module cooling under tropical climate conditions. *J. Clean. Prod.* **2021**, *293*, 126065. [CrossRef]
6. Hasan, A.; Hejase, H.; Abdelbaqi, S.; Assi, A.; Hamdan, M.O. Comparative effectiveness of different phase change materials to improve cooling performance of heat sinks for electronic devices. *Appl. Sci.* **2016**, *6*, 226. [CrossRef]
7. Vanaga, R.; Blumberga, A.; Freimanis, R.; Mols, T.; Blumberga, D. Solar facade module for nearly zero energy building. *Energy* **2018**, *157*, 1025–1034. [CrossRef]
8. Cabeza, L.F.; Castellón, C.; Nogués, M.; Medrano, M.; Leppers, R.; Zubillaga, O. Use of microencapsulated PCM in concrete walls for energy savings. *Energy Build.* **2007**, *39*, 113–119. [CrossRef]
9. Experimental Investigation of PCM Transient Performance in Free Cooling of the Fresh Air of Air Conditioning Systems-ScienceDirect. Available online: https://www.sciencedirect.com/science/article/pii/S2352710219316882?casa_token=g7ablpICE6gAAAAA:6Nrl-B3yqM1LoHvOHoUDVfu-YJ5T1EBy_NIgMqWZqZY-rCHtQZkx_PhEyTcnPJ353bYXgTu6gC8z (accessed on 17 September 2023).
10. Energy Saving in Building with PCM Cold Storage-Stritih-2007-International Journal of Energy Research-Wiley Online Library. Available online: https://onlinelibrary.wiley.com/doi/10.1002/er.1318 (accessed on 17 September 2023).
11. Liu, S.; Iten, M.; Shukla, A. Numerical study on the performance of an air—Multiple PCMs unit for free cooling and ventilation. *Energy Build.* **2017**, *151*, 520–533. [CrossRef]
12. Gado, M.G.; Hassan, H. Energy-saving potential of compression heat pump using thermal energy storage of phase change materials for cooling and heating applications. *Energy* **2023**, *263*, 126046. [CrossRef]
13. Chaiyat, N.; Kiatsiriroat, T. Energy reduction of building air-conditioner with phase change material in Thailand. *Case Stud. Therm. Eng.* **2014**, *4*, 175–186. [CrossRef]
14. Huang, B.; Zheng, Z.; Lu, G.; Zhai, X. Design and experimental investigation of a PCM based cooling storage unit for emergency cooling in data center. *Energy Build.* **2022**, *259*, 111871. [CrossRef]
15. Akbari, A.D.; Talati, F.; Mahmoudi, S.M.S. New solution method for latent energy storage and thermoeconomic optimization for an air conditioning system. *Int. J. Refrig.* **2020**, *109*, 12–24. [CrossRef]
16. Said, M.A.; Hassan, H. Parametric study on the effect of using cold thermal storage energy of phase change material on the performance of air-conditioning unit. *Appl. Energy* **2018**, *230*, 1380–1402. [CrossRef]
17. Nagano, K.; Takeda, S.; Mochida, T.; Shimakura, K.; Nakamura, T. Study of a floor supply air conditioning system using granular phase change material to augment building mass thermal storage—Heat response in small scale experiments. *Energy Build.* **2006**, *38*, 436–446. [CrossRef]
18. A Comparative Study on PCM and Ice Thermal Energy Storage Tank for Air-Conditioning Systems in Office Buildings-ScienceDirect. Available online: https://www.sciencedirect.com/science/article/abs/pii/S1359431115013551?via=ihub (accessed on 17 September 2023).
19. Vakilaltojjar, S.M.; Saman, W. Analysis and modelling of a phase change storage system for air conditioning applications. *Appl. Therm. Eng.* **2001**, *21*, 249–263. [CrossRef]
20. Mosaffa, A.H.; Farshi, L.G.; Ferreira, C.A.I.; Rosen, M.A. Advanced exergy analysis of an air conditioning system incorporating thermal energy storage. *Energy* **2014**, *77*, 945–952. [CrossRef]
21. Mosaffa, A.H.; Farshi, L.G. Exergoeconomic and environmental analyses of an air conditioning system using thermal energy storage. *Appl. Energy* **2016**, *162*, 515–526. [CrossRef]
22. Maccarini, A.; Hultmark, G.; Bergsøe, N.C.; Afshari, A. Free cooling potential of a PCM-based heat exchanger coupled with a novel HVAC system for simultaneous heating and cooling of buildings. *Sustain. Cities Soc.* **2018**, *42*, 384–395. [CrossRef]
23. Hai, T.; Sajadi, S.M.; Zain, J.M.; El-Shafay, A.S.; Sharifpur, M. The effect of using tubes filled with phase change materials in the air conditioning system of a residential building. *J. Build. Eng.* **2022**, *49*, 104079. [CrossRef]
24. Farah, S.; Liu, M.; Saman, W. Numerical investigation of phase change material thermal storage for space cooling. *Appl. Energy* **2019**, *239*, 526–535. [CrossRef]
25. Nie, B.; She, X.; Du, Z.; Xie, C.; Li, Y.; He, Z.; Ding, Y. System performance and economic assessment of a thermal energy storage based air-conditioning unit for transport applications. *Appl. Energy* **2019**, *251*, 113254. [CrossRef]
26. Hasan, A.; McCormack, S.J.; Huang, M.J.; Norton, B. Characterization of phase change materials for thermal control of photovoltaics using Differential Scanning Calorimetry and Temperature History Method. *Energy Convers. Manag.* **2014**, *81*, 322–329. [CrossRef]

27. Lewis, R.W.; Morgan, K.; Thomas, J.R.; Seetharamu, I.N. *The Finite Element Method in Heat Transfer Analysis*; John Wiley & Sons: Hoboken, NJ, USA, 1999.
28. Reese, T.G.; Heid, O.; Weisskoff, R.M.; Wedeen, V.J. Reduction of eddy-current-induced distortion in diffusion MRI using a twice-refocused spin echo. *Magn. Reson. Med. Off. J. Int. Soc. Magn. Reson. Med.* **2003**, *49*, 177–182. [CrossRef]
29. Bang, D.; Lee, J.H.; Lee, E.S.; Lee, S.; Choi, J.S.; Kim, Y.K.; Cho, B.K.; Koh, J.K.; Won, Y.H.; Kim, N.I.; et al. Epidemiologic and clinical survey of Behcet's disease in Korea: The first multicenter study. *J. Korean Med. Sci.* **2001**, *16*, 615–618. [CrossRef]
30. Lamberg, P.; Lehtiniemi, R.; Henell, A.M. Numerical and experimental investigation of melting and freezing processes in phase change material storage. *Int. J. Therm. Sci.* **2004**, *43*, 277–287. [CrossRef]

buildings

MDPI

Article

Economic Feasibility of PV Mounting Structures on Industrial Roofs

Alicia Perdigones [1], José L. García [1], Isaac García [1], Fátima Baptista [2] and Fernando R. Mazarrón [1,*]

[1] Escuela Técnica Superior de Ingeniería Agronómica, Alimentaria y de Biosistemas, Departamento de Ingeniería Agroforestal, Universidad Politécnica de Madrid, 28040 Madrid, Spain; alicia.perdigones@upm.es (A.P.); joseluis.garciaf@upm.es (J.L.G.); i.gmoreno@alumnos.upm.es (I.G.)

[2] Escola de Ciências e Tecnologia, MED—Instituto Mediterrâneo para a Agricultura, Ambiente e Desenvolvimento, Departamento de Engenharia Rural, Universidade de Évora, 7000-849 Évora, Portugal; fb@uevora.pt

* Correspondence: f.ruiz@upm.es; Tel.: +34-910-671-001

Abstract: This study determines the viability and profitability of photovoltaic (PV) mounting structures on industrial roofs. For this purpose, more than 656,000 different cases have been analyzed, combining different consumption patterns, energy prices, locations, inclinations, azimuths, capacity installed, and excess income. The results show that the industry's consumption pattern is a key factor, leading to significant reductions in the available assembly budget for inclined structures compared to the coplanar option when the pattern is seasonal and/or irregular. The increase in energy prices experienced in the last 2 years represents a substantial change in the viability of the structures. The budget for inclined structures increases by hundreds of euros compared to the coplanar option. Depending on the azimuth and inclination of the roof, the maximum available budget can vary by more than a thousand euros per kWp, being highly profitable in orientations close to the east and west and on roofs partially inclined to the north. Differences between low-irradiation and high-irradiation locations can mean variations in the average budget of more than 1 k€/kWp, especially with high electricity prices.

Keywords: photovoltaic; mounting structures; azimuth; inclination; industrial roofs; seasonality

Citation: Perdigones, A.; García, J.L.; García, I.; Baptista, F.; Mazarrón, F.R. Economic Feasibility of PV Mounting Structures on Industrial Roofs. *Buildings* **2023**, *13*, 2834. https://doi.org/10.3390/buildings13112834

Academic Editors: Yaolin Lin and Wei Yang

Received: 3 October 2023
Revised: 30 October 2023
Accepted: 9 November 2023
Published: 12 November 2023

1. Introduction

The increase in energy prices experienced since mid-2021 has led to a significant rise in the energy bill, compromising the viability and competitiveness of many industries [1]. In this context, photovoltaic energy has been the main measure to mitigate the increase in costs in many sectors. The new context has significant effects on the viability and profitability of photovoltaic (PV) systems and their mounting structures, which, in many cases, are not taken into consideration. The adaptation to the new context requires adjustments in the sizing of PVs, which is also associated with changes in the return on investment [2]. In many cases, the design is carried out following established paradigms, with approximate calculations, without analyzing in detail the consumption pattern of the industry.

Load matching is an inevitable problem that restricts the development of PV systems used in buildings [3]. Optimal PV orientation should not only be based on maximizing energy production but also on expected demand patterns and market prices. In residential buildings, providing electricity not only around solar noon but also in the morning and late afternoon, when demand increases, helps to maximize self-consumption/-sufficiency and reduces costs for end-users and utilities [4]. Seasonality strongly conditions the optimal size of PV installations, the return on the investment, and the potential savings [2]. Differences between demand patterns caused large variations in optimal PV orientations [5].

Therefore, the optimization of PVs based on the consumption pattern has been analyzed in numerous specific cases, such as in residential buildings [5–7], commercial buildings [5,8], and military facilities [9]. Within the industrial sector, the causality is immense, with subsectors with many seasonal production industries, such as the agro-industrial [1]. For example, Jimenez-Castillo et al. [10] demonstrated the importance of optimizing the orientation and inclination in olive mills, based on the uniqueness of its consumption pattern, adapting it to seasonality. In wineries, seasonality was key in the profitability of the system for self-consumption, related to the optimum power to be installed in PVs [11].

The electricity tariff, which varies with the time of day and time of year, also impacts the optimal orientation of non-tracking PV modules [12]. Fitting the solar PV power generation curve according to the consumption curve is not always the most profitable way to optimize the orientation but also taking the electricity price curve into account is very significant when assessing profitability [13]. Therefore, a correct design of the PV must take into account the variations in the price of energy, carrying out a precise calculation of the resulting bill.

The optimization of azimuth and inclination on rooftops arouses great interest in the scientific community, with different approaches and criteria for decision-making depending on the application case. Maximizing the energy generated throughout the year is a common design criterion [14–21]. For a given location, the best-fixed orientation of a PV panel can be determined by achieving the maximum incident solar irradiance throughout the year [22]. A precise estimation of solar irradiation on a solar panel is critical for an optimal PV design [23].

Non-ideal inclination and azimuth angles can also lead to acceptable levels of electric energy generation [24]. A range for the optimal panel orientation according to the target loss can be used in decision-making for practical optimization of the PV panel [25].

Beyond the fixed PV mounting, several studies have analyzed the periodic change in tilt: daily, monthly, and/or seasonally optimum [26–31]. The monthly optimum tilt angle for a PV panel changes throughout the year [32]. Solar radiation energy is lost when using the yearly average fixed optimal tilt angle as compared with the monthly optimal tilt angle [33]. Although the daily optimal tilt angle produces the maximum energy generated, it is costly and impractical [34].

Optimization is also common taking into account economic and environmental criteria (through life cycle cost analysis) [9,35–39]. Cases with an inclination other than the one that maximizes energy can have economic and environmental advantages [35].

Optimization to minimize self-shading while accounting for practical considerations (ensuring rooftop walkability, distances required for maintenance, curtail the cooling load, etc.) has been analyzed in some papers [24,40–42]. There are also studies that have analyzed the optimal deployment of photovoltaic module rows installed on irregular flat roof shapes [43] and vertical installation of PV modules [44]. In modern structures, photovoltaic panels are being installed on different parts of buildings, resulting in panels mounted at various angles [45].

Although most studies focus on a specific city, several analyzed different locations, obtaining variations in the optimal inclinations [18,24,27,33,46–49]. On the contrary, Beringer et al. pointed to a reduced importance of panel tilt in mid-latitudes [50]. Therefore, locations with very different irradiance must be analyzed if a global view of PV optimization is to be obtained.

Despite the advantages that azimuth and inclination optimization may present, coplanar mounting is still frequently used on industrial roofs. This is due to its obvious advantages, including lower investment, greater simplicity, less weight, little visual impact, and less wind load. It is also attributed to the paradigms established in years of lower energy prices. However, structures that modify the orientation and/or inclination of the solar panels could be a much more profitable solution in the new energy context. The wide range of industrial roofs, from flat roofs to sawtooth roofs, with all possible azimuths, complicates the optimization of the PV. The optimization valid for one industry may not be suitable for another. In addition, in many cases, the limited space available raises the need to also take advantage of the most unfavorable roofs. A correct design of azimuth and inclination is key to the maximum profitability of the PVs.

Despite the numerous investigations on the subject, the extra cost and profitability of inclined structures versus coplanar structures have not been exhaustively analyzed. Most studies analyze the PV as a whole, without focusing on the best mounting option and the budget available for it based on the azimuth, inclination, and industrial demand pattern. This knowledge gap, coupled with the changes brought about by the new energy context, raises the need to carry out a study that provides a global vision of the economic viability of using PV mounting structures to change the inclination and azimuth of the panels on industrial roofs.

2. Materials and Methods

The study has quantified the available budget and the profitability of using mounting structures that improve the inclination and/or azimuth of the PVs, compared to a coplanar scheme. Specifically, two alternatives have been analyzed (Figure 1):

- Inclined structures on the roof to install solar modules in the optimal plane to maximize production, changing azimuth and inclination.
- Inclined structures on the roof, with the base of the panels parallel to the ridge. Only the tilt to reach the optimum inclination is modified, maintaining the roof azimuth.

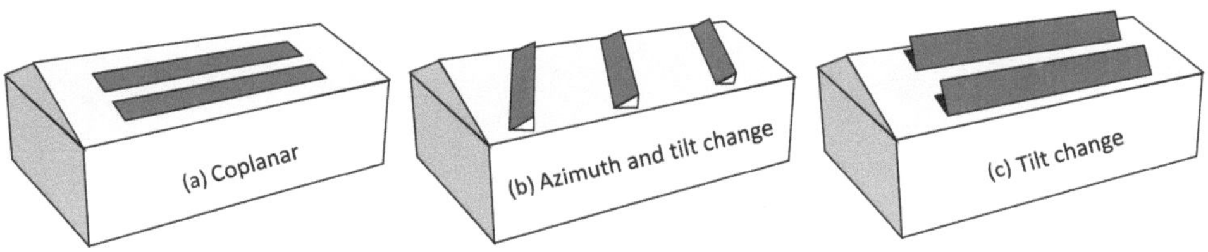

Figure 1. Alternatives analyzed for mounting the photovoltaic panels: (**a**) coplanar, (**b**) change in azimuth and inclination, and (**c**) change in inclination while maintaining the azimuth of the roof.

The suitability of optimizing the collection plane of the modules instead of using a coplanar structure is conditioned by numerous factors, such as the consumption pattern, the size of the PV (capacity installed), the azimuth and inclination of the industry's roofs, the solar radiation, electricity prices, or the income received from the production of electricity injected into the grid. All these factors have been analyzed through a precise calculation, reproducing a typical billing system of industries, carrying out a quarter-hourly energy calculation (Figure 2).

2.1. *Quarter-Hourly Energy Analysis*

The actual electricity billing system is usually based on data energy consumption collected every 15 min in almost every industry. Therefore, the energy analysis was carried out on a quarter-hourly basis. Industrial electricity consumption was compared with the electricity supplied by a PV system according to a certain installed capacity, with the objective of quantifying the reduction in grid energy consumption and the possibility of injecting electricity into the grid (as an upside). Variations in the Flash test report (test measuring the module's current–voltage (I–V) curve) are not taken into account, considering that the marked power corresponds to that of the photovoltaic module.

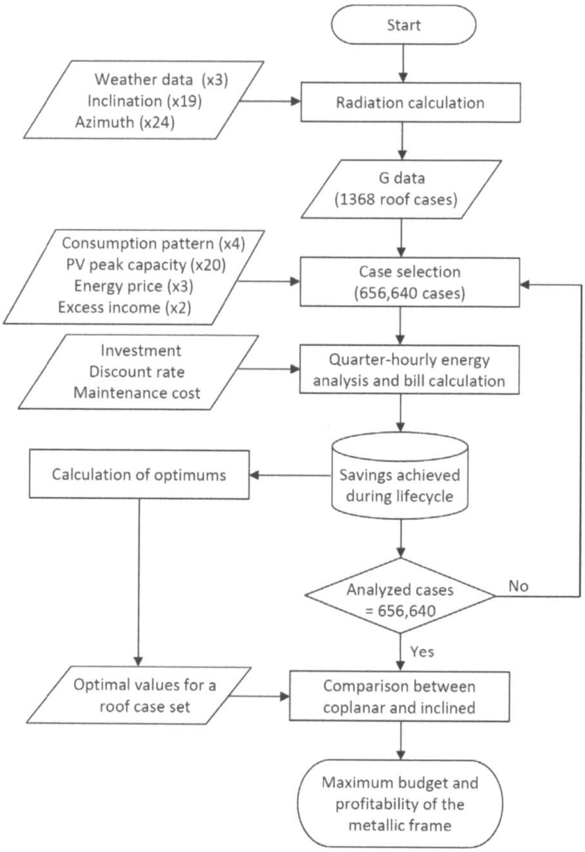

Figure 2. Methodology flowchart.

The equations used in the energy calculation have been previously evaluated and validated [2]. For each scenario, the energy supplied by the PV ($E_{PV},\ kWh$) during each 15 min interval was calculated by multiplying the peak power of the system (P_{PV}, kWp) by the radiation on the inclined plane ($G, kW\ m^{-2}$), the duration of the interval (t, h), and by an overall efficiency of the system (μ_T , −) of 0.8 (Equation (1)). A similar level of efficiency (similar to the performance ratio) has been assumed in all cases with the objective of working under the same conditions. In a later section, the calculation of radiation on the inclined plane using the EnergyPlus software (https://energyplus.net/) is detailed.

$$E_{PV} = P_{PV} \cdot t \cdot G \cdot \mu_T \tag{1}$$

At any interval, the electricity produced by the PV system is compared with electricity demanded by the industry ($E_D,\ kWh$), calculating the electricity savings achieved by the PV on the bill to be paid ($E_s,\ kWh$), electricity acquired from the grid due to capacity limitations ($E_G,\ kWh$), and the excess energy produced by the PV ($E_X,\ kWh$) that could generate a certain level of income if country regulations allow for such an upside (Equations (2) and (3)).

$$If\ E_D \geq E_{PV}:\ E_s = E_{PV};\ E_G = E_D - E_{PV};\ E_X = 0 \tag{2}$$

$$If\ E_{PV} > E_D:\ E_s = E_D;\ E_G = 0;\ E_X = E_{PV} - E_D \tag{3}$$

In addition, in each quarter-hour interval, the new power demanded from the grid (P_G, kW) has been calculated by the difference between the power demanded by the industry (P_D, kW) and the effective power of the PV ($\frac{E_{PV}}{t}$, kW) (Equation (4)). According to this, any possible penalty applicable for excess power is calculated in a precise and realistic way.

$$P_G = P_D - \frac{E_{PV}}{t} \tag{4}$$

After calculating the electricity balances for each quarter of an hour of the year, the summation of energy supplied by the PV system (reduction in network consumption), energy demanded from the network that the PV could not supply, and energy injected into the grid (when exceeding demand) has been carried out in each of the tariff periods involved, reproducing the methodology used by the electricity providers.

2.2. Actual Impact on the Electricity Bill

In order to carry out a rigorous analysis of the savings generated by a photovoltaic system, it is necessary to reproduce an actual electricity billing system. The Spanish electricity pricing system has been used as a basis for calculation, with a fixed term for capacity and a variable term for energy, subject to a 21% value added tax (VAT) and excise tax of 5.11%. Electricity prices vary throughout each day according to 3 periods. Those periods are modified monthly basis, creating 6 periods with different price levels. Therefore, the impact on the PV system over the final bill depends on the time and month in which it occurs. Due to this fact, the methodology is based on an analysis with a quarter-hourly interval, calculating the annual amount of the expected bill based on 35,040 data.

A calculation tool based on "Visual Basic for Application" (VBA) was developed. This application is an evolution of a previously developed one to reproduce the billing of electricity companies, used as the basis for a recent publication [1]. The tool has been validated in scenarios without PV, using prices and consumption data from real invoices, verifying that the total amounts match.

An energy calculation module based on the previously described methodology has been added to the billing tool, which compares at each interval the real demand of the industry and the production of the photovoltaic system. From Equations (1)–(4), every 15 min, the electricity supplied by the PV, the electricity acquired from the grid, the energy savings achieved (compared to demand without PV), the excess energy produced by the PV injected to the grid (when the demand is less than the energy produced), and the new power registered are calculated. The tool has been validated with PVsyst in several cases on a random basis, selecting a PV system with an overall efficiency equivalent to that assumed and using the same radiation data.

After performing the energy calculations for each of the 35,040 intervals of the year, the new monthly bills and the resulting annual amount are calculated using the new values of energy and power demanded from the grid. The tool also provides monthly and annual summaries of the aforementioned energy variables. To do this, the billing tool reproduces the calculation of the electricity companies, taking into account price variations according to the period, the fixed term power, penalties for capacity excess, taxes, etc. In scenarios that include income from excess energy injected into the grid, the monthly and annual amounts are reduced by the profits generated.

2.3. Maximum Budget and Profitability of the Metallic Frame

A total lifecycle of the PV installation of 20 years has been assumed. This conservative hypothesis, with a shorter useful life than those considered by many investors, reduces the uncertainty of assuming an excessively long period in which substantial changes in the energy context are foreseeable.

The total updated savings achieved by the PV during the lifecycle have been calculated through the net present value (NPV) from the annual savings generated. For this, a discount rate of 2% has been applied, a value close to the interest rate on 20-year bonds in recent months in countries like Germany. A 1% of the investment was assumed as annual maintenance cost. Annual cash flows have been calculated by subtracting maintenance costs from the annual savings generated on the invoice.

In order to compare systems of different sizes, the total savings have been divided by the peak capacity installed in each case (k€/kWp). To compare the suitability of using inclined mounting structures versus the coplanar option, the maximum budget available for the inclined structure has been calculated based on the savings generated during the useful life of the PV system (calculated NPV) compared to that obtained without changing the azimuth and inclination of the industrial roof (coplanar).

In the case of structures that change the azimuth and inclination, the budget has been calculated as the difference between the savings achieved in the particular case of the roof (azimuth and inclination) and the optimal case that maximizes the energy, taking into account all possible azimuths and inclinations. In the case of structures that only change the inclination, the calculation has been equivalent, but taking as reference the maximum value of each azimuth instead of the total of the set of scenarios. The optimal values—those with the highest energy production—have been identified through VBA programming among the set of cases defined in each analysis. The results have been stored in a database based on the characteristics of each specific scenario.

The available budget values thus calculated are the maximum possible, considering that panel installation does not create any shading. In some cases, especially those close to north orientation and with great inclination, the installation could involve shading if it is not carried out in the upper area and be limited to a single line of panels.

The casuistry of the mounting structures is very wide, from low-priced coplanars to complex and expensive structures that elevate the panels. The calculated budget quantifies the extra cost of the inclined structure compared to the coplanar, regardless of the type of material, its cost, etc., providing very useful information for the design of the PVs. However, to have a reference to compare profitabilities, the updated payback has also been calculated, assuming an extra cost of the inclined structure of 0.2 k€/kWp compared to the coplanar one, considering an inclined structure with a certain installation complexity.

2.4. Analyzed Scenarios

A total of 656,640 different cases have been analyzed, corresponding to the combination of 4 consumption patterns, 3 energy prices, 3 locations, 20 sizes of the PV, 19 inclinations, 24 azimuths, and 2 excess incomes.

2.4.1. Inclination and Azimuth

The aim is to quantify the impact of the inclination and azimuth of industrial roofs on the savings generated by the PV, determining the margin to incorporate structures that optimize the system. For this, 19 different inclinations were analyzed, from 0° (flat roof) to 90° (assuming structures attached to vertical enclosures on sawtooth roofs), with an interval of 5°. In turn, each of the previous inclinations was analyzed assuming different azimuths of the roof and/or the structure of the PV. Specifically, the 360° divided into intervals of 15°.

The incident radiation on each inclined plane has been calculated using the EnergyPlus program. For this, a nearly cylindrical construction was designed, similar to a mill, with the base being a polygon of 24 sides. The roof of the construction is also divided into 24 polygons, each oriented to a specific azimuth. When carrying out an annual energy simulation from this basic model and the climatological file of a given location, the program calculated the incident irradiation for each of the 24 surfaces that make up the roof, with a quarter-hourly interval. The rest of the variables that can be calculated with the energy simulation were discarded. To calculate radiation as a function of inclination, 19 different input data files were generated. In each one, the inclination of the polygons that make up the roof was modified, adapting them to the specific case. By carrying out the 19 simulations, which include 24 roof orientations each, a total of 456 different radiation scenarios are obtained for the same location. Furthermore, the simulations were calculated using 3 different climatological files, obtaining a total of 1368 cases.

2.4.2. Radiation

The calculations have been carried out in three locations with very different radiation. For the selection, the horizontal plane radiation data available in the EnergyPlus climatological files [51] have been used. After calculating the peak solar hours (PSH) of more than 200 locations, 3 cities were selected with values of 3.0 (Stuttgart), 4.0 (Rome), and 5.0 (Los Angeles). Specifically, the following climatological files (*epw) were used to carry out the simulations:

- DEU_Stuttgart.107380_IWEC.epw
- ITA_Rome.162420_IWEC.epw
- USA_CA_Los.Angeles.Intl.AP.722950_TMY.epw

2.4.3. Consumption Pattern

The seasonality and homogeneity of the demand pattern of industries could strongly influence the profitability of mounting structures. Therefore, the study has been based on four industries with very different consumption patterns:

- Autu_7d_Peak (Winery): It has a strongly seasonal demand, with large peaks of energy consumption and high power in August, September, and October. During these months, the processing of grapes for wine production takes place from Monday to Sunday. The rest of the year the energy demand is very small, mainly from Monday to Friday.
- Wint_7d_Peak (Olive mill): The demand pattern is very similar to that of the first industry, but concentrated in December, January, and February. In addition, the power required by the machinery used in olive milling and processing is somewhat lower than in the previous case.
- Year_7d_Unif (Fruits and vegetables): The industry presents a very stable and uniform pattern throughout all months of the year. The cooling equipment demands energy every day of the week, with a stable maximum power, much lower than in the previous cases.
- Year_5d_Peak (Feed factory): As in the previous case, it has very stable energy consumption throughout the months. However, its energy demand is concentrated during a few hours a day, from Monday to Friday, registering large peaks of power.

The selected industries, partners in projects such as "Transferring Energy Save Laid on Agroindustry"(TESLA) and "Saving COOPerative Energy"(SCOOPE), have very similar annual energy consumption. Specifically, adding the quarter-hourly values for a typical year, they show a consumption of 124 MWh, 107 MWh, 122 MWh, and 109 MWh. The consumption data of these 4 industries have been the subject of a previous study to evaluate the impact of the new energy context on the electricity bill [1].

To eliminate any bias caused by differences in consumption, the values of power and energy demanded every 15 min have been adjusted by a small correction factor. Thus, an overall factor of 0.9292 has been applied to the first industry, 1.0753 to the second, 0.9447 to the third, and 1.0512 to the fourth. As a result, the 4 patterns analyzed have exactly the same annual consumption, 115 MWh, maintaining the energy demand behavior (Figure 3).

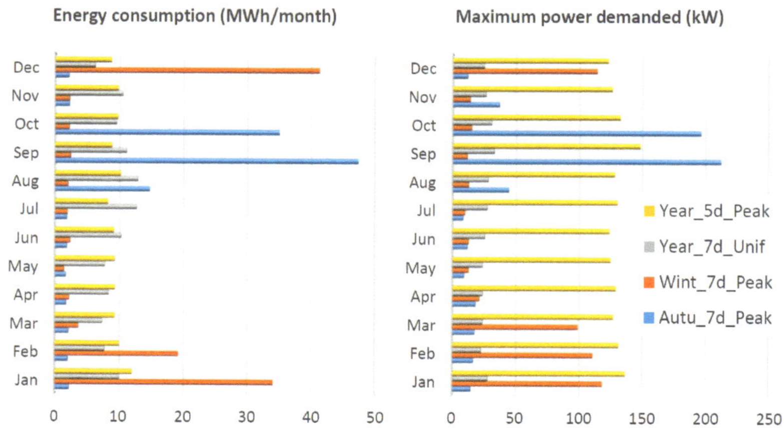

Figure 3. Monthly energy consumption (**left**) and maximum power demanded (**right**) in each of the industries analyzed.

2.4.4. Remaining Parameters

In order to analyze the impact of the size of the PV system, PVs with different peak capacities have been analyzed, from 5 kWp to 100 kWp, with an interval of 5 kWp.

Given the great variation in energy prices, 3 scenarios have been considered. The assumed values are those offered by a marketing company in Spain (Som Energia), specifically:

- Low prices before the energy crisis (June 21): 0.135, 0.119, 0.099, 0.085, 0.073, and 0.069 €/kWh for each of the 6 existing tariff periods.
- Maximum prices reached during the energy crisis (February 22): 0.34, 0.31, 0.291, 0.267, 0.243, and 0.231 €/kWh.
- Intermediate prices close to those in force in 2023, calculated as the average of the 2 previous scenarios: 0.238, 0.215, 0.195, 0.176, 0.158, and 0.150 €/kWh.

To simplify the analyses carried out, the same price has been assumed for the power term in the three scenarios: 24.732, 21.529, 12.320, 9.897, 2.834, 1.571, and 72.884 €/kW year.

The sale of excess energy presents great differences and legal nuances even within the same country. For example, there may be unavailability of discharge points by the distribution companies, as the nodes of the network are saturated; there may be limitations on compensation for energy consumed in certain periods, etc. In cases where this injected energy is paid, it presents great variations depending on the area, the company, the year, etc. Therefore, both the assumption that injecting into the grid is not allowed and the assumption that a third of the cost of the purchased energy is received for the excess energy will be analyzed (value close to the offer of the last months of the reference company).

This work does not quantify the reduction in greenhouse gases due to the increase in energy supplied that is achieved with inclined structures. If the extra financial benefits by carbon credit were taken into account, the budget available for the structure would be greater than the obtained values.

3. Results and Discussions

In order to analyze the wide causality, scenarios of intermediate values have been taken as a reference framework, carrying out a subsequent sensitivity analysis, specifically, the intermediate value of the energy price (close to the current reality) without injection income and intermediate radiation (PSH 4.0).

3.1. Optimization by Modifying Azimuth and Inclination

3.1.1. Influence of the Uniformity and Seasonality of Demand

The consumption pattern strongly conditions the potential savings of the PV and, as a consequence, the financial viability to optimize the orientation and azimuth of the PV system. In those industries with strong seasonality (Autu_7d_Peak and Wint_7d_Peak), the increase in installed capacity is associated with a rapid reduction in the useful energy used per kWp, for any azimuth (Figure 4). This is due to the increase in excess energy that needs to be injected into the grid or in the worst case, curtailed, due to the impossibility of matching electrical supply and demand. The reduction in energy demand on weekends and the lack of uniformity from Monday to Friday (Year_5d_Peak case) also penalize the performance of the PV when increasing the installed capacity. On the contrary, in the industry with relatively uniform and stable consumption throughout the year (Year_7d_Unif), the increase in installed capacity causes a limited loss of energy used for each kWp until its usual demand is exceeded (considering 25 kW as the calculation base).

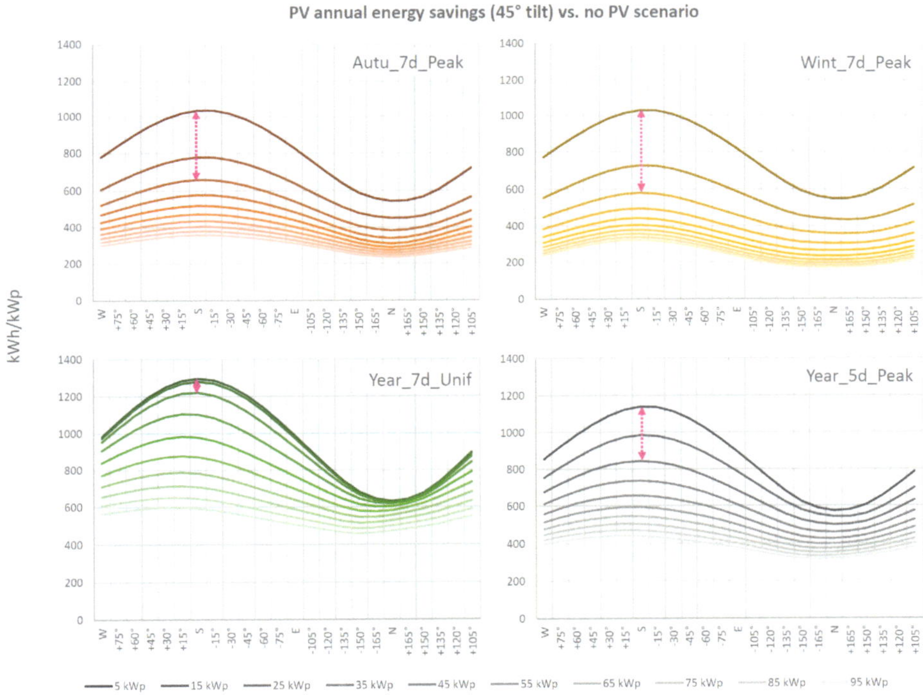

Figure 4. Annual energy savings achieved by a PV installation compared to the scenario without PV, depending on the installed peak capacity and the azimuth of the panel. PV installed with a 45° inclination at the location of 4.0 PSH.

Reduction in the industry's electricity bill is directly linked with the behavior of the energy used, with small differences due to variations in the price of energy and penalties for excess power demanded (Figure 5). Therefore, the greater the seasonality and irregularity of demand, the smaller the differences when changing the azimuth (€/kWp), and therefore, the lower the budget available for an optimization of the PV plane.

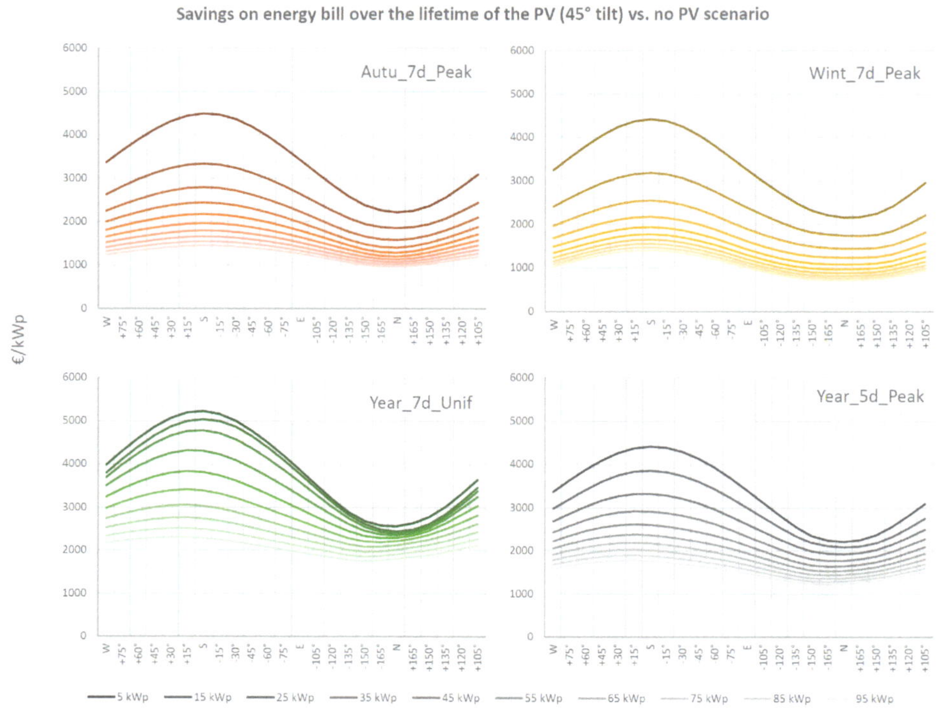

Figure 5. Bill reduction achieved during the lifecycle of a PV installation compared to the scenario without PV, depending on the installed peak capacity and the azimuth of the panel; 45° PV inclination, intermediate energy price, no excess income, and HSP 4.0.

3.1.2. PV with Limited Excess

The suitability of each case has been analyzed through the maximum budget available for the extra cost of an inclined structure that optimizes the plane compared to the budget of a coplanar structure (k€/kWp).

With a small installed capacity where the excesses are reduced (5 kW in the analyzed industries), the consumption pattern has limited importance (Figure 6). In this case, maximum use of the captured energy is ensured, increasing the available margin. Except in cases of azimuth and inclination very close to the optimum, there is a wide budget to optimize the structure. Thus, the average budget of the set of scenarios analyzed for the reference framework is 1.2 ± 0.8 k€/kW in the most unfavorable pattern and 1.5 ± 1.0 k€/kW in the most favorable one. On flat roofs, the wide margin for the extra cost of the structure, between 0.6 and 07 k€/kW, supports the use of inclined structures in all cases. On vertical sawtooth roofs, the budget starts at 1.1–1.4 k€/kW for south orientation, increasing as the azimuth varies, as long as shading can be avoided. Considering only azimuths between 90° and −90°, the margin for the structure is reduced to averages between 0.7 ± 0.5 k€/kW and 0.9 ± 0.7 k€/kW.

Assuming an extra cost of the inclined structure of 0.2 k€/kWp compared to the coplanar one, in 92–94% of the cases, the structure would be amortized over the lifetime of the PV (between 85% and 89% for azimuths between 90° and −90°). The average payback of the different patterns ranges between 4.6 ± 5.0 and 5.5 ± 5.4 years. Scenarios with a payback of fewer than 5 years range between 63% in the most unfavorable pattern and 70% in the uniform pattern. This percentage is reduced to 42–50% with azimuths between 90° and −90°.

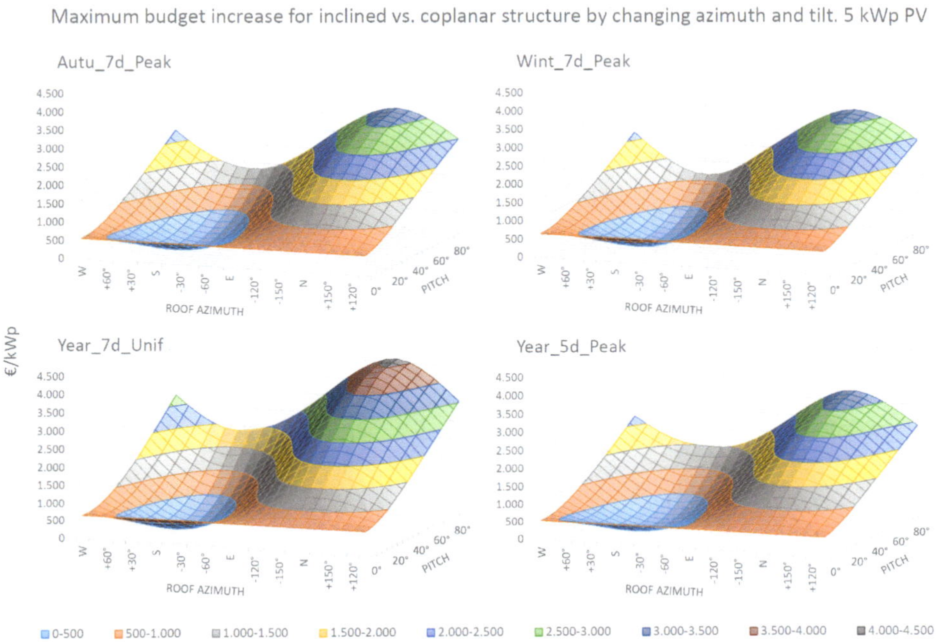

Figure 6. Maximum budget for the extra cost of a structure that changes the orientation and inclination to the optimal case, compared to the cost of the coplanar option; 5 kWp PV, intermediate energy price, PSH 4.0, and no excess income.

However, a low installed capacity leads to a limited reduction in the electricity bill, giving up the potential benefits of an optimized sizing PV system. In the most favorable cases where 5 kWp is installed, the PV would achieve a reduction in the industry's annual consumption of 6–7 MWh compared to the scenario without PV, just 5% of the total (115 MWh).

3.1.3. PV Designed to Increase Savings on the Energy Bill

The increase in installed peak capacity, seeking to maximize the industry's savings, leads to a reduction in the available budget for the extra cost of an inclined structure compared to a coplanar one. This reduction is more pronounced the greater the seasonality and irregularity of demand (Figure 7). Thus, for example, in the industry with the most uniform and constant pattern, the increase in installed capacity to 25 kWp (close to the maximum recorded in it) hardly produces variations in the available budget for the extra cost of the structure, with an average of 1.3 $\pm$ 0.9 k€/kW. On the contrary, in the other three patterns, there is a drastic cut, with an average of 0.6 $\pm$ 0.4 k€/kW, 0.6 $\pm$ 0.5 k€/kW, and 0.8 $\pm$ 0.6 k€/kW for Wint_7d_Peak, Autu_7d_Peak, and Year_5d_Peak, respectively. Considering azimuths between 90° and $-90°$, these values would fall to 0.8 $\pm$ 0.6 k€/kW and 0.4 $\pm$ 0.3 k€/kW. In the most unfavorable pattern, the budget for the extra cost of the inclined structure is reduced to 0.25 €/kWp on flat roofs.

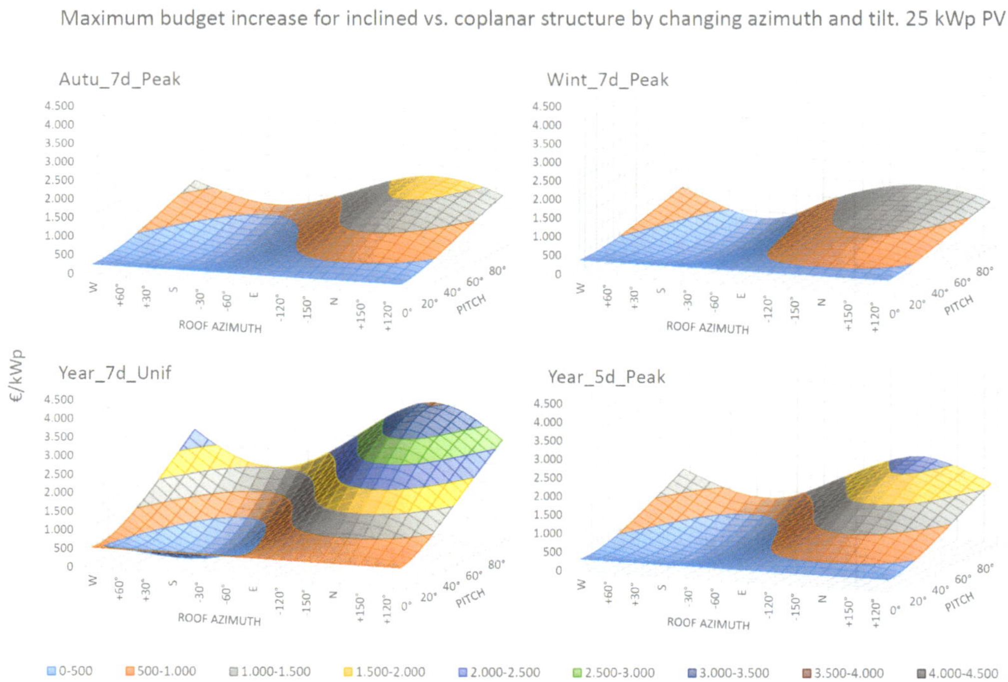

Figure 7. Maximum budget for the extra cost of a structure that changes the orientation and inclination to the optimal case, compared to the cost of the coplanar option; 25 kWp PV, intermediate energy price, PSH 4.0, and no excess income.

The percentage of cases in which the inclined structure would be amortized (cost of the structure of 0.2 k€/kWp) would remain almost the same in the uniform pattern but would drop between 83% and 86% in the other more irregular patterns (67–74% between east and west), with a high number of cases below 0.3 k€/kWp. The payback increases several years in the most irregular patterns, with averages between 8.1 $\pm$ 6.3 and 9.3 $\pm$ 6.6 years. In the uniform pattern, the increase is more moderate, with an average payback of 5.4 $\pm$ 5.4 (an increase of less than one year). Scenarios with a payback of fewer than 5 years would be reduced to 39–65% (11–44% between 90° and −90°).

Therefore, in seasonal and/or irregular consumption patterns, the suitability against the coplanar structure loses strength in a large number of scenarios around the optimal case. On the contrary, in an industry with seasonal and uniform consumption, the increase in installed capacity to a value close to its maximum demand hardly reduces the available budget and the profitability of the inclined structure.

The analyzed scenario of 25 kWp would imply a reduction in annual consumption close to 16 MWh in the most seasonal patterns and 31 MWh in the uniform one, which represents percentages of 14% and 27% of the total consumption (115 MWh). The increase in installed peak capacity to further reduce the bill would lead to greater reductions in the available budget for the structure

3.2. *Optimization by Modifying Only the Inclination (Same Azimuth)*

The suitability of each case has been analyzed through the maximum budget available for the extra cost of an inclined structure that optimizes the panel's inclination compared to the budget of a coplanar structure (k€/kWp), maintaining the azimuth.

The option to change only the inclination, using structures with the base of the panels parallel to the ridge of the roof, further reduces the available budget compared to the coplanar option (Figures 8 and 9). In the case of a low-capacity PV system (5 kWp) with reduced excesses, the average budget of the set of scenarios analyzed for the reference framework falls to 1.0 ± 0.9 k€/kW in the most favorable pattern and values close to 0.8 ± 0.8 k€/kW in the other three (Figure 8). Considering only azimuths between 90° and −90°, the margin for the structure is much smaller, with averages between 0.4 ± 0.4 k€/kW and 0.6 ± 0.6 k€/kW.

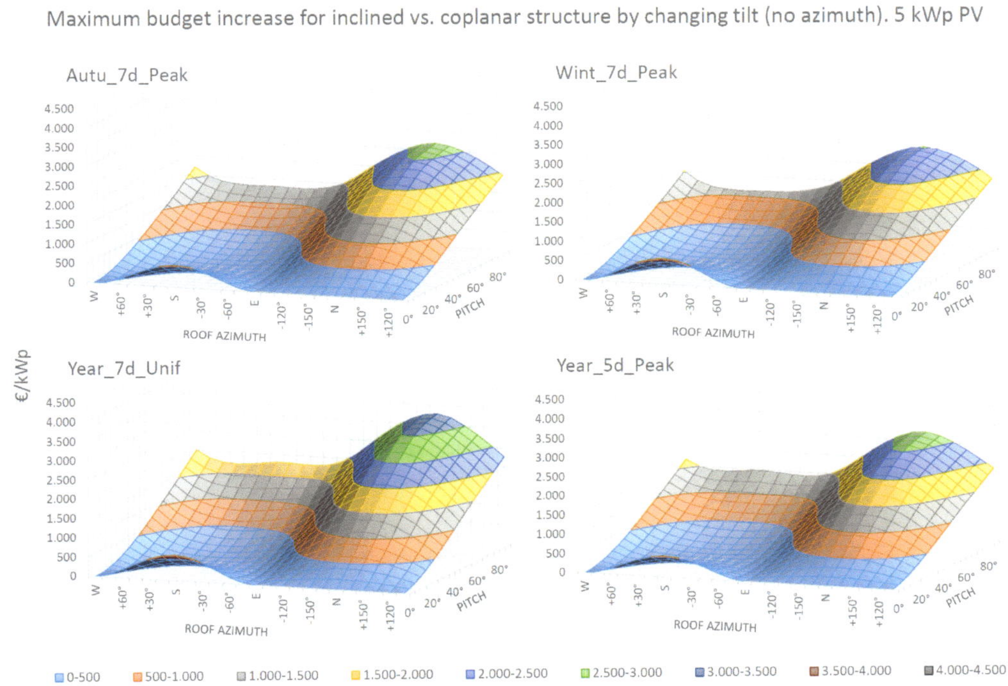

Figure 8. The maximum available budget for the extra cost of a structure that achieves the optimal inclination (but not the azimuth), compared to the coplanar option; 5 kWp PV, intermediate energy price, PSH 4.0, and no excess income.

Assuming the same extra cost of the inclined structure (0.2 k€/kWp), the percentage of cases in which the structure would be amortized in 20 years is reduced to 70–74% of the analyzed cases (between 56% and 62% for azimuths between 90° and −90°). The payback of the different patterns would increase to averages between 8.5 ± 7.7 and 9.7 ± 7.6 years. Scenarios with a payback of fewer than 5 years would fall to 41% in the most unfavorable pattern and 52% in the uniform pattern (23–34% with azimuths between 90° and −90°).

When increasing the installed capacity to 25 kWp, the differences between the irregular patterns and the uniform pattern increase (Figure 9). The average threshold budget is 0.3 ± 0.3 k€/kW for the winter seasonal pattern, 0.5 ± 0.4 k€/kW for the autumn one, 0.5 ± 0.5 k€/kW for the weekday one, and 0.9 ± 0.9 k€/kW for the uniform one. Limiting the scenarios to azimuths between east and west, these values are reduced to 0.2 ± 0.2 k€/kW, 0.2 ± 0.2 k€/kW, 0.3 ± 0.3 k€/kW, and 0.5 ± 0.5 k€/kW.

Assuming an additional cost of the inclined structure of 0.2 k€/kWp compared to the coplanar one, the percentage of cases in which the structure would be amortized in 20 years is reduced to 53–72% of the cases analyzed (between 31% and 60% for azimuths between 90° and −90°). The payback of the different patterns would increase to averages between 9.0 ± 7.7 and 13.7 ± 6.8 years. Scenarios with a payback of fewer than 5 years remain around 50% in the most favorable pattern but fall to 18–27% in the rest. Considering azimuths between 90° and −90°, in the winter pattern, no case presents a payback of fewer than 5 years.

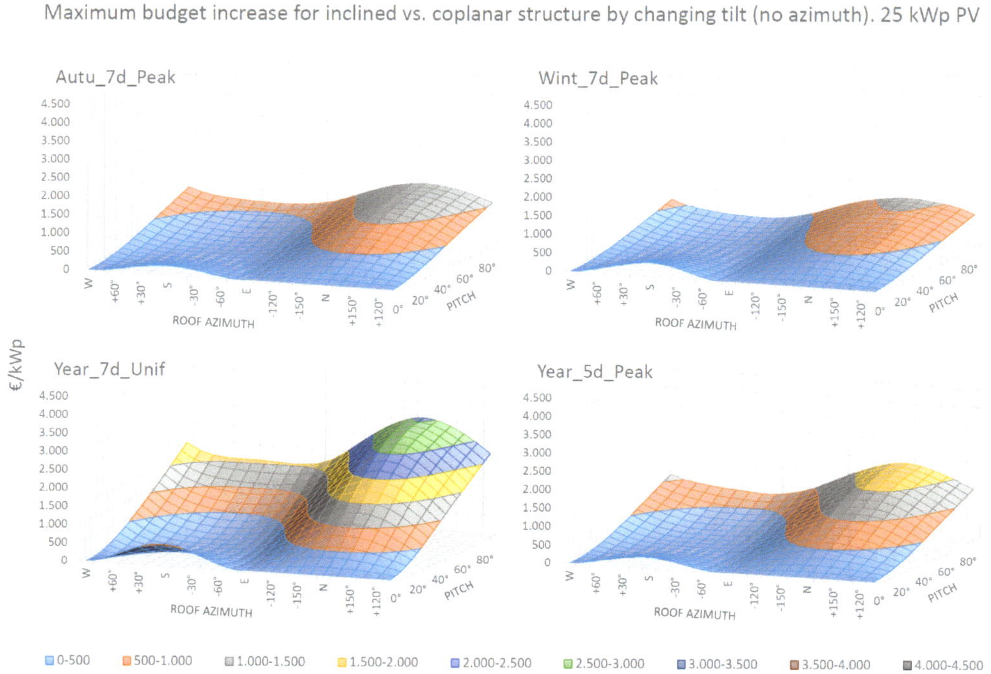

Figure 9. Maximum budget available for the additional cost of a structure that achieves the optimal inclination (but not the azimuth), compared to the coplanar option; PV of 25 kWp, intermediate energy price, PSH 4.0, and no excess income.

Therefore, in irregular and/or seasonal consumption patterns, when seeking to maximize energy bill savings (with associated energy excesses), the design of structures by modifying only the inclination presents disadvantages compared to azimuth optimization.

3.3. *Sensitivity Analysis*

The maximum budgets available for an inclined structure compared to the coplanar one experience significant variations when modifying the values assumed in the reference framework. Table 1 summarizes the variations of the global average compared to the reference framework, both in the case of optimization with azimuth and inclination and in the case of only inclination.

Table 1. Average variation in the maximum available budget (k€/kWp) compared to the reference framework (PSH 4.0, without income from energy surpluses, and intermediate energy price).

Peak Capacity	Energy Price	Excess Income	Azimuth and Inclination						Inclination					
			PSH 3.0		PSH 4.0		PSH 5.0		PSH 3.0		PSH 4.0		PSH 5.0	
			$\dot{x}$	σ	$\dot{x}$	σ	$\dot{x}$	σ	$\dot{x}$	σ	$\dot{x}$	σ	$\dot{x}$	σ
5 kWp	High	X	0.0	0.1	0.6	0.4	1.2	0.9	0.0	0.0	0.4	0.4	0.8	0.8
		√	0.1	0.1	0.7	0.5	1.5	1.0	0.1	0.1	0.5	0.5	1.0	0.9
	Med.	X	−0.4	0.3	0.0	0.0	0.5	0.4	−0.3	0.2	0.0	0.0	0.3	0.3
		√	−0.3	0.2	0.1	0.1	0.6	0.5	−0.2	0.2	0.1	0.1	0.4	0.4
	Low	X	−0.8	0.6	−0.6	0.4	−0.4	0.3	−0.5	0.5	−0.4	0.4	−0.2	0.3
		√	−0.8	0.6	−0.5	0.4	−0.3	0.3	−0.5	0.5	−0.4	0.4	−0.2	0.2

Table 1. *Cont.*

Peak Capacity	Energy Price	Excess Income	Azimuth and Inclination						Inclination					
			PSH 3.0		PSH 4.0		PSH 5.0		PSH 3.0		PSH 4.0		PSH 5.0	
			ẋ	σ	ẋ	σ	ẋ	σ	ẋ	σ	ẋ	σ	ẋ	σ
25 kWp	High	X	0.0	0.1	0.4	0.3	0.8	0.7	0.0	0.1	0.3	0.3	0.5	0.6
		√	0.3	0.2	0.8	0.5	1.3	0.9	0.2	0.2	0.5	0.5	0.9	0.8
	Med.	X	−0.3	0.2	0.0	0.0	0.3	0.3	−0.2	0.2	0.0	0.0	0.2	0.2
		√	−0.1	0.2	0.3	0.2	0.6	0.5	0.0	0.2	0.2	0.2	0.4	0.4
	Low	X	−0.5	0.4	−0.4	0.3	−0.2	0.3	−0.3	0.4	−0.3	0.3	−0.2	0.2
		√	−0.4	0.4	−0.3	0.3	−0.1	0.3	−0.3	0.4	−0.2	0.3	0.0	0.2

3.3.1. Energy Price

Increasing electricity prices has translated into a substantial change in the viability of structures and subsequently in the collection area. The change in values from 2021 (low price analyzed) to the values of mid-2023 (intermediate price analyzed), if maintained in the future, implies an increase in the budget available for the additional cost of the structures of several hundred euros in large part of the scenarios analyzed (Table 1).

Thus, for example, considering a 5 kW PV in the reference framework, the average budget available for the structure would go from 0.6 ± 0.4–0.8 ± 0.6 k€/kW with low prices prior to the crisis (low price) to 1.2 ± 0.9–1.5 ± 1.0 k€/kW with intermediate prices close to the present. The increase would be much more spectacular if prices in the future stabilized around the maximums recorded during 2022 (1.8 ± 1.3–2.2 ± 1.6 k€/kW).

3.3.2. Radiation Site

Solar radiation is another parameter that leads to significant variations in the budget available for inclined versus coplanar structures. The differences between locations with low irradiation (PSH 3.0) and high irradiation (PSH 5.0) can mean variations in the average budget of more than 1 k€/kWp, especially with high energy prices (Table 1). Thus, for example, in the case of a 5 kW system and intermediate energy price, the average for the budget would go from 0.8 ± 0.6–1.1 ± 0.7 k€/kW for PSH 3.0 to 1.6 ± 1.2–2.0 ± 1.4 k€/kW for PSH 5.0. Considering azimuths between 90° and −90°, they would go from 0.5 ± 0.4–0.6 ± 0.5 k€/kW (PSH 3.0) to 1.0 ± 0.7–1.3 ± 0.9 k€/kW (PSH 5.0).

3.3.3. Optimization with Income from Surplus Energy Injected into the Grid

The possibility of being paid for the electricity injected into the grid reduces the differences between consumption patterns as the PV capacity increases. For example, with a 25 kWp PV system in the reference framework, assuming income for each kWh injected into the grid of 1/3 of the price of the energy purchased, the average budget for the structure would increase from 0.6 ± 0.4 k€/kW without income to 1.0 ± 0.7 k€/kW with income for the most seasonal patterns (Wint_7d_Peak and Autu_7d_Peak). Considering azimuths between 90° and −90°, such levels would go from 0.4 ± 0.3 to 0.6 ± 0.4 k€/kW.

In all cases, income from electricity injected into the grid improves results, ensuring a higher profitability of the installation. However, its importance is limited to low energy prices, even in cases where the installed capacity is high (Table 1). On the contrary, with high energy prices, such a scenario allows increasing the maximum budget by several hundred euros.

4. Conclusions

The industry consumption pattern is a key factor in ensuring the viability and profitability of any mounting structures of solar photovoltaic installations. In industries with a seasonal and/or irregular consumption pattern, the increase in installed peak capacity, with the aim of maximizing savings on the energy bill, leads to a significant reduction in the available mounting budget for inclined structures compared to the coplanar option. Thus, for example, with a strongly seasonal pattern, going from a reduction in annual consumption of 4% to 16% leads to a halving of the mounting budget; in an industry with relatively uniform consumption, going from 5% to 31% barely implies a reduction of 15% of the budget.

Depending on the azimuth and inclination of the industry's rooftop, the maximum budget available for an inclined mounting structure that optimizes the collection plane can vary by more than a thousand euros per kWp. The further the azimuth and inclination move from the optimal values, the higher the available budget will be; therefore, it is especially profitable in orientations close to the east and west and in roofs partially inclined to the north, as long as the panels are arranged avoiding shading. For example, with a PV designed to increase savings on the energy bill (installed capacity of 25 kW) and assuming intermediate energy prices and radiation, the average budget for optimizing the system on a west-facing roof (considering all the inclinations) was 1.3 k€/kWp for the uniform consumption pattern and 0.7 k€/kWp for the seasonal and/or irregular consumption pattern.

The design of structures by modifying only the inclination presents disadvantages compared to azimuth optimization, especially in irregular and/or seasonal consumption patterns. In the previous example (optimizing the west-facing roof), the mean budget was reduced to 0.8 k€/kWp (uniform consumption pattern) and 0.4 k€/kWp (seasonal patterns). The cost of the mounting structure would have to differ little from the coplanar option and be less than that of azimuth-modifying structures to make the choice worthwhile.

In the last 2 years, the increase in electricity prices has led to a substantial change in the viability and profitability of structures optimized for the plane compared to the coplanar option. If the prices at the beginning of 2023 are maintained in the future, the budget available for the structures would increase by several hundred euros compared to the prices in mid-2021, modifying accepted design paradigms until now. In the example of optimizing a west orientation, the budget would be halved considering pre-crisis prices and would almost double if the prices were increased to crisis highs.

The differences between locations with low irradiation and high irradiation can mean variations in the average budget of more than 1 k€/kWp, especially with high energy prices. The average budget doubles from 3.0 HSP to 5.0 HSP.

The possibility of selling electricity injected into the grid reduces the differences between consumption patterns. In the previous example (optimizing the west-facing roof), the mean budget was increased to 1.4 k€/kWp for the uniform consumption pattern and 1.0 k€/kWp for the seasonal patterns. However, its importance is limited to low electricity prices, even in cases where the installed capacity is high.

Author Contributions: Conceptualization, J.L.G., F.B. and F.R.M.; methodology, J.L.G. and F.R.M.; software, A.P., F.R.M. and I.G.; validation, A.P.; formal analysis, F.R.M., A.P. and I.G.; investigation, A.P.; resources, J.L.G. and F.B.; writing—original draft preparation, A.P., F.R.M. and I.G.; writing—review and editing, J.L.G. and F.B.; visualization, A.P. and I.G.; supervision, J.L.G. and F.B.; project administration, J.L.G. and F.B.; funding acquisition, J.L.G. and F.B. All authors have read and agreed to the published version of the manuscript.

Funding: This research received no external funding.

Data Availability Statement: The data presented in this study are available on request from the corresponding author.

Acknowledgments: The authors would like to thank the participating agroindustries for their collaboration in the research and also the R+D+i 2023 Program of the UPM for the support received to carry out an international research stay.

Conflicts of Interest: The authors declare no conflict of interest.

References

1. Garcia, J.L.; Perdigones, A.; Benavente, R.M.; Mazarron, F.R. Influence of the new energy context on the spanish agri-food industry. *Agronomy* **2022**, *12*, 977. [CrossRef]
2. García, J.L.; Perdigones, A.; Benavente, R.M.; Álvarez, J.; Baptista, F.; Mazarrón, F.R. Feasibility of photovoltaic systems for the agrifood industry in the new energy and climate change context. *Agronomy* **2023**, *13*, 2620. [CrossRef]
3. Liu, C.; Xu, W.; Li, A.; Sun, D.; Huo, H. Analysis and optimization of load matching in photovoltaic systems for zero energy buildings in different climate zones of china. *J. Clean. Prod.* **2019**, *238*, 117914. [CrossRef]
4. Freitas, S.; Brito, M.C. Non-cumulative only solar photovoltaics for electricity load-matching. *Renew. Sustain. Energy Rev.* **2019**, *109*, 271–283. [CrossRef]
5. Litjens, G.; Worrell, E.; van Sark, W. Influence of demand patterns on the optimal orientation of photovoltaic systems. *Sol. Energy* **2017**, *155*, 1002–1014. [CrossRef]
6. Shin, H.; Geem, Z.W. Optimal design of a residential photovoltaic renewable system in south korea. *Appl. Sci.* **2019**, *9*, 1138. [CrossRef]
7. Zubair, M.; Awan, A.B. Optimization of photovoltaic energy systems for residential customers in hot climate areas based on seasonal and average daily load profile. *Energy Technol.* **2021**, *9*, 2100036. [CrossRef]
8. Kurdi, Y.; Alkhatatbeh, B.J.; Asadi, S. The role of demand energy profile on the optimum layout of photovoltaic system in commercial buildings. *Energy Build.* **2022**, *271*, 112320. [CrossRef]
9. Jeong, K.; Hong, T.; Ban, C.; Koo, C.; Park, H.S. Life cycle economic and environmental assessment for establishing the optimal implementation strategy of rooftop photovoltaic system in military facility. *J. Clean. Prod.* **2015**, *104*, 315–327. [CrossRef]
10. Jimenez-Castillo, G.; Munoz-Rodriguez, F.J.; Martinez-Calahorro, A.J.; Tina, G.M.; Rus-Casas, C. Impacts of array orientation and tilt angles for photovoltaic self-sufficiency and self-consumption indices in olive mills in spain. *Electronics* **2020**, *9*, 348. [CrossRef]
11. Perdigones, A.; Garcia, I.; Porras-Prieto, C.J.; Gomez-Villarino, M.T.; Baptista, F.; Garcia, J.L. Evaluation of the cost-effectiveness of grid-connected photovoltaic solar energy in wineries. *Cienc. E Tec. Vitivinic.* **2021**, *36*, 45–54. [CrossRef]
12. Haysom, J.E.; Hinzer, K.; Wright, D. Impact of electricity tariffs on optimal orientation of photovoltaic modules. *Prog. Photovolt.* **2016**, *24*, 253–260. [CrossRef]
13. Merilainen, A.; Puranen, P.; Kosonen, A.; Ahola, J. Optimization of rooftop photovoltaic installations to maximize revenue in finland based on customer class load profiles and simulated generation. *Sol. Energy* **2022**, *240*, 422–434. [CrossRef]
14. Mitiu, M.A.; Olteanu, M.V.; Deak, G.; Raischi, N.S.; Cociorva, D. Determination the Optimal Tilt and Azimuth Angle of Photovoltaic System for Maximum Energy in Bucharest Region, Romania—Case Study. In *Proceedings of the International Conference on Renewable Energy and Environment Engineering (REEE), Paris, France, 29–31 October 2018*; E D P Sciences: Paris, France, 2018.
15. Hailu, G.; Fung, A.S. Optimum tilt angle and orientation of photovoltaic thermal system for application in greater toronto area, canada. *Sustainability* **2019**, *11*, 6443. [CrossRef]
16. Zhong, Q.; Tong, D.Q. Spatial layout optimization for solar photovoltaic (pv) panel installation. *Renew. Energy* **2020**, *150*, 1–11. [CrossRef]
17. Bozikova, M.; Bilcik, M.; Madola, V.; Szaboova, T.; Kubik, L.; Lendelova, J.; Cviklovic, V. The effect of azimuth and tilt angle changes on the energy balance of photovoltaic system installed in the southern slovakia region. *Appl. Sci.* **2021**, *11*, 8998. [CrossRef]
18. Dhimish, M.; Silvestre, S. Estimating the impact of azimuth-angle variations on photovoltaic annual energy production. *Clean Energy* **2019**, *3*, 47–58. [CrossRef]
19. Gwesha, A.O.; Li, P.W.; Alfulayyih, Y.M. Optimization of fixed photovoltaic panel "tilt" angles for maximal energy harvest considering year-around sky coverage conditions. *J. Sol. Energy Eng. Trans.-ASME* **2021**, *143*, 021002. [CrossRef]
20. Nfaoui, M.; El-Hami, K. Optimal tilt angle and orientation for solar photovoltaic arrays: Case of settat city in morocco. *Int. J. Ambient Energy* **2020**, *41*, 214–223. [CrossRef]
21. Hussein, H.M.S.; Ahmad, G.E.; El-Ghetany, H.H. Performance evaluation of photovoltaic modules at different tilt angles and orientations. *Energy Convers. Manag.* **2004**, *45*, 2441–2452. [CrossRef]
22. Masili, M.; Ventura, L. Local tilt optimization of photovoltaic solar panels for maximum radiation absorption. *Int. J. Photoenergy* **2019**, *2019*, 3254780. [CrossRef]
23. Diez, F.J.; Martinez-Rodriguez, A.; Navas-Gracia, L.M.; Chico-Santamarta, L.; Correa-Guimaraes, A.; Andara, R. Estimation of the hourly global solar irradiation on the tilted and oriented plane of photovoltaic solar panels applied to greenhouse production. *Agronomy* **2021**, *11*, 495. [CrossRef]
24. Barbon, A.; Bayon-Cueli, C.; Bayon, L.; Rodriguez-Suanzes, C. Analysis of the tilt and azimuth angles of photovoltaic systems in non-ideal positions for urban applications. *Appl. Energy* **2022**, *305*, 117802. [CrossRef]
25. Oh, M.; Kim, J.Y.; Kim, B.; Yun, C.Y.; Kim, C.K.; Kang, Y.H.; Kim, H.G. Tolerance angle concept and formula for practical optimal orientation of photovoltaic panels. *Renew. Energy* **2021**, *167*, 384–394. [CrossRef]
26. Kwak, Y.; Mun, S.H.; Park, C.D.; Lee, S.M.; Huh, J.H. Statistical analysis of power generation of semi-transparent photovoltaic (stpv) for diversity in building envelope design: A mock-up test by azimuth and tilt angles. *Renew. Energy* **2022**, *188*, 651–669. [CrossRef]

27. Melhem, R.; Shaker, Y. Optimum tilt angle and solar radiation of photovoltaic modules for gulf collaboration council countries. *Int. J. Energy Res.* **2023**, *2023*, 8381696. [CrossRef]
28. Mukisa, N.; Zamora, R. Optimal tilt angle for solar photovoltaic modules on pitched rooftops: A case of low latitude equatorial region. *Sustain. Energy Technol. Assess.* **2022**, *50*, 101821. [CrossRef]
29. Yildirim, E.; Aktacir, M.A. Investigation of azimuth and tilt angle effects on building integrated photovoltaic system. *J. Fac. Eng. Archit. Gazi Univ.* **2019**, *34*, 609–619.
30. Calabro, E. Determining optimum tilt angles of photovoltaic panels at typical north-tropical latitudes. *J. Renew. Sustain. Energy* **2009**, *1*, 033104. [CrossRef]
31. Guo, M.; Zang, H.; Gao, S.; Chen, T.; Xiao, J.; Cheng, L.; Wei, Z.; Sun, G. Optimal tilt angle and orientation of photovoltaic modules using hs algorithm in different climates of china. *Appl. Sci.* **2017**, *7*, 1028. [CrossRef]
32. Kacira, M.; Simsek, M.; Babur, Y.; Demirkol, S. Determining optimum tilt angles and orientations of photovoltaic panels in sanliurfa, turkey. *Renew. Energy* **2004**, *29*, 1265–1275. [CrossRef]
33. Ashetehe, A.A.; Gessesse, B.B.; Shewarega, F. Development of optimal tilt angle models of a photovoltaic module for maximum power production: Ethiopia. *Int. J. Photoenergy* **2022**, *2022*, 8729570. [CrossRef]
34. Abdelaal, A.K.; El-Fergany, A. Estimation of optimal tilt angles for photovoltaic panels in egypt with experimental verifications. *Sci Rep.* **2023**, *13*, 3268. [CrossRef]
35. Annibaldi, V.; Condemi, A.; Cucchiella, F.; Rotilio, M. Solar photovoltaic optimal tilt angles in public building. *Environ. Clim. Technol.* **2020**, *24*, 265–277. [CrossRef]
36. Ban, C.; Hong, T.; Jeong, K.; Koo, C.; Jeong, J. A simplified estimation model for determining the optimal rooftop photovoltaic system for gable roofs. *Energy Build.* **2017**, *151*, 320–331. [CrossRef]
37. Tripathy, M.; Joshi, H.; Panda, S.K. Energy payback time and life-cycle cost analysis of building integrated photovoltaic thermal system influenced by adverse effect of shadow. *Appl. Energy* **2017**, *208*, 376–389. [CrossRef]
38. Saedpanah, E.; Asrami, R.F.; Sohani, A.; Sayyaadi, H. Life cycle comparison of potential scenarios to achieve the foremost performance for an off-grid photovoltaic electrification system. *J. Clean. Prod.* **2020**, *242*, 118440. [CrossRef]
39. Sherwani, A.F.; Asjad, M.; Haleem, A. Performance evaluation of solar photovoltaic electricity-generating systems: An indian perspective. *Int. J. Sustain. Eng.* **2019**, *12*, 70–75. [CrossRef]
40. Alharbi, A.; Awwad, Z.; Habib, A.; de Weck, O. Economical sizing and multi-azimuth layout optimization of grid-connected rooftop photovoltaic systems using mixed-integer programming. *Appl. Energy* **2023**, *335*, 120654. [CrossRef]
41. Yang, H.; Lu, L. The optimum tilt angles and orientations of pv claddings for building-integrated photovoltaic (bipv) applications. *J. Sol. Energy Eng. Trans.-ASME* **2007**, *129*, 253–255. [CrossRef]
42. Awan, A.B. Optimization and techno-economic assessment of rooftop photovoltaic system. *J. Renew. Sustain. Energy* **2019**, *11*, 033501. [CrossRef]
43. Barbon, A.; Ghodbane, M.; Bayon, L.; Said, Z. A general algorithm for the optimization of photovoltaic modules layout on irregular rooftop shapes. *J. Clean. Prod.* **2022**, *365*, 132774. [CrossRef]
44. Zimmerman, R.; Panda, A.; Bulovic, V. Techno-economic assessment and deployment strategies for vertically-mounted photovoltaic panels. *Appl. Energy* **2020**, *276*, 115149. [CrossRef]
45. Bempah, K.O.; Kwon, K.; Kim, K.A. Experimental study of photovoltaic panel mounting configurations for tube-shaped structures. *Appl. Energy* **2019**, *240*, 754–765. [CrossRef]
46. Liu, C.P.; Xu, W.; Li, A.G.; Sun, D.Y.; Huo, H.M. Energy balance evaluation and optimization of photovoltaic systems for zero energy residential buildings in different climate zones of china. *J. Clean. Prod.* **2019**, *235*, 1202–1215. [CrossRef]
47. Obiwulu, A.U.; Erusiafe, N.; Olopade, M.A.; Nwokolo, S.C. Modeling and estimation of the optimal tilt angle, maximum incident solar radiation, and global radiation index of the photovoltaic system. *Heliyon* **2022**, e09598. [CrossRef]
48. Hartner, M.; Ortner, A.; Hiesl, A.; Haas, R. East to west—The optimal tilt angle and orientation of photovoltaic panels from an electricity system perspective. *Appl. Energy* **2015**, *160*, 94–107. [CrossRef]
49. Chang, Y.-P. Optimal the tilt angles for photovoltaic modules in taiwan. *Int. J. Electr. Power Energy Syst.* **2010**, *32*, 956–964. [CrossRef]
50. Beringer, S.; Schilke, H.; Lohse, I.; Seckmeyer, G. Case study showing that the tilt angle of photovoltaic plants is nearly irrelevant. *Sol. Energy* **2011**, *85*, 470–476. [CrossRef]
51. Energyplus Weather Data. Available online: https://energyplus.net/weather (accessed on 11 July 2023).

buildings

MDPI

Article

Effect of Orientation and Skylight Area Ratio on Building Energy Efficiency in the Qinghai–Tibet Plateau

Yingmei Wang [1,2], Haosen Qin [1,2], Yan Wang [3], Ji Chen [4,*], Xin Hou [4], Pengfei Rui [4], Shouhong Zhang [5] and Hanyu Song [6]

1 School of Energy and Power Engineering, Lanzhou University of Technology, Lanzhou 730050, China
2 Key Laboratory of Complementary Energy System of Biomass and Solar Energy, Lanzhou 730050, China
3 School of Materials Science and Engineering, Wuhan University of Technology, Wuhan 430070, China
4 Beiluhe Observation and Research Station of Frozen Soil Engineering and Environment, State Key Laboratory of Frozen Soil Engineering, Northwest Institute of Eco-Environmental and Resources, Chinese Academy of Sciences, Lanzhou 730000, China
5 China Railway Qinghai–Tibet Group Co., Ltd., Xining 810007, China
6 Shangqiu Branch of China Tower Co., Ltd., Shangqiu 476000, China
* Correspondence: chenji@lzb.ac.cn

Abstract: The Qinghai–Tibet plateau, with an average altitude of over 4000 m, has low annual average temperatures and a high demand for building heating. This region's abundant solar energy resources hold substantial practical significance for improving the indoor heat environment and reducing building energy consumption. This paper investigates the impact of orientation and skylight area ratio on building heat load and indoor temperature, using both actual measurement and simulation methods, with a case study of the comprehensive building at Beiluhe Observation and Research Station of Frozen Soil Engineering and Environment (Beiluhe Station), located in the Qinghai–Tibet Plateau region. Initially, a model was established using the EnergyPlus 9.4 software, with orientation variables set from east to west in 15° increments, to simulate the variations in building heat load resulting from orientation changes; simulations were then conducted for three different skylight area ratios under the optimal orientation to evaluate their influence on heat load and indoor temperature. The results show that for the architectural style examined in this paper, the optimal building orientation within the region is 30° south by east, with the optimal orientation range spanning from 45° south by east to due south. Heating load is negatively correlated with the skylight area ratio, and beyond a certain threshold, the rate of decrease in heat load diminishes or even stabilizes. The conclusions of this paper offer guidance for the orientation and skylight design of new buildings on the Qinghai–Tibet Plateau.

Keywords: building orientation; skylight area ratio; Qinghai–Tibet plateau; Beiluhe station

Citation: Wang, Y.; Qin, H.; Wang, Y.; Chen, J.; Hou, X.; Rui, P.; Zhang, S.; Song, H. Effect of Orientation and Skylight Area Ratio on Building Energy Efficiency in the Qinghai–Tibet Plateau. *Buildings* **2024**, *14*, 755. https://doi.org/10.3390/buildings14030755

Academic Editor: Rafik Belarbi

Received: 17 February 2024
Revised: 4 March 2024
Accepted: 8 March 2024
Published: 11 March 2024

1. Introduction

As the global energy crisis intensifies and environmental issues become more severe, Carbon Neutrality and Carbon Peak have become a common focus of attention for the international community. The building sector is one of the largest energy-consuming industries in terms of total energy consumption [1], and its proportion is increasing yearly [2,3]. According to statistics [2], from 2005 to 2018, the energy consumption of the entire lifecycle of buildings in China rose from 934 million tons of coal equivalent (tce) in 2005 to 2.147 billion tce in 2018, with an average annual growth of 6.6%. In 2018, the lifecycle energy consumption of buildings accounted for 46.5% of total energy consumption [2]. Therefore, building energy efficiency is significant for carbon reduction and global sustainable development [4,5].

The Qinghai–Tibet Plateau, known as the "Water Tower of Asia", is the source of several major rivers in Asia [6]. However, with an average altitude of over 4000 m and low average annual temperatures, the region has a prolonged heating period and high energy consumption for heating [7,8]. The traditional fossil fuel heating methods pose challenges to the region's ecological environment and water resource security. Located in the mid to low latitudes with high atmospheric transparency, the Qinghai–Tibet Plateau has abundant solar energy resources with tremendous potential for utilization [9]. Fully tapping into the region's rich solar energy resources can effectively reduce building energy consumption and decrease reliance on external traditional fossil fuels [10], promote green development, and accelerate progress towards the global dual-carbon goals.

During the design stage of a building, optimizing the building's orientation, structure, and layout can significantly enhance the thermal utilization efficiency of solar energy and reduce the building's heat load [11–13]. Orientation is an essential factor affecting building energy consumption [14]. Choosing the building orientation wisely can effectively increase the acquisition of solar energy for the building [15]. Skylights not only utilize natural light sources but also make use of solar radiation for indoor heating, which improves building energy efficiency [16]. The orientation of buildings and the use of skylights as highly effective and low-cost energy-saving strategies have attracted the attention of numerous researchers.

Numerous studies have demonstrated that optimizing building orientation can significantly reduce energy consumption [15]. In New Minia in Egypt, the difference in energy consumption between residential buildings' best and worst orientation can reach 7.5% [17]. In Iran, orienting a building to the south can reduce energy consumption by 11–39% [18]. In India, the optimal orientation of a building can save an additional 3.41% of energy compared with a reference building [19]. In China's hot summer and warm winter regions, the north–south orientation of residential buildings is 15% more energy-efficient than the east–west orientation [20]. Other studies also indicate that optimizing building orientation can reduce energy consumption by 20–36% [21,22]. In winter, the orientation of a building can significantly reduce its heating energy consumption [23], and an appropriate orientation can help reduce about 35% of a building's thermal losses during the winter season [24]. Regarding building energy consumption, the best orientation varies with geographical location and climatic conditions [25]. In the UK, the optimal residential orientation is south, while the least efficient is northwest [26]. The best building orientation is northward, and the worst is southward, in New Minia, Egypt [17]. In the Shandong region of China, the optimal range for building orientation is south to 15° south by east [27].

In addition to building orientation, skylight design is a critical factor in optimizing the use of solar energy [28]. As atriums serve as hubs connecting indoor and outdoor spaces, skylights are an essential part of their envelope structures [16,29]. In cold regions, large skylights can introduce solar radiation to reduce the heat load during the day [30], but they may also lead to significant heat loss at night due to poor thermal performance [31]. The area ratio (AR), which is the ratio of skylight area to roof area, is an important parameter affecting the thermal performance of buildings [32,33], and determining the appropriate skylight area is crucial for ensuring thermal comfort in atriums. Simulation studies on energy consumption indicate that reducing the skylight area can improve the thermal environment [34], achieving optimal daylighting and thermal comfort conditions when the AR is 1/4 [35]. However, other studies suggest increasing the area ratio can improve daylighting performance [36]. Furthermore, the skylight area-to-floor area ratio also significantly affects building energy consumption. When the skylight-to-floor area ratio is between 5.5% and 6%, it can reduce energy demand by 19% [28].

These studies above demonstrate clearly that orientation and AR have significant and complex impacts on building energy consumption. Especially in the Qinghai–Tibet Plateau region, characterized by unique geographical conditions and severe climate, the research field regarding building orientation and the skylight area ratio has not been sufficiently explored, which suggests a clear research void in this region. The Qinghai–Tibet plateau is the world's most extensive plateau with abundant solar thermal resources and is home to nearly ten million people with a strong demand for building heating. Exploring the effect of orientation and AR on building energy consumption in this region has significant practical value. This article takes the comprehensive building of the Beiluhe Station on the Qinghai–Tibet plateau as the research subject. It validates the applicability of the EnergyPlus 9.4 software based on monitoring data. It simulates the heat load of buildings at different orientations from 90° (due east) to 270° (due west), as well as the heat load and room temperature under skylights with different ARs, providing reasonable building orientation and skylight AR for the region. This research can support the energy-saving design of buildings in the Qinghai–Tibet plateau and similar cold regions.

2. Overview of the Study Region

2.1. Study Subject

This paper selects the comprehensive building of the Beiluhe Station in the hinterland of the Source Region of the Yangtze River, Yellow River, and Lancang River on the Tibetan Plateau as the research subject. The station is located at 34°51′ N and 92°56′ E, with an altitude of 4628 m. The annual average temperature is −3.1 °C, and it is classified as a severe cold A zone according to the "Standard of Climatic Regionalization for Architecture" [37]. The building was constructed in 2017. It is a two-story ventilated concrete frame structure with a floor height of 3.3 m and covers an area of 631.24 m^2, with a shape coefficient of 0.31. The building is oriented due south, and at its center is an atrium covering 137.45 m^2. The total roof area is 650.67 m^2, and the southern slope of the roof has a skylight covering 316.73 m^2, resulting in an AR of 48.68%. The building's exterior wall facades are equipped with windows configured differently, with the window-to-wall ratios of each facade listed in Table 1 and the building's floor plan presented in Figure 1.

Table 1. Window-to-wall ratio.

	Total	North	East	South	West
Total Wall Area (m^2)	734.36	242.22	124.96	242.22	124.96
Window Area (m^2)	84.64	6.60	0.90	76.24	0.90
Overall Window-to-Wall Ratio (%)	11.53	2.72	0.72	31.28	0.72

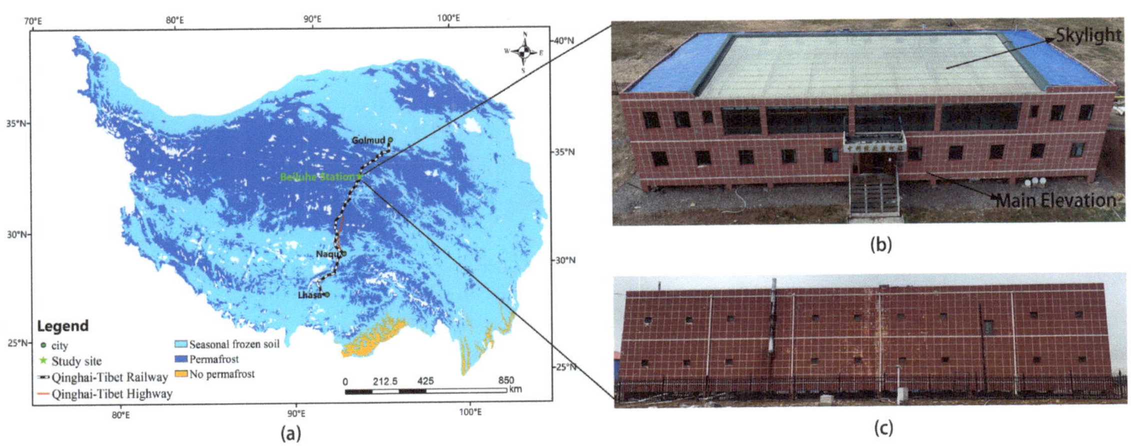

Figure 1. (**a**) Location of the study site; (**b**) main elevation of the building; (**c**) back elevation of the building.

2.2. Monitoring Method

Temperature and humidity sensors were installed in typical rooms within the building. The model of the sensors is RC-4HC, produced by Jiangsu Jingchuang Electric Co., Ltd. in Xuzhou, Jiangsu, China. The recording interval is 1 h, the accuracy of temperature recording is ±0.5 °C, and the resolution is 0.1 °C. The sensors are mounted on brackets fixed to the wall 3 m above the ground. A layout diagram of the building's internal monitoring setup and a schematic of the monitoring equipment are shown in Figure 2.

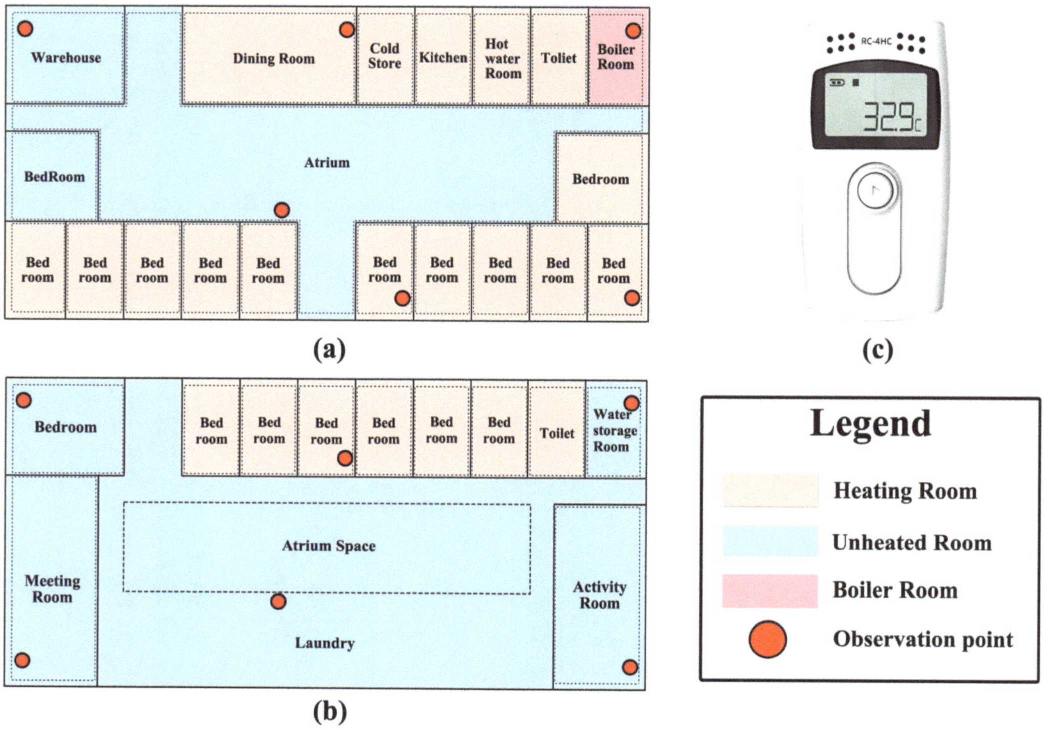

Figure 2. Room function, heating conditions, and monitoring point locations. (**a**) First floor; (**b**) second floor; (**c**) monitoring equipment.

In addition, a weather station is located 200 m west of the building to monitor the meteorological conditions of the region. The weather station has sensors for temperature and humidity, atmospheric pressure, four-component radiation, wind direction, and wind speed.

2.3. Local Climate Characteristics

Figure 3 displays the temperature distribution, radiation distribution, and wind frequency diagrams around the Beiluhe Station on the Qinghai–Tibet Plateau. The average annual temperature at Beiluhe Station is −3.1 °C, with monthly average temperatures below 0 °C for seven months. The coldest month is January, with an average monthly temperature of −14.32 °C, while the warmest month is July, with an average monthly temperature of 7.31 °C. According to the observational data, the maximum solar irradiance is 1155.00 Wh/m^2. The average solar radiation before noon, at 12 p.m. local time, is 37.8% higher than in the afternoon. The wind frequency diagram indicates that the prevailing winds in the region are from the west, with maximum wind speeds reaching up to 20 m/s.

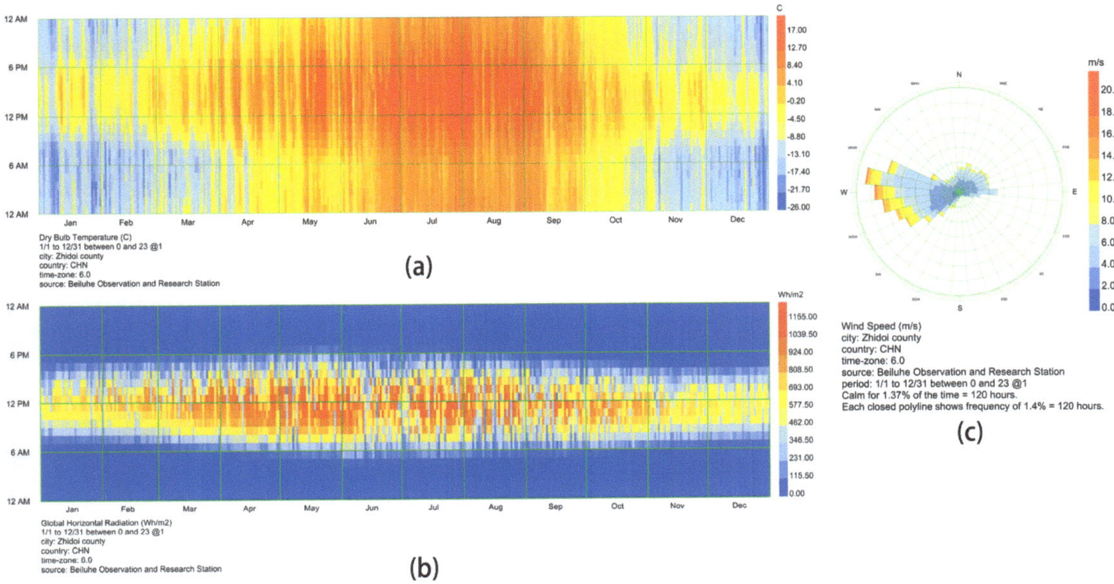

Figure 3. (**a**) Annual temperature distribution graph; (**b**) total radiation distribution graph; (**c**) wind frequency graph.

3. Study Methods

3.1. Energy Consumption Simulation Process

3.1.1. Simulation Software

The software EnergyPlus 9.4 used in this paper was co-developed by the U.S. Department of Energy and Lawrence Berkeley National Laboratory. It is widely used in building design and research and can perform comprehensive energy consumption simulation and analysis for heating, cooling, lighting, ventilation, and other energy uses in buildings [38]. EnergyPlus is recognized by numerous institutions and enterprises worldwide; hence, its simulation results are widely accepted.

3.1.2. Meteorological Data

The meteorological data mainly comes from the Beiluhe Meteorological Station. The separation of the total radiation into direct and diffuse radiation using the Gompertz function method [39], as shown in Equations (1) and (2).

$$K_n = A_1 {A_2}^{A_3 {A_2}^{-A_4 K_t}} \quad (1)$$

In the formula:

$$\begin{aligned} A_1 &= -0.1556 \sin^2 h + 0.1028 \sin h + 0.3748 \\ A_2 &= 0.7973 \sin^2 h + 0.1509 \sin h + 3.035 \\ A_3 &= 5.4307 \sin h + 7.2182 \\ A_4 &= 2.990 \\ K_t &= \frac{E}{E_0} \end{aligned} \quad (2)$$

In the formula, E represents the solar irradiance on the surface of the earth, E_0 corresponds to solar irradiance outside the atmosphere on a horizontal plane, and h denotes the solar altitude angle, and the calculation formulas for E, E_0, h, and local time, etc., can be found in the literature [40,41].

3.1.3. Setting of Building Envelope and Thermal Parameters

The construction of each building envelope structure and the thermal parameters are shown in Tables 2 and 3. The thermal parameters are derived from the Chinese standard "Thermal Design Code for Civil Buildings" [42], and the data for transparent envelope structures come from the International Glass Database (IGDB) [43]. This model is also based on these fixed settings.

Table 2. Building envelope details and thermal parameters.

Building Envelope	Construction	Heat Conductivity (W/m·K) [42]	Thickness (m)	Heat-Transfer Coefficient (W/m²·K)
Exterior wall	Cement mortar	0.930	0.005	0.230
	Cement lime plaster mortar	0.870	0.011	
	Autoclaved aerated concrete block	0.140	0.300	
	XPS insulation board	0.030	0.060	
	Cement mortar	0.930	0.018	
Interior wall	Cement mortar	0.930	0.005	0.580
	Cement lime plaster mortar	0.870	0.011	
	Autoclaved aerated concrete block	0.140	0.200	
	Cement lime plaster mortar	0.870	0.011	
	Cement mortar	0.930	0.005	
External floor	Cement mortar	0.930	0.020	0.221
	Reinforced concrete slab	1.740	0.200	
	Polyurethane rigid foam	0.024	0.100	
	Cement mortar	0.930	0.020	
Internal floor	Reinforced concrete slab	1.740	0.200	2.460
	Cement mortar	0.930	0.010	
Roof	SBS sheet	0.170	0.008	0.776
	Cement mortar	0.930	0.025	
	Coal ash	0.230	0.080	
	Cement cinder	0.760	0.510	
	Reinforced concrete slab	1.740	0.120	

Table 3. Thermal performances of skylight and external window.

Building Envelope	Construction	Solar Heat Gain Coefficient	Heat-Transfer Coefficient (W/m²·K)
External window	Insulating glass (6 + 12A + 6)	0.749	2.695
Skylight	PC board	0.701	2.014

3.1.4. HVAC System

The heating control temperature is set at 18 °C (24 h), with a startup temperature of 12 °C, and the heating period extends from January to June and from August to December. The lighting power density is 5 W/m^2, and the equipment power density is 3.8 W/m^2. The operation schedules for lighting, equipment, and personnel occupancy are detailed in the referenced literature [44].

3.1.5. Model Validation

This paper verifies the model by comparing the temperature data obtained from the atrium measurements with the data obtained from simulations, using the correlation coefficient (R^2) to evaluate the model's accuracy. The larger the R^2 value, the closer the simulation values are to the observed values. The formula for calculating the evaluation index is as shown in Equation (3):

$$R^2 = 1 - \sum_{i=1}^{n} (O_i - S_i)^2 / \sum_{i=1}^{n} (O - \overline{O})^2 \tag{3}$$

In the formula, O_i, S_i represent the observed and simulated values, respectively, and $\overline{O}$ is the average of the observed values.

Figure 4 displays the comparison between the simulated values and the monitored values of the atrium temperature, with an R^2 value of 0.954. The simulation curve slightly differs from the curve of the observed values, but, overall, the trends are consistent. The reasons for the simulation error include discrepancies between the actual thermal properties of the building envelope materials and their parameters in the model. There are also errors in the process of separating direct and diffuse radiation. The comparative analysis of the simulation and measurement values indicates that the temperature simulation results of the EnergyPlus 9.4 software for the object of this study are accurate.

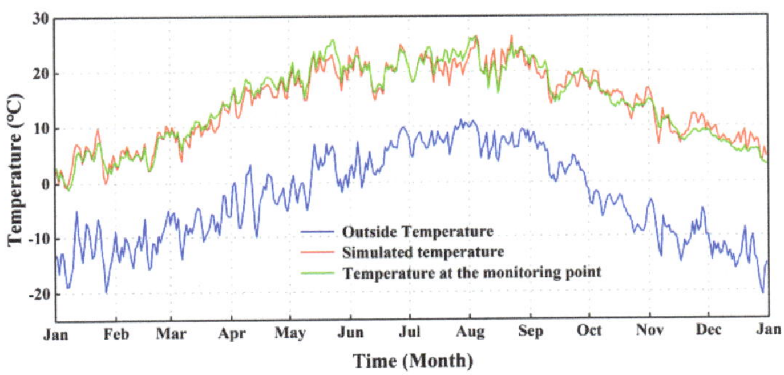

Figure 4. Comparison between simulated and measured values of atrium temperature.

The heating coal for Beiluhe Station comes from the Mulei Coal Mine in Xinjiang, with an ash content of 3.83% and a calorific value of 23.97 MJ/kg, classified as Class II anthracite. Per the "specification for small type boilers and atmospheric hot water boilers" [45], the boiler's thermal efficiency is 60%. According to the "general principles for the calculation of the comprehensive energy consumption" [46], the comprehensive energy consumption is calculated using formula (4), and the verification of energy consumption uses formula (5) for determination.

$$E = \sum_{i=1}^{n} (E_i \times k_i) \tag{4}$$

In the formula, E represents the comprehensive energy consumption, n represents the number of types of energy consumed, E_i represents the quantity of the i type of energy actually consumed during production or service processes, and k_i represents the standard coal conversion coefficient for the i type of energy.

$$ERR_{month} = \left[\frac{(M - S)_{month}}{M_{month}} \right] \times 100\% \tag{5}$$

In the formula, ERR_{month} represents the monthly error between the simulated and actual energy consumption, M represents the actual measured value of energy consumption, and S represents the simulated value of energy consumption.

Through the analysis of meteorological data, December and January are identified as the coldest months of the year. Based on the survey of coal usage at Beiluhe Station during this period, the average daily coal consumption is about 70 kg. Hence, December and January are selected as the key verification months for assessing heating energy consumption. According to the "general principles for the calculation of the comprehensive energy consumption" [46], the conversion coefficient for thermal value to standard coal is 0.03412 kgce/MJ for anthracite, and 0.82 kgce/MJ for smokeless coal. Through simulation, the heating amounts for December and January are 8149.92 kW·h and 9226.57 kW·h, respectively. Converting these amounts to kilograms of standard coal equivalent (kgce) using formula (4), they are 1001.07 kgce and 1133.32 kgce, respectively. Ignoring the thermal loss from the heating system's pipeline network, the actual heating amounts for December and January are 1067.64 kgce. Comparing the actual energy consumption with the simulated values, the deviations for December and January are 6.65% and −6.15%, respectively. According to the "technical code for the retrofitting of public building on energy efficiency" [47], a deviation between calculations and simulations within ±15% is an acceptable error range. Therefore, the energy consumption data simulated by the EnergyPlus 9.4 software are accurate and reliable.

Based on the dual verification of temperature and energy consumption, EnergyPlus 9.4 software can be used for the analysis of energy consumption and temperature at the comprehensive building of Beiluhe station.

3.2. Operation Setting

As shown in Figure 5. This paper defines north as 0°, increasing clockwise by 15° increments from 90° (east) to 270° (west), and uses EnergyPlus 9.4 software to simulate heat loads for 13 different orientations, as well as the heat load and atrium temperature under various ARs.

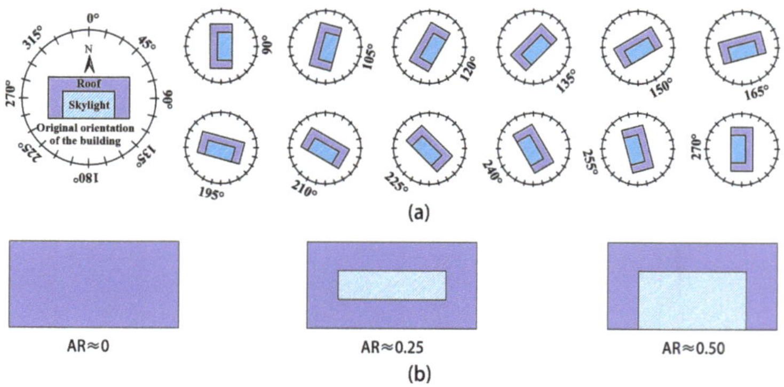

Figure 5. Simulation method. (**a**) Orientation; (**b**) area ratio.

4. Simulation Results and Analysis

4.1. Analysis of Heat Load Energy Saving for Different Building Orientations

Figure 6 shows the variation curve of the annual total heat load of a building under different orientations. As shown in Figure 6, the overall annual heat load presents a parabolic shape that opens upward, showing a trend of decreasing first and then increasing. Specifically, from 90° to 150°, the heat load decreases. Starting from the 150° orientation, the heat load begins to rise until it reaches 180°. This is because the general weather pattern of cloud cover changes in the plateau region: there is less stable cloud cover in the morning, and it increases rapidly in the afternoon [48–50]. Cloud cover blocks the solar radiation reaching the ground, leading to the asymmetry in the diurnal variation of total solar radiation on the plateau, with higher levels in the morning and lower in the afternoon [51]. This result is consistent with the radiation data measured by the Beiluhe weather station. Therefore, when the building's orientation is between 90° and 270°, under the condition that the absolute value of the angle difference between the building's orientation and the due south direction is the same, the main facade and the skylight of the building facing southeast receive more solar radiation than those facing southwest. The influence of wind on the primary facade orientation is minor within this range (Figure 2).

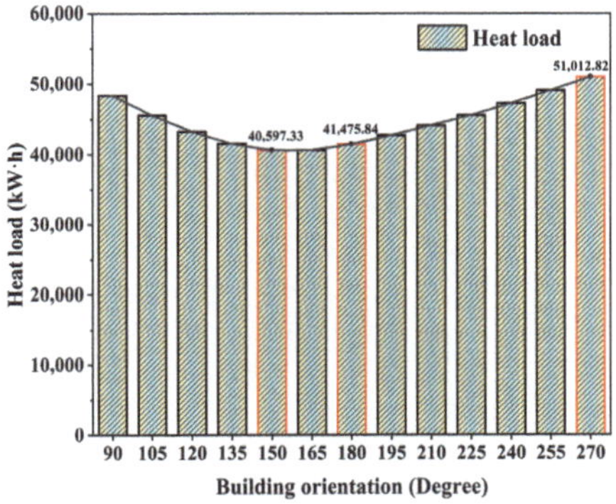

Figure 6. Changes in annual heat load of building under different orientations.

At 150°, the building's heat load is the smallest, at 40,597.33 kWh, which is an energy savings of 2.12% compared with the current actual orientation (180°) of the Beiluhe Station comprehensive building. The maximum heat load occurs at 270°, at 51,012.82 kWh, which is 22.99% higher than the energy consumption of the current actual orientation (180°). The difference in heat load between the two is 25.66%. To further clarify the relationship between orientation and heat load, a study was conducted on the monthly values of building heat load.

Figure 7 shows the variation in building heat load from January to December under different orientations. As depicted, with the progression of months, the overall heat load for each orientation exhibits a parabola shape that opens upward, showing a trend of decreasing first and then increasing. This is due to the change in outdoor temperature affecting the building's heating demand. From January to June, the outdoor temperature gradually increases, leading to a decrease in heat load. The outdoor temperature gradually decreases from August to December, and the heat load increases accordingly. From January to March and August to December, the heat loads are lower between 135° and 180°. These months belong to the colder seasons, and due to the strong diurnal variation characteristics of solar radiation in the Qinghai–Tibet Plateau region, the solar radiation within the range of 135° to 180° contributes more to the building's heat gain, reducing the heating requirements of the building.

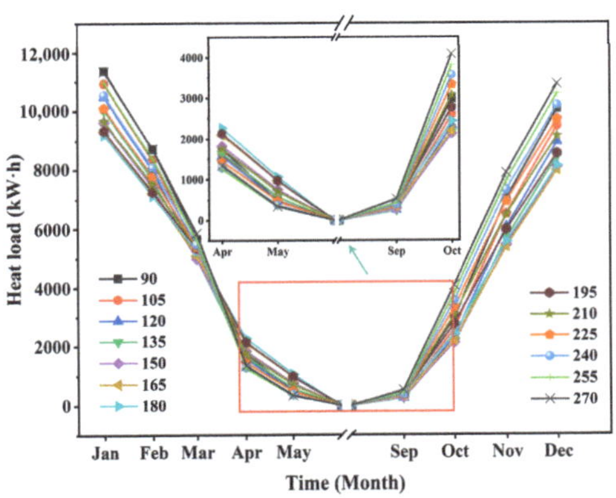

Figure 7. Monthly variation curve of building heat load.

In plateau regions, where temperatures are relatively low throughout the four seasons, maintaining indoor temperatures requires significant energy. Observing the variation curve of building heat loads from January to December, it can be seen that the heat load between 135° and 180° remains at a lower level for most of the year. Considering the annual total heat load of the building, the heat load at 150° is the lowest.

4.2. Analysis of Heat Load with Varied Skylight Areas

Figure 8 presents the annual total thermal load of buildings with different ARs at a 150° orientation, demonstrating significant variations in heat load with different skylight areas. When the AR is 0, the building's heat load is the highest at 50,535.93 kW·h; with an AR of 0.25, the heat load decreases to 44,619.69 kW·h; and the smallest heat load is found at an AR of 0.5, with 40,597.32 kW·h. This indicates that the heat load is inversely correlated with the area of the skylight. That is, the larger the skylight area, the smaller the building's heat load. Adopting skylights in building roofs on the plateau can significantly save energy. With an AR of 0.5, there is an energy saving of 19.67% compared with buildings with an AR of 0. The larger the skylight area, the more solar radiation enters the atrium. After the sun's short-wave radiation enters the atrium through the skylight, the indoor temperature rises, and the skylight blocks the long-wave radiation emitted from the atrium, reducing energy release and thus leading to an increase in atrium temperature [52]. Meanwhile, the existence of the atrium's buffering effect [53] also avoids direct contact of the surrounding rooms with the external environment, creating a suitable daytime activity space for personnel at workstations, while also providing auxiliary heating to surrounding rooms.

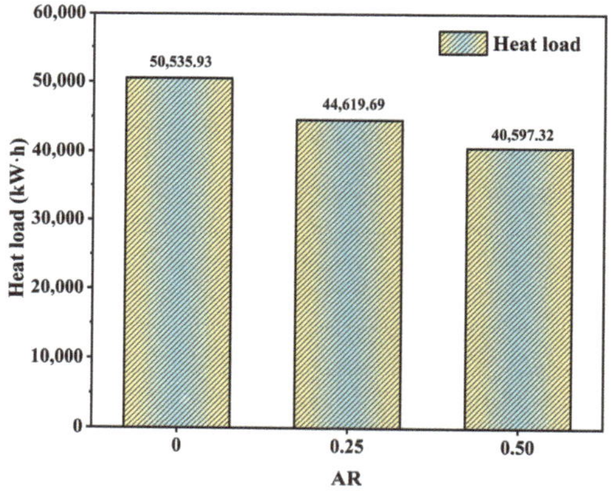

Figure 8. Building heat load diagram with different AR under 150° orientation.

Figure 9 depicts the skylight heat gains and the curve for solar radiation entering indoors through the skylight at a 150° orientation with different ARs. The graph shows that the heat losses through the skylight and the solar radiation energy entering indoors positively correlate with the AR. During the summer months of June, July, and August, skylights contribute significant heat gains to the interior regardless of AR, peaking in July. In March, April, May, and September, the solar radiation introduced by the skylight far exceeds its heat losses, and this difference becomes more pronounced as the AR increases. This indicates that more enormous skylights in these seasons can more effectively utilize solar radiation to provide heat indoors. In winter, more enormous skylights can lead to more significant heat loss through the skylight. However, the heat losses can be partially or entirely offset by the solar radiation energy they introduce, reducing or eliminating the building's reliance on traditional heating systems and lowering energy consumption.

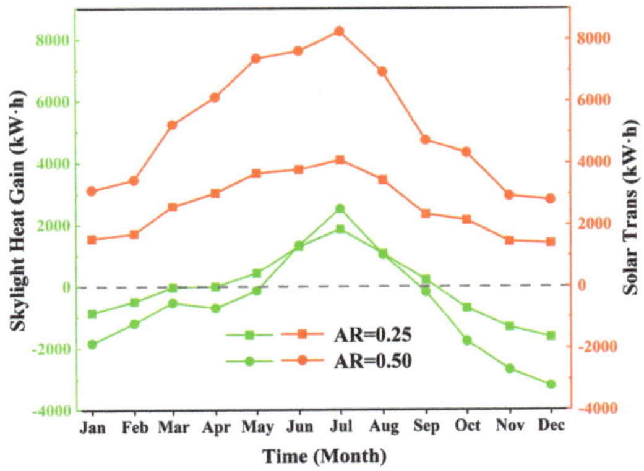

Figure 9. Skylight heat gain and solar radiation curves for skylight with different ARs under 150° orientation.

4.3. Indoor Temperature Analysis under Different Skylight Areas

Skylights can effectively utilize daytime solar radiation to improve the indoor thermal environment, but due to their poor thermal performance, there are significant heat losses at night. The solar radiation received by the skylight directly affects the atrium, and since the warehouse is less affected by external disturbances compared with other rooms, it can serve as an indicator of how changes in the skylight area influence temperature variations in other rooms. Figure 10 shows the daily temperature variation curves for the atrium and warehouse under different ARs at a 150° orientation. From the graph, it is evident that for both the warehouse and the atrium, the overall trend of the temperature curves is consistent across the three ARs, with larger ARs resulting in more pronounced temperature fluctuations. The temperature positively correlates with the AR for the atrium and warehouse from March to October. From November to February of the following year, whether for warehouses or atriums, the temperature difference under the three area ratios is not significant. During this period, the skylight maintains a balance between the energy gained during the day and the energy lost at night.

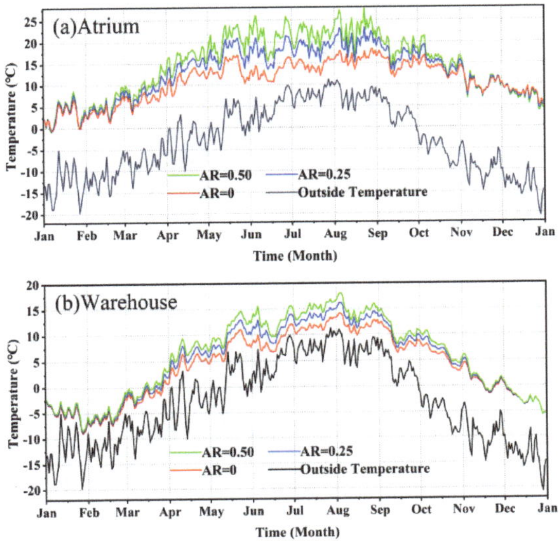

Figure 10. Daily average temperature curve of atrium and warehouse with different ARs under 150° orientation.

5. Discussions

5.1. Optimal Building Orientation in the Qinghai–Tibet Plateau Region

This paper utilized EnergyPlus 9.4 software and relied on actual meteorological data to explore the impact of building orientation on heat load in the cold region of the Qinghai–Tibet Plateau. The results show that building orientation significantly impacts the heat load in this region. However, the optimal orientation differs from previous studies, with 150° (30° south by east) being the orientation with the smallest heat load, and the heat loads between orientations of 135° to 180° differing very minimally (Figure 7).

Building energy consumption exhibits differences in heating or cooling demand under varying climatic conditions. In regions with high cooling energy demands, building orientations that reduce solar radiation are preferred, such as the best orientation for Bandung in Indonesia, which is south of the equator, being southwest [54], while for New Minia in Egypt, which is north of the equator, the orientation is towards the north [17]. For the Iranian Plateau, which is at a similar latitude to the Tibetan Plateau, there is a need for heating in the winter. However, due to the relatively arid climate of the region, the difference in radiation between morning and afternoon is small, making a southward orientation optimal for buildings in this area [18].

The optimal orientation for the comprehensive building of Beiluhe Station is skewed towards the east. This result aligns with the radiation mechanism of the Yin–Yang slope along the Qinghai–Tibet Railway, where solar radiation on the plateau is stronger in the morning than in the afternoon. For the north–south Qinghai–Tibet line, the morning solar radiation primarily affects the eastern slope, while the afternoon solar radiation primarily affects the western slope [50]. Therefore, the eastern slope inevitably receives more solar radiation than the western slope [50]. Thus, for buildings in the Qinghai–Tibet Plateau region, an eastward orientation allows them to receive more solar radiation. As Morrissey has pointed out, buildings with many windows on their façades usually benefit from higher levels of solar radiation inside [55], which can even replace air conditioning systems in high solar radiation areas, like the Qinghai–Tibet Plateau during winter [10]. The main façade of the comprehensive building of Beiluhe Station has a large window-to-wall ratio and a large skylight on the roof. Given the strong total solar radiation in the region and the particularly intense solar radiation in the morning, this leads to the best orientation range being between 135° (45° south by east) to 180° (due south).

5.2. The Impact of Varied ARs on Heat Load

The simulation results presented in this paper indicate that, as the skylight AR increased from 0 to 0.25 and 0.50, the heating load of the building decreased by 11.71% and 19.67%, respectively (Figure 8). Studies in Austin and Boston suggest that heat load would increase regardless of the variation in skylight area ratio [56,57].

Austin and Boston, being low-altitude regions with less winter solar radiation, found that the positive impact of solar radiation indoors cannot offset the negative effect of heat dissipation through skylights. Hence, any skylight area can increase the thermal load in these regions [56,57]. However, for the Qinghai–Tibet Plateau region, the intense winter solar radiation can partially or completely compensate for the energy loss through skylights (Figure 9). Consequently, the skylight of the Beiluhe Station's comprehensive building still plays a significant role in the warming effect during winter.

Skylight introduces solar radiation indoors, which is a beneficial factor for reducing energy consumption in the frigid winter months. For low-altitude regions with cold winters and hot summers, skylights increase the demand for cooling energy in summer [30,58]. However, in the Source Region of the Yangtze River, Yellow River, and Lancang River on the Qinghai–Tibet Plateau, the average temperature during the warmest month in summer is below 10 °C. According to China's "Indoor Air Quality Standards" [59], heating is mandatory for buildings when the temperature falls below 16 °C. Hence, even the summer season requires heating in this region. Under such climatic conditions, buildings in the region have year-round heating demands. Solar radiation entering indoors will only positively impact the indoor thermal environment and will not cause overheating in summer (Figure 11). An overall analysis for both winter and summer suggests that skylight design in the Qinghai–Tibet Plateau region can reduce the building's annual energy consumption.

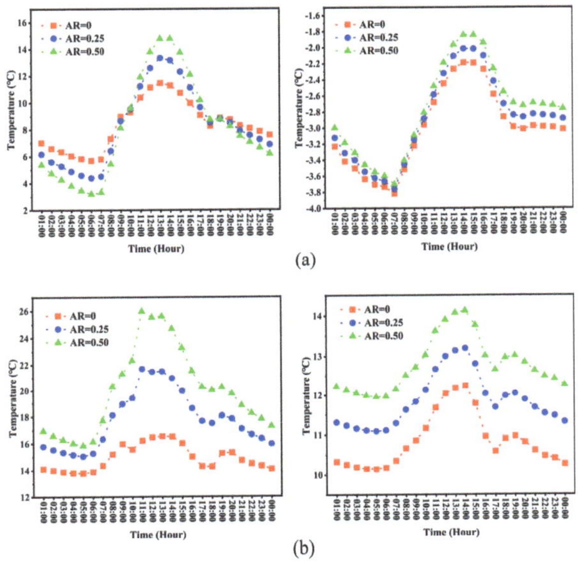

Figure 11. (**a**) Temperature of atrium and warehouse on the winter solstice facing 150; (**b**) temperature of atrium and warehouse at 150 towards the lower summer solstice.

While skylight can reduce the heating load of buildings on the Qinghai–Tibet Plateau, the energy savings are only 9.01% when the skylight AR increases from 0.25 to 0.50, which is less than the decrease from 0 to 0.25. This indicates no linear negative correlation between the heating heat load and the skylight area ratio in the Qinghai–Tibet Plateau region. Once the skylight area ratio reaches a certain threshold, the rate of decline in the building's heat load will slow down or even stabilize.

6. Conclusions

This paper takes the comprehensive building of Beiluhe Station in the Source Region of the Yangtze River, Yellow River, and Lancang River, located in the hinterland of the Qinghai–Tibet Plateau, as a case study. Based on monitoring and simulation results, the paper discusses the impact of building orientation and the area ratio of skylights on the heating energy consumption of buildings in the Qinghai–Tibet Plateau area and draws the following conclusions:

(a) In regions with intense solar radiation like the Qinghai–Tibet Plateau, the orientation and design of skylights significantly affect the building's energy efficiency and the quality of the indoor environment.

(b) For the building style discussed in this paper, the optimal orientation of the building is 30° south by east, with the best orientation ranging from 45° south by east to due south.

(c) The skylight design optimization can reduce buildings' annual energy consumption in the Qinghai–Tibet Plateau region. There is a negative correlation between the building's heat load and the area ratio of the skylight. When the area ratio of the skylight reaches a certain value, the rate of decline in building heat load will slow down or even tend to stabilize.

The conclusions of this paper can provide valuable guidance for the architectural design and construction in the Qinghai–Tibet Plateau region, especially when considering the orientation of buildings and the area of skylights. It is recommended that future design practices take these factors into comprehensive consideration to enhance the energy efficiency of buildings and the comfort of occupants. This paper provides empirical support for maximizing building energy efficiency in the region and offers valuable references for architectural professionals in designing for similar geographic and climatic conditions. Future research will focus on the optimal window-to-wall ratio for the external fenestration of building envelopes in the Qinghai–Tibet Plateau region. The expansion of this research is expected to further improve the buildings' ability to harness solar radiation and contribute to the development of more refined design solutions that can adapt to the constantly changing climate conditions.

Author Contributions: Conceptualization, Y.W. (Yingmei Wang); Methodology, H.Q., S.Z. and H.S.; Software, Y.W. (Yingmei Wang); Validation, Y.W. (Yingmei Wang), H.Q. and J.C.; Formal Analysis, H.Q. and Y.W. (Yan Wang); Investigation, Y.W. (Yan Wang), X.H. and P.R.; Resources, Y.W. (Yingmei Wang) and J.C.; Data Curation, H.Q.; Writing—Original Draft, H.Q.; Writing—Review and Editing, Y.W. (Yingmei Wang); Visualization, X.H. and P.R.; Supervision, Y.W. (Yingmei Wang); Project Administration, S.Z. and H.S.; Funding Acquisition, Y.W. (Yingmei Wang) and J.C. All authors have read and agreed to the published version of the manuscript.

Funding: This research was supported by the National Key Research and Development Program of China (2022YFF1302600), the State Key Laboratory of Frozen Soil Engineering Funds (SKLFSE-ZY-19), and the program of the Research and Development of Science and Technology of China State Railway Group Co., Ltd. (K2022G017).

Data Availability Statement: The original contributions presented in the study are included in the article, further inquiries can be directed to the corresponding author.

Conflicts of Interest: Author Shouhong Zhang was employed by the company China Railway Qinghai-Tibet Group Co., Ltd. Author Hanyu Song was employed by the company Shangqiu Branch of China Tower Co., Ltd. The remaining authors declare that the research was conducted in the absence of any commercial or financial relationships that could be construed as a potential conflict of interest.

Abbreviations

AR	area ratio
ARs	area ratios
IGDB	International Glass Database

References

1. Conti, J.; Holtberg, P.; Diefenderfer, J.; LaRose, A.; Turnure, J.T.; Westfall, L. *International Energy Outlook 2016 with Projections to 2040*; USDOE Energy Information Administration (EIA): Washington, DC, USA, 2016.
2. China Association of Building Energy Efficiency. China Building Energy Consumption Annual Report 2020. *J. Build. Energy Effic.* **2021**, *49*, 1–6. (In Chinese)
3. Jiang, Y. Current building energy consumption in china and effective energy efficiency measures. *Heat. Vent. Air Cond.* **2005**, *35*, 30–40. (In Chinese) [CrossRef]
4. Fang, Y.; Cho, S. Design optimization of building geometry and fenestration for daylighting and energy performance. *Solar Energy* **2019**, *191*, 7–18. [CrossRef]
5. Mirrahimi, S.; Mohamed, M.F.; Haw, L.C.; Ibrahim, N.L.N.; Yusoff, W.F.M.; Aflaki, A. The effect of building envelope on the thermal comfort and energy saving for high-rise buildings in hot–humid climate. *Renew. Sustain. Energy Rev.* **2016**, *53*, 1508–1519. [CrossRef]
6. Yao, T.; Bolch, T.; Chen, D.; Gao, J.; Immerzeel, W.; Piao, S.; Su, F.; Thompson, L.; Wada, Y.; Wang, L. The imbalance of the Asian water tower. *Nat. Rev. Earth Environ.* **2022**, *3*, 618–632. [CrossRef]
7. Yihao, L. Improve the Indoor Thermal Environment of Alpine Region Building Technical Measures and Research in Tibet. Master's Thesis, Southwest Jiaotong University, Chengdu, China, 2012. (In Chinese)
8. Wang, M. Study on Indoor Thermal Environment of Buildings in Cold Plateau Area. Master's Thesis, Chongqing University, Chongqing, China, 2019. (In Chinese)

9. Wang, Q.; Qiu, H.-N. Situation and outlook of solar energy utilization in Tibet, China. *Renew. Sustain. Energy Rev.* **2009**, *13*, 2181–2186. [CrossRef]
10. Gong, X.; Akashi, Y.; Sumiyoshi, D. Optimization of passive design measures for residential buildings in different Chinese areas. *Build. Environ.* **2012**, *58*, 46–57. [CrossRef]
11. Baharvand, M.; Ahmad, M.H.B.; Safikhan, T.; Mirmomtaz, S.M.M. Thermal performance of tropical atrium. *Environ. Clim. Technol.* **2013**, *12*, 34–40. [CrossRef]
12. Boukli Hacene, M.; Chabane Sari, N. Energy efficient design optimization of a bioclimatic house. *Indoor Built Environ.* **2020**, *29*, 270–285. [CrossRef]
13. Su, Y.; Zhao, Q.; Zhou, N. Improvement strategies for thermal comfort of a city block based on PET Simulation—A case study of Dalian, a cold-region city in China. *Energy Build.* **2022**, *261*, 111557. [CrossRef]
14. Wong, K.d.; Fan, Q. Building information modelling (BIM) for sustainable building design. *Facilities* **2013**, *31*, 138–157. [CrossRef]
15. Pacheco, R.; Ordóñez, J.; Martínez, G. Energy efficient design of building: A review. *Renew. Sustain. Energy Rev.* **2012**, *16*, 3559–3573. [CrossRef]
16. Acosta, I.; Varela, C.; Molina, J.F.; Navarro, J.; Sendra, J.J. Energy efficiency and lighting design in courtyards and atriums: A predictive method for daylight factors. *Appl. Energy* **2018**, *211*, 1216–1228. [CrossRef]
17. Elhadad, S.; Baranyai, B.; Gyergyák, J. The impact of building orientation on energy performance: A case study in new Minia, Egypt. *Pollack Period.* **2018**, *13*, 31–40. [CrossRef]
18. Kohansal, M.E.; Akaf, H.R.; Gholami, J.; Moshari, S. Investigating the simultaneous effects of building orientation and thermal insulation on heating and cooling loads in different climate zones. *Archit. Eng. Des. Manag.* **2022**, *18*, 410–433. [CrossRef]
19. Prasad, A.; Anchan, S.; Kamath, M.; Akella, V. Impact of building orientation on energy consumption in the design of green building. *Int. J. Emerg. Res. Manag. Technol.* **2017**, *6*, 2278–9359.
20. Hu, D.; Chen, D.; Shan, P.; Huang, F.; Huang, H. Influence of Residential Building Orientations to Energy Consumption in Hot Summer and Warm Winter Zone. *J. Build. Energy Effic.* **2017**, *45*, 57–60. (In Chinese) [CrossRef]
21. Spanos, I.; Simons, M.; Holmes, K.L. Cost savings by application of passive solar heating. *Struct. Surv.* **2005**, *23*, 111–130. [CrossRef]
22. Aksoy, U.T.; Inalli, M. Impacts of some building passive design parameters on heating demand for a cold region. *Build. Environ.* **2006**, *41*, 1742–1754. [CrossRef]
23. Albatayneh, A.; Alterman, D.; Page, A.; Moghtaderi, B. The significance of the orientation on the overall buildings thermal performance-case study in Australia. *Energy Procedia* **2018**, *152*, 372–377. [CrossRef]
24. Albatayneh, A.; Mohaidat, S.; Alkhazali, A.; Dalalah, Z.; Bdour, M. The influence of building's orientation on the overall thermal performance. *Environ. Sci. Sustain. Dev.* **2018**, *3*, 63–69. [CrossRef]
25. Ren, W.; Zhao, J.; Chang, M. Energy-saving optimization based on residential building orientation and shape with multifactor coupling in the Tibetan areas of western Sichuan, China. *J. Asian Archit. Build. Eng.* **2023**, *22*, 1476–1491. [CrossRef]
26. Abanda, F.; Byers, L. An investigation of the impact of building orientation on energy consumption in a domestic building using emerging BIM (Building Information Modelling). *Energy* **2016**, *97*, 517–527. [CrossRef]
27. Ji, W.; Wang, J.; Xie, Y. Influence of Dormitory Building Orientation on Energy Consumption in Cold Zones: Taking the Dormitory of Shandong Jianzhu University as an Example. *J. Build. Energy Effic.* **2022**, *50*, 34–41. (In Chinese) [CrossRef]
28. Motamedi, S.; Liedl, P. Integrative algorithm to optimize skylights considering fully impacts of daylight on energy. *Energy Build.* **2017**, *138*, 655–665. [CrossRef]
29. Liu, X. The Design of Atrium and Architectural Daylight Roof. Master's Thesis, Hebei University of Technology, Wuhan, China, 2006.
30. Yang, Y.; Wu, H.; Yang, L.; Xu, T.; Ding, Y.; Fu, P. Thermal and day-lighting performance of aerogel glazing system in large atrium building under cooling-dominant climates. *Energy Procedia* **2019**, *158*, 6347–6357. [CrossRef]
31. Sher, F.; Kawai, A.; Güleç, F.; Sadiq, H. Sustainable energy saving alternatives in small buildings. *Sustain. Energy Technol. Assess.* **2019**, *32*, 92–99. [CrossRef]
32. Sudan, M.; Mistrick, R.G.; Tiwari, G. Climate-Based Daylight Modeling (CBDM) for an atrium: An experimentally validated novel daylight performance. *Sol. Energy* **2017**, *158*, 559–571. [CrossRef]
33. Fan, Z.; Yang, Z.; Yang, L. Daylight performance assessment of atrium skylight with integrated semi-transparent photovoltaic for different climate zones in China. *Build. Environ.* **2021**, *190*, 107299. [CrossRef]
34. Wu, P.; Zhou, J.; Li, N. Influences of atrium geometry on the lighting and thermal environments in summer: CFD simulation based on-site measurements for validation. *Build. Environ.* **2021**, *197*, 107853. [CrossRef]
35. Nasrollahi, N.; Abdolahzadeh, S.; Litkohi, S. Appropriate geometrical ratio modeling of atrium for energy efficiency in office buildings. *J. Build. Perform.* **2015**, *6*, 95–104.
36. Li, J.; Ban, Q.; Chen, X.; Yao, J. Glazing sizing in large atrium buildings: A perspective of balancing daylight quantity and visual comfort. *Energies* **2019**, *12*, 701. [CrossRef]
37. *GB 50178-1993*; Standard of Climatic Regionalization for Architecture. Ministry of Construction of the People's Republic of China: Beijing, China, 1993.
38. Pan, Y. *Building Energy Simulation Handbook*; China Architecture & Building Press: Beijing, China, 2013; ISBN 9787112155101.

39. Zhang, Q.; Yang, H. *Manual of Standard Meteorological Data for Buildings*; China Architecture & Building Press: Beijing, China, 2012; ISBN 9787112137701.
40. Wang, B. Solar Radiation Calculation Lecture one: Calculation of Astronomical Parameters in Solar Energy. *Sol. Energy* **1999**, 8–10. (In Chinese)
41. Wang, B. Solar Radiation Calculation Lecture three: Calculation of Extraterrestrial Horizontal Radiation. *Sol. Energy* **1999**, 12–13. (In Chinese)
42. *GB 50176-2016*; Thermal Design Code for Civil Buildings. Ministry of Construction of the People's Republic of China: Beijing, China, 2016.
43. Available online: https://windows.lbl.gov/software/igdb (accessed on 18 December 2023).
44. *JGJ 26-2010*; Design Standard for Energy Efficiency of Residential Buildings in Severe Cold and Cold Zones. Ministry of Construction of the People's Republic of China: Beijing, China, 2010.
45. *JB/T 7985-2002*; Specification for Small Type Boilers and Atmospheric Hot Water Boilers. State Economic and Trade Commission of the People's Republic of China: Beijing, China, 2002.
46. *GB/T 2589-2020*; General Principles for the Calculation of the Comprehensive Energy Consumption. Energy Fundamentals and Management: Beijing, China, 2020.
47. *JGJ 176-2009*; Technical Code for the Retrofitting of public Building on Energy Efficiency. Ministry of Construction of the People's Republic of China: Beijing, China, 2009.
48. Ye, D.; Gao, Y. *Meteorology of the Qinghai—Tibet Plateau*; Science Press: Beijing, China, 1979; ISBN 130311072.
49. Zhang, J. *Advances in Qinghai—Tibet Plateau Meteorology*; Science Press: Beijing, China, 1988; ISBN 7030011066.
50. Sun, L.; Dong, X.; Zhou, Y.; Zhao, X.; Wang, G.; Chen, J. The Effect of Embankment Slope Orientation along the Qinghai-Tibet Routes and Routes and Related Radiation Mechanism. *J. Glaciol. Geocryol.* **2008**, *30*, 610–616. (In Chinese)
51. Sun, H. *The Formation and Evolution of the Qinghai—Tiebt Plateau*; Shanghai Scientific & Technical Publishers: Shanghai, China, 1996; ISBN 9787532340231.
52. Cui, T.; Wang, L.; Zhao, J.; Jiang, L.; Zhang, A.; Long, E. Analysis of greenhouse effect on greenhouse buildings. *Refrig. Air-Cond.* **2005**, *z1*, 280–282. (In Chinese) [CrossRef]
53. Liang, R. Research on the Optimization Strategy of Solar Thermal Performance of Existing Building Atriums in Universities in Cold Regions. Master's Thesis, Shandong Jianzhu University, Jinan, China, 2022. (In Chinese)
54. Mulyani, R.; Kholidasari, I. Wardi, The impact of building orientation on energy use: A case study in Bung Hatta University, Indonesia. *Int. J. Real Estate Stud.* **2017**, *11*, 43–48.
55. Morrissey, J.; Moore, T.; Horne, R.E. Affordable passive solar design in a temperate climate: An experiment in residential building orientation. *Renew. Energy* **2011**, *36*, 568–577. [CrossRef]
56. Motamedi, S. Energy analysis of different toplights for office buildings in austin. In Proceedings of the World Renewable Energy Forum Conference, Denver, CO, USA, 13–17 May 2012.
57. Ghobad, L.; Place, W.; Cho, S. *Design Optimization of Square Skylights in Office Buildings*; AIVC: Ghent, Belgium, 2013.
58. Vujošević, M.; Krstić-Furundžić, A. The influence of atrium on energy performance of hotel building. *Energy Build.* **2017**, *156*, 140–150. [CrossRef]
59. *GB/T 18883-2022*; Indoor Air Quailty Standard. National Health Commission of the People's Republic of China: Beijing, China, 2022.

buildings

MDPI

Article

The Evaluation National Green Building Index Based on a Survey of Personnel Satisfaction: The Case of Hubei Province, China

Shengda Ouyang [1], Xiaofang Shan [1], Qinli Deng [1,*], Zhigang Ren [1], Wenyu Wu [1], Tingwei Meng [1] and Yinguang Wu [2]

1 School of Civil Engineering and Architecture, Wuhan University of Technology, Wuhan 430070, China; ouyangsd@whut.edu.cn (S.O.); xfshan@whut.edu.cn (X.S.); renzg@whut.edu.cn (Z.R.); wwenyu@whut.edu.cn (W.W.); tingweimeng@whut.edu.cn (T.M.)

2 Central-South Architectural Design Institute Co., Ltd. (CSADI), Wuhan 430071, China; wuyinguang6065@csadi.cn

* Correspondence: deng4213@whut.edu.cn

Abstract: With the rapid development of China's urbanization process and the promotion of the 'double carbon' strategy, green buildings will become an inevitable trend in the future development of the construction industry. Among the various building evaluation criteria, it is important to discuss how to promote the development of green buildings more efficiently and adaptively according to the characteristics of personnel needs. This study constructed a questionnaire to assess building use satisfaction based on China's national standards. Field research was conducted on 23 projects in six cities in Hubei Province, China, and a total of 2251 questionnaires were collected. The survey evaluated satisfaction with the current use of green buildings across different age groups and genders. A new satisfaction evaluation model is constructed through fuzzy comprehensive evaluation to provide guidance for the differentiated development of green buildings in different cities.

Keywords: green building; personnel satisfaction; evaluation indicators; survey questionnaire; fuzzy comprehensive evaluation

Citation: Ouyang, S.; Shan, X.; Deng, Q.; Ren, Z.; Wu, W.; Meng, T.; Wu, Y. The Evaluation National Green Building Index Based on a Survey of Personnel Satisfaction: The Case of Hubei Province, China. *Buildings* **2024**, *14*, 868. https://doi.org/10.3390/buildings14040868

Academic Editor: Antonio Caggiano

Received: 1 March 2024
Revised: 15 March 2024
Accepted: 19 March 2024
Published: 22 March 2024

1. Introduction

The global urban population is expected to reach 68% by 2050 due to the acceleration of the urbanization process [1]. The construction industry is a rapidly growing sector, accounting for 40% of total global energy consumption [2,3]. In 2020, China pledged to 'peak carbon emissions by 2030 and achieve carbon neutrality by 2060' [4]. China's construction sector alone emits close to 2 billion tonnes of CO_2 annually [5]. According to reference [6], green buildings consume 25% to 30% less energy than traditional buildings. Therefore, developing green buildings is an effective way to reduce carbon emissions [7,8]. In 1990, the United Kingdom released the world's first green building assessment method, which brought green building assessment to the public's attention [9,10]. In comparison to traditional energy-saving buildings, green buildings offer greater advantages in energy conservation and emission reduction [11]. Currently, many countries have established a green building sustainable development evaluation system that aligns with their national conditions [12]. Mature green building technologies can effectively reduce the energy consumption of buildings [13,14]. Although green buildings offer more advantages than traditional buildings, they still face challenges such as technology, costs, and benefits [15,16]. Each national evaluation system has its own focus, but generally, it aims to promote environmental protection, energy efficiency, and sustainable development [9,17]. China's Assessment Standard for Green Building GB/T 50378-2019 (ASGB 2019) focuses on adapting to the country's environmental conditions, resources, and energy conditions, as well as residents' living habits [18]. It also emphasizes the applicability of residential and public buildings. Table 1 compares green building evaluation standards across different countries.

Table 1. Current status of green building evaluation standards in various countries.

Nation	Evaluation Criteria	Certification Level	Evaluation Aspect
China	ASGB 2019	Basic, One Star, Two Stars, Three Stars	The evaluation method is based on five green performance indices: safety and durability, health and comfort, convenience of life, resource saving, and livable environment.
America	LEED	Certified, Silver, Gold, Platinum	Evaluation criteria include sustainable sites, water efficiency, energy and atmosphere, materials and resources, and indoor environmental quality.
Britain	BREEAM	Pass, Good, Very Good, Excellent, Outstanding	The assessment covers management, health and well-being, energy, transport, water, materials, waste, land use and ecology, and pollution.
Japan	CASBEE	C, B−, B+, A, S	The evaluation indicators include the building's energy efficiency, material efficiency, indoor environment, etc.
Australia	NABERS	Between 0.5 and 6 stars, with 6 representing market-leading performance	The score is based on aspects such as energy consumption, water use, waste management, and indoor environmental quality.

Compared with developed countries, China's demand for green buildings remains high, with about 2 billion square meters of new commercial buildings added every year [19,20]. While learning from international experience, China's green building development also emphasizes the integration of traditional Chinese architectural culture, regionalism, adaptability, and the protection of cultural heritage [21]. In 2005, China's former Ministry of Construction and Ministry of Science and Technology jointly issued the 'Green Building Technical Guidelines'. In accordance with China's national conditions and international building evaluation standards, the first edition of the Assessment Standard for Green Building (GB/T50378-2006) was promulgated in 2006. This standard provides a clear definition of 'green building' in China [22]. The standard evaluates the environmental performance of buildings in six categories: land use and outdoor environment, energy efficiency, water efficiency, material efficiency, indoor environmental quality, and operation management. It also establishes basic indicators [23,24]. In 2009 and 2010, two evaluation standards were implemented: Evaluation Standards for Green Industrial Buildings and Evaluation Standards for Green Office Buildings [25]. The 2014 second edition of the Assessment Standard for Green Building extended the scope of the standards to include all types of civil buildings. It also optimized and supplemented specific requirements [26]. In 2019, a new evaluation method was proposed based on five green performance indicators: safety and durability, health and comfort, convenience of life, resource saving, and livable environment [18]. China has developed a comprehensive system for green building standards that covers three levels: applicable object, applicable stage, and standard type [27]. Table 2 illustrates the development process of China's Assessment Standard for Green Building.

In 2019, China's total building area exceeded 500 billion square meters, with only 10% of it certified for green building [28]. By 2020, more than 77% of new civil buildings in cities and towns were green buildings [29]. In 2021, China's green buildings increased by 2.362 billion square meters, and the proportion of new green buildings reached 84.22% of the annual new buildings [30]. In accordance with green building evaluation requirements, buildings must meet green building grade standards while fulfilling their general use functions throughout their entire life cycle [31]. Despite the rapid development of China's green building industry in recent years, there are still issues with meeting expected outcomes [32].

Therefore, it is important to consider the specific needs of occupants when designing the architecture comprehensively [33].

Table 2. Development process of China's Assessment Standard for Green Building.

Edition	Evaluation Opportunity	Evaluation Content	Evaluation Principle	Technical Requirement	Individuation
2006	Post-construction	Operation evaluation	/	/	/
2014	After the design, after the construction is completed	Design evaluation, operation evaluation	/	/	/
2019	After the construction drawing design is completed (pre-evaluation), upon completion of the building	Pre-evaluation, operational evaluation	Four sections and one environmental protection, people-oriented	More detailed and comprehensive	Pay attention to regional differences

At present, whether the comfort of Chinese buildings and the surrounding living environment meet people's growing needs for a better life has become the focus of current green building development. However, in ASGB 2019, all evaluation indicators are evaluated using the same system, which inevitably reduces the pertinence of building use functions and fails to reflect the impact of evaluation indicators on life satisfaction [34]. Therefore, it is necessary to evaluate and improve the indicators in the green evaluation standard according to the needs and satisfaction of personnel.

Hubei Province is situated in the heartland of China and is a typical central region [35]. The winter temperature in most areas ranges from 3 °C to 5 °C, while the summer temperature ranges from 26 °C to 29.5 °C. The average annual precipitation is 1200.7 mm, and the climate is characterized by hot summers and cold winters. Therefore, buildings require both cooling in summer and heating in winter, resulting in high energy consumption demands. It is important to understand the needs of users [36]. With the development of the economy, Hubei Province has become an important transportation hub in China, leading to a sharp increase in population and demand for buildings. This paper aims to comprehensively assess people's use and demand for green building functions in Hubei Province, China. To achieve this, we use the satisfaction degree of existing green building users in Hubei Province as an indicator, following the guidelines set out in ASGB 2019. This study focuses on public and residential buildings. A satisfaction evaluation model was developed using the analytic hierarchy process, based on questionnaire surveys and statistical analysis. The weight of the satisfaction index of residential buildings was calculated, and the weight ratio of each index in ASGB 2019 was compared. The research findings provide a valuable theoretical foundation for the current direction of green building development. The questionnaire survey uses a five-level scale, with the first- and second-level index questions compiled based on the scoring items in ASGB 2019. This paper focuses on two main issues: (1) This report evaluates the satisfaction levels of the first-level index for both residential and public buildings, as well as their current functional status. (2) Additionally, it examines the weight distribution and satisfaction characteristics of secondary indicators for these types of buildings. The purpose of this evaluation is to identify indicators based on satisfaction and provide technical guidance and suggestions for the development of green buildings and the green transformation of existing buildings.

2. Investigation and Implementation

2.1. Survey Area

This survey involved research on a total of 6 prefecture-level cities in Hubei Province, including 9 public buildings and 14 residential buildings. Figure 1 shows the project distribution map, while Table 3 provides specific research project information and green building assessment time.

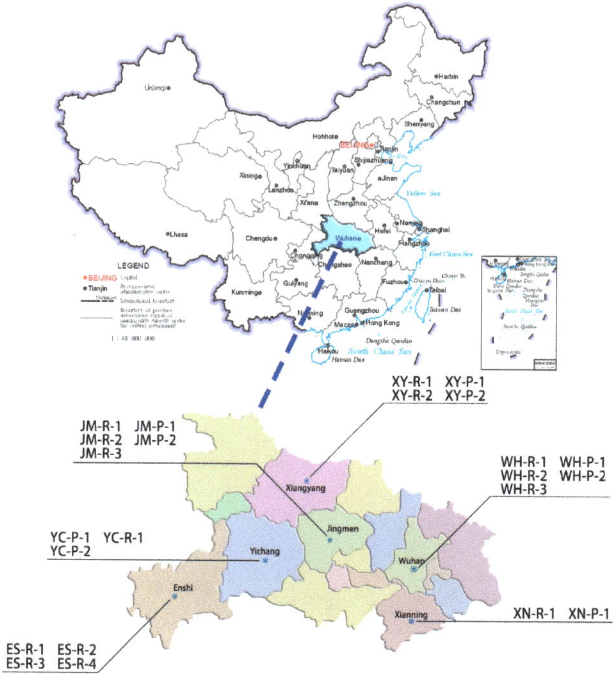

Figure 1. Project layout.

Table 3. Field research project information.

Project Location	Item Number	Project Name	Building Area (10^4 m^2)	Type of Building	Green Building Rating Time
Wuhan	WH-R-1	Building 1–10, Phase 8, Shimaolong Bay, Wuhan	25.65	Residential building	2020
	WH-R-2	Huafa Bund capital	11.99	Residential building	2020
	WH-R-3	Building 1~3, 5, Zhongjian Yi Pin LAN Hui, Wuhan	9.46	Residential building	2019
	WH-P-1	The new Hubei Science and Technology Museum	7.03	Public building	2019
	WH-P-2	Zhongcarbon Deng Building, Zhongbei Road, Wuchang District	7.85	Public building	2019

Table 3. *Cont.*

Project Location	Item Number	Project Name	Building Area (10^4 m^2)	Type of Building	Green Building Rating Time
Xiangyang	XY-R-1	Residential buildings No. 1~3, 7~8, 11~13, 15~19, Lot A1 "Dongjin Century City", Xiangyang	11.41	Residential building	2020
	XY-R-2	Residential building No. 1~3, 5~10, 15~17, 19~23, Lot A, PanggongBiyuan Project, Xiangyang	14.79	Residential building	2020
	XY-P-1	Laohekou public service comprehensive functional area	4.78	Public building	2019
	XY-P-2	Xiangyang Hubei Free Trade Zone Xiangyang Area Comprehensive Service Center project	0.99	Public building	2020
Enshi	ES-R-1	Blocks 1–3, 5–13, 15–17, Yubin Mansion, Country Garden	15.67	Residential building	2019
	ES-R-2	Building 1–3, 5 to 11, No. 9 Zhongliang Courtyard, Enshi	9.68	Residential building	2019
	ES-R-3	Building 21, Yipin Phase II, Enshi Capital	1.85	Residential building	2020
	ES-R-4	Building 6, Phase 3, Guangyin Haitangwan, Enshi	1.04	Residential building	2020
Yichang	YC-R-1	Building 1 to 3, 5 to 13, and 16, Area C, Country Garden Phoenix City, Yichang	16.47	Residential building	2020
	YC-P-1	Law enforcement and case handling site of Yichang Municipal Supervisory Commission	3.49	Public building	2019
	YC-P-2	Yichang Women's and Children's Activity Center	2.37	Public building	2019

Table 3. *Cont.*

Project Location	Item Number	Project Name	Building Area (10^4 m^2)	Type of Building	Green Building Rating Time
Xianning	XN-R-1	Building 1–3, 5, 7, 10, 11, Block A6, Ziwan International Resort, Xianning Greenland	11.22	Residential building	2020
	XN-P-1	Tongcheng County Yinshan Culture and Art Center (six halls, one hospital, and one venue) planning comprehensive hall	3.43	Public buildings	2020
Jingmen	JM-R-1	Residential building 1–7, Block 1, Longshan Central Business District, Jingmen	12.69	Residential building	2020
	JM-R-2	Commercial building No. 1–12, Lot 5, Longshan Central Business District, Jingmen	7.15	Residential building	2020
	JM-R-3	Building 1–17, Silver Lake City, Zhong Xiang	40.96	Residential building	2020
	JM-P-1	Jingmen Huijin Center Hotel Project	13.38	Public building	2020
	JM-P-2	Jingmen CPC Party School—Comprehensive building, lecture hall, lecture theatre, activity center	1.75	Public building	2019

2.2. Research Method

This study conducted a questionnaire survey on green building users in Hubei Province to assess their satisfaction with building use. A questionnaire survey was used as a research method:

(1) Determine research objectives and samples:

The survey aimed to evaluate satisfaction with the use of green buildings in Hubei Province. To ensure representative samples from different regions and building types, the investigation focused on users of residential and public buildings that have been rated as green buildings in the past five years in Hubei Province.

(2) Questionnaire design:

Questionnaire content: The design included basic user information (age, gender) and questions related to satisfaction with green building use. Questions were compiled according to key indicators in ASGB 2019.

Types of questions: The questionnaires were all compiled using a five-level satisfaction scale (e.g., on a scale of 1 to 5, from "very dissatisfied" to "very satisfied") to rate different aspects of satisfaction.

(3) Distribution method:

A paper questionnaire and an online questionnaire were used to conduct the survey. Participants were selected for the survey using random sampling at the green building site. To ensure sample diversity and representativeness, project property management staff were authorized to issue online questionnaires to owners on their behalf. The deadline for online questionnaire collection was determined, and the data were organized and checked for integrity after collection. The photograph below displays the field research conducted on green buildings. For detailed information on the survey questionnaire, please refer to Appendix A. Figure 2 illustrates the research scenario, with Figure 2a–c representing typical buildings and Figure 2d–f depicting the distribution of survey questionnaires offline. Meanwhile, Figure 2g–i show the distribution of online questionnaires to property management personnel.

(4) Data analysis:

Figure 2. Architectural survey photos.

Statistical software was utilized to analyze data both descriptively and inferentially. The satisfaction results were analyzed, highlighting areas of high satisfaction and identifying areas for improvement. Based on the findings, specific recommendations were made to enhance the design and management of green buildings in China in the future.

2.3. Questionnaire Design

The questionnaire was compiled based on the ASGB 2019 in China, and some scoring items with high attention were selected. The survey was conducted through field visits to the star-rated buildings that passed the green building assessment in Hubei Province in 2019–2020, and the questionnaire was released in online and offline forms. Tables 4 and 5 show the questionnaire questions and corresponding index numbers.

Table 4. Questionnaire information for residential buildings.

Level 1 Index	Topic Number	Corresponding Specification Number	Level 2 Index
Safe and durable	SD-1-1	4.2.2	Balconies, outside windows, windowsills, protective railings, etc., have strengthened anti-fall design to reduce the risk of falling objects hurting people
	SD-2-1	4.2.4	Put anti-slip measures on the ground or pavement outside
	SD-3-1	4.2.5	Human–vehicle diversion, outdoor lighting effects
	SD-4-1	4.2.7	Original hardware fittings, pipe valves, switch taps, etc., are good-quality and easy to replace
	SD-5-1	4.2.8	Interior wall cracking
	SD-6-1	4.2.9	Building exterior paint fading, water seepage condition
Healthy and comfortable	HC-1-1	5.2.1	The indoor public space has a good ventilation effect and no odor
	HC-2-1	5.2.3	Outdoor landscape water is clean and pollution-free
	HC-3-1	5.2.4	Domestic water is clean and pollution-free
	HC-4-1	5.2.7	Indoor sound insulation effects
	HC-5-1	5.2.8	Indoor and basement lighting conditions
	HC-6-1	5.2.10	Indoor natural ventilation comfort
Live comfortably	LC-1-1	6.2.1	Accessibility to public transportation
	LC-2-1	6.2.2	The indoor and outdoor public areas of the building meet the requirements of all-age design
	LC-3-1	6.2.3	The surrounding area is equipped with corresponding public service facilities, such as hospitals and cultural centers
	LC-4-1	6.2.4	Open spaces such as urban green spaces, squares and public sports fields within walking distance
	LC-5-1	6.2.5	Set up a centralized outdoor fitness activity area
	LC-6-1	6.2.8	Real-time information on water use available online

Table 4. *Cont.*

Level 1 Index	Topic Number	Corresponding Specification Number	Level 2 Index
Environmental habitability	EH-1-1	8.2.2	Conditions for standing water outside during rainy weather
	EH-2-1	8.2.3	Comfort of greenery around this building
	EH-3-1	8.2.4	Reasonable setting of outdoor smoking areas
	EH-4-1	8.2.6	The noise level of the outdoor site
	EH-5-1	8.2.7	Outdoor light pollution levels
	EH-6-1	8.2.8	Comfort with natural ventilation outside
	EH-7-1	8.2.9	Outdoor shade comfort

This survey included public and residential buildings. Public buildings encompassed shopping malls, hospitals, and office buildings. The survey participants were of varying genders and ages, including teenagers (under 18 years old), youth (18–30 years old), prime age (31–50 years old), and elderly (over 51 years old).

The researchers distributed the offline questionnaire, and the project owners filled it out. On the other hand, the property management staff distributed the online questionnaire to the community owners or users of public buildings. This distribution strategy aimed to ensure that all participants were actual users of the current green buildings, thus ensuring the authenticity and reliability of the questionnaire results.

Table 5. Questionnaire information for public buildings.

Level 1 Index	Topic Number	Corresponding Specification Number	Level 2 Index
Safe and durable	SD-1-2	4.2.2	Balconies, exterior windows, windowsills, protective railings, etc., have strengthened anti-fall design to reduce the risk of falling objects hurting people
	SD-2-2	4.2.4	Put anti-slip measures on the ground or pavement outside
	SD-3-2	4.2.5	Human–vehicle diversion, outdoor lighting effects
	SD-4-2	4.2.7	Durability of equipment such as lighting, faucets, and plumbing in public areas
	SD-5-2	4.2.8	Use durable steel, anticorrosive wood, and other building materials
	SD-6-2	4.2.9	Indoor floor tiles have good wear resistance and low breakage rate
Healthy and comfortable	HC-1-2	5.2.1	Indoor public spaces are well ventilated and odor-free
	HC-2-2	5.2.3	Outdoor landscape water is clean and pollution-free
	HC-3-2	5.2.4	Domestic water is clean and pollution-free
	HC-4-2	5.2.7	Indoor sound insulation effects
	HC-5-2	5.2.8	Indoor and basement lighting
	HC-6-2	5.2.10	Natural ventilation comfort in the building during the transition season

Table 5. *Cont.*

Level 1 Index	Topic Number	Corresponding Specification Number	Level 2 Index
Live comfortably	LC-1-2	6.2.1	Accessibility to public transportation
	LC-2-2	6.2.2	The indoor and outdoor activity areas of the building meet the requirements of all-age design
	LC-3-2	6.2.3	The building has shared meeting facilities, fitness facilities, dining facilities, etc.
	LC-4-2	6.2.4	Open spaces such as urban green spaces, squares, and public sports fields within walking distance
	LC-5-2	6.2.5	Set up outdoor walking slow lanes
	LC-6-2	6.2.9	Mobile data network connection effect
Environmental habitability	EH-1-2	8.2.2	Conditions for standing water outside during rainy weather
	EH-2-2	8.2.3	Comfort of greenery around this building
	EH-3-2	8.2.4	Reasonable setting of outdoor smoking areas
	EH-4-2	8.2.6	Noise levels in outdoor activity areas
	EH-5-2	8.2.7	Outdoor light pollution levels
	EH-6-2	8.2.8	Comfort in natural ventilation outside
	EH-7-2	8.2.9	Outdoor shade comfort

2.4. Reliability and Validity Analysis

Upon completion of the survey, the reliability of the collected questionnaires was calculated to determine the reliability of the questionnaire data. The reliability of the questionnaire was determined using Cronbach's Alpha coefficient. A coefficient higher than 0.8 indicates high reliability, while a value between 0.7 and 0.8 indicates appropriate reliability. If the value falls between 0.6 and 0.7, it indicates that the data are available only after the questionnaire is modified. If the value is less than 0.6, it indicates low reliability and should be rejected. Equation (1) shows the reliability coefficient:

$$\alpha = \frac{K}{K-1}\left(1 - \frac{\sum S_i^2}{s_x^2}\right) \tag{1}$$

where α is the reliability coefficient; K is the number of test questions; S_i^2 is the variation of all subjects' scores on question i; s_x^2 is the variance of the total score of all subjects [37].

2.5. Correlation Analysis of Satisfaction

This questionnaire analyzes the preferences of different groups regarding the use needs of residential and public buildings based on their satisfaction. Factors such as gender and age of building users affect the demand for building use. The report proposes an analysis of the use needs of different groups of buildings based on their satisfaction and in accordance with ASGB 2019.

This study analyzed the correlation between sample variables and factors and investigated the relationship between gender and age and satisfaction and demand of each index using a *T*-test, one-way ANOVA, and chi-square test. The language used is clear, objective, and value-neutral, with a formal register and precise word choice. The text adheres to conventional structure and formatting features, including consistent citation and footnote style. The grammar, spelling, and punctuation are correct. No changes in content were made.

2.5.1. Independent Sample *T*-Test

The independent sample *T*-test is employed to compare the significance of the mean difference in continuous variables between two groups. In the case of two groups, the difference between gender and satisfaction is calculated, and the sample size and variance are taken into account to determine whether there is a significant difference in the mean.

The purpose of the independent sample *T*-test is to determine the probability of a difference occurring between two averages. Equation (2) is used to compare the difference between a sample average and a known population average to test for significance.

$$t = \frac{\overline{X}_1 - \overline{X}_2}{\sqrt{\frac{(n_1-1)S_1^2+(n_2-1)S_2^2}{n_1+n_2-2}\left(\frac{1}{n_1}+\frac{1}{n_2}\right)}} \quad (2)$$

where S_1^2 and S_2^2 are the sample variances; n_1 and n_2 are the sample sizes; $\overline{X}_1$ and $\overline{X}_2$ are the sample averages.

2.5.2. One-Way Analysis of Variance

One-way analysis of variance is used to compare the means of continuous variables between multiple groups for statistically significant differences when there are multiple levels (groups) of an independent variable (the level of independent variable should be $\geq$3). By comparing the variation of age and satisfaction with the size of the variation within the group, it can be determined whether there is a significant difference in the mean value. The correlation test of personnel satisfaction involves the application of one-way ANOVA, which is calculated according to the following steps:

(1) The sum of the squares of the deviation of the overall data is SS_T. The specific calculation formula is as follows:

$$SS_T = \sum_{i=1,j=1}^{m,n}\left(x_{ij} - \overline{\overline{x}}\right)^2 \quad (3)$$

where x_{ij} is the result of any test; $\overline{\overline{x}}$ is the total average value of the test results.

(2) The sum of the squares of the difference between the groups is SS_A, indicating the degree of difference between the groups:

$$SS_A = \sum_{i=1}^{m}\left(\overline{x}_i - \overline{\overline{x}}\right)^2 \quad (4)$$

where $\overline{x}_i$ is the average value of test results in any group; the meanings of other parameters are the same as the preceding ones.

(3) The sum of the squares of the intra-group deviation SS_E represents the degree of difference within the group:

$$SS_E = \sum_{i=1}^{m}\sum_{j=1}^{n}\left(x_{ij} - \overline{x}_i\right)^2 \quad (5)$$

(4) To eliminate the effect of the number of samples on the sum of squares of deviation, divide the sum of squares of deviation by the corresponding number of degrees of freedom. The sum of the squared deviations between groups A is transformed into the variance between groups SS_A, as shown in the equation for MS_A:

$$MS_A = \frac{SS_A}{m-1} \quad (6)$$

where $m-1$ is the degree of freedom of variance between groups; the meanings of other parameters are the same as the preceding ones.

(5) The sum of the squares of the intra-group deviations is converted to the intra-group variance MS_E:

$$MS_E = \frac{SS_E}{N-m} \tag{7}$$

where N is the total number of samples; that is, $m \times n$ and $n-m$ are the degrees of freedom of variance within the group; the meanings of other parameters are the same as the preceding ones.

(6) Finally, a statistic is used to test the significant influence of factors on the results, and F-distribution is used to test and analyze the results, as shown in Equation (8):

$$F = \frac{MS_A}{MS_E} \tag{8}$$

If the difference between samples has little effect on the detection results, then only random error affects the intra-group and inter-group variance, and the ratio will be close to 1.

If the difference between samples has a large impact on the detection result, then the inter-group variance will be greater than the intra-group variance, and the ratio will be greater than 1.

When this ratio is greater than a certain degree (F-test critical value), it indicates that there is a significant difference in the levels of different factors, or that factors have a significant impact on the results; in this case, the differences between different samples are too large and the samples are not uniform.

2.5.3. Chi-Square Test

The chi-square test is employed to compare whether there is a significant difference between the observed and expected values. It is suitable for analyzing categorical variables to test the correlation and independence between gender, age, and demand for each dimension. The test calculates the difference between the observed frequency and the expected frequency to determine whether the variables are correlated or independent. Once the theoretical and actual values have been obtained, the chi-square test can be performed. The significance of the difference is indicated by the chi-square value, with a smaller *p*-value indicating a more significant difference. A *p*-value of less than 0.05 (α level) indicates a significant difference. The calculation is as follows:

$$\chi^2 = \sum \frac{(A-T)^2}{T} \tag{9}$$

where χ^2 is the chi-square value; A is the actual value; T is the theoretical value.

2.6. Analytic Hierarchy Process

2.6.1. Construct Hierarchical Structure Model

AHP is an analytical method that converts qualitative data into quantitative data based on fuzzy comprehensive evaluation [38]. The method of fuzzy solution is used for calculation, but better results can be obtained with a large number of data samples [39]. The hierarchical structure model is constructed based on the original questionnaire design classification. Using the questionnaire as a hierarchical structure, identify the target layer, criterion layer, and index layer.

The judgment matrix is constructed, and the specific formula is as follows:

$$A = \left(b_{ij}\right)_{n\times n} \tag{10}$$

In the formula, b_{ij} is the indicator, and the importance scale of a_i and a_j relative to first-level indicators adopts the classical 1–9 scale method, and the quantization value k is shown in Table 6.

Table 6. The 1~9 scale method.

Factor a_i over Factor a	Quantization Value k
Equally important	1
Slightly important	3
Stronger important	5
Strongly Important	7
Extremely important	9
The middle of two adjacent judgments	2, 4, 6, 8
Count backwards	$b_{ij} = 1/b_{ji}$

The judgment matrix is computed by row $\overline{b}_j$, and the specific calculation formula is as follows:

$$\overline{b}_j = \prod_{i=1}^{n} b_{ij} \tag{11}$$

with quadrature heel $\overline{\omega}_i$.

$$\overline{\omega}_i = \sqrt[4]{\overline{b}_j} \tag{12}$$

Find the weight coefficient $\overline{\omega}_i$, which will be normalized to obtain ω_i.

$$\omega_i = \overline{\omega}_i / \sum_{i=1}^{n} \overline{\omega}_i \tag{13}$$

Solve for maximum feature root.

$$\lambda_{\max} = \sum_{i}^{n} ((A \times \omega)_i / n\omega_i) \tag{14}$$

where $(A \times \omega)_i = a_{i1}\omega_1 + a_{i2}\omega_2 + \ldots + a_{in}\omega_n$.

The weight vector is equal to the eigenvector ω corresponding to the largest eigenroot $\lambda_{\max}$ of matrix A.

2.6.2. Consistency Check

To construct the judgment matrix, a consistency test is needed to make the judgment result meet the basic consistency and order consistency.

To calculate the consistency index, see Equation (15):

$$CI = (\lambda_{\max} - n)/(n-1) \tag{15}$$

where CI is a consistency index.

The matrix average consistency index RI value is the query value used in the process of analytic hierarchy process consistency test and the average value obtained after 500 sampling tests by scientists, which is generally applicable to the judgment matrix consistency test. Table 7 shows the consistency indicator RI values.

Table 7. Average random consistency index.

Rank	**4**	**5**	**6**	**7**	**8**	**9**
RI value	0.89	1.12	1.26	1.36	1.41	1.46

To calculate the consistency ratio, see Equation (16):

$$CR = CI/RI \tag{16}$$

When $CR < 0.1$, the judgment matrix is consistent and acceptable. On the contrary, when $CR \geq 0.1$, it is considered that the judgment matrix does not meet the consistency test and should be adjusted to maintain a certain degree of consistency [40].

3. Analysis and Discussion

3.1. Questionnaire Basic Information

A total of 2251 questionnaires were collected, comprising 1108 for residential buildings and 1143 for public buildings.

The needs of different gender and age groups for building use vary. For instance, the provision of commercial and social service facilities can help alleviate transportation issues for the elderly, enabling them to live more comfortably in the community [41]. Therefore, it is necessary to have a more comprehensive understanding of the satisfaction of various types of people with the existing buildings and the needs of the current buildings in terms of safety and durability, health and comfort, convenience of life, and livable environment. Table 8 presents the basic information of the respondents.

Table 8. Construction satisfaction survey.

Residential Buildings				Public Buildings			
Variables	**Options**	**Frequency**	**Percentage**	**Variable**	**Options**	**Frequency**	**Percentage**
Gender	male	479	43%	Gender	male	497	44%
	female	629	57%		female	646	57%
Age	Under 18	32	3%	Age	Under 18	35	3%
	18–30	701	63%		18–30	725	63%
	31–50	346	31%		31–50	356	31%
	51+	29	3%		51+	27	2%
1108 in total				1143 in total			

The survey was conducted both online and offline. The respondents were predominantly middle-aged and young, with a slightly higher proportion of women than men.

To ensure the reliability of the results, measures were taken to reduce the impact of accidental factors, such as measurement errors, that could cause the respondents' actual scores to deviate from their true scores. To evaluate the reliability of the questionnaire, reliability and validity coefficients are used.

Figure 3 shows the reliability coefficients for both residential and public buildings. Based on the reliability analysis results, it is evident that the standardized reliability coefficients of both residential and public buildings are greater than 0.7 for safety and durability, health and comfort, convenience of life, livable environment, and overall sub-items, indicating good relative reliability.

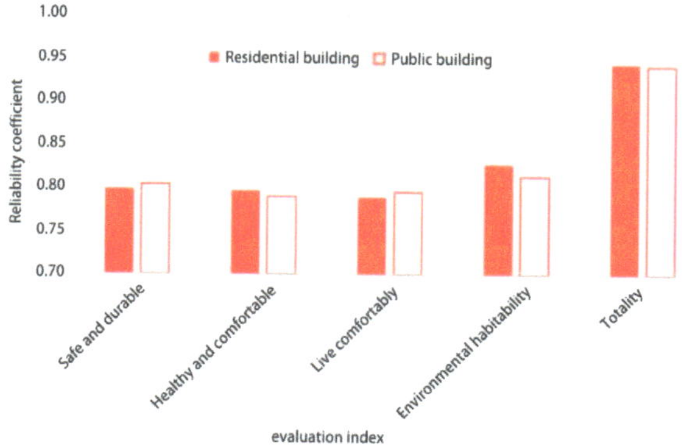

Figure 3. Reliability coefficient.

Table 9 displays the validity calculation outcomes for both types of buildings.

Table 9. Analysis of KMO and Bartlett *T*-test.

Residential Buildings			Public Buildings	
KMO sampling appropriateness measure		0.971		0.972
Bartlett's sphericity test	Approximate chi-square Degree of freedom Salience	12,237.473 300 0	Approximate chi-square Degrees of freedom Salience	12,130.045 300 0

According to the results of the above exploratory analysis, it can be seen that the coefficient results of the KMO test are all greater than 0.9, and the value of the coefficient of the KMO test ranges from 0 to 1. The closer the coefficient is to 1, the better the validity of the questionnaire.

According to the significance of the sphericity test, it can also be seen that the significance of this time is infinitely close to 0, so the null hypothesis is rejected. Therefore, the questionnaire has good validity.

3.2. Evaluation Index Satisfaction Difference Analysis

3.2.1. Level 1 Index Satisfaction

Table 10 shows the gender differences in satisfaction.

Table 10. Analysis of gender differences in satisfaction (independent sample *T*-test).

Variables	Gender	Number of Cases		t		sig	
		Residential Building	Public Building	Residential Building	Public Building	Residential Building	Public Building
Safety and durability satisfaction	male female	479 629	497 646	0.355	1.197	0.723	0.231
Health comfort satisfaction	male female	479 629	497 646	0.210	1.338	0.834	0.181
Convenience satisfaction	male female	479 629	497 646	0.151	1.156	0.880	0.248
Environmental habitability satisfaction	male female	479 629	497 646	0.166	0.815	0.868	0.415

According to the results of the *T*-test of the above independent samples, it can be seen that the gender difference in the satisfaction of all dimensions of residential buildings and public buildings is greater than 0.05, indicating that there is no gender difference in the satisfaction of safety and durability, health and comfort, convenience of life, and livable environment.

Based on the results of multiple comparisons presented in Table 11, it is evident that individuals aged 31–50 report higher levels of satisfaction with regard to safety and durability, health and comfort, convenience of life, and livable environment compared to those aged 18–30. Young people may prioritize quality of life, leading to higher standards for residential buildings. In contrast, middle-aged individuals tend to prioritize functional demand for green buildings and may not prioritize quality and functionality as much as younger individuals. Compared to residential buildings, public buildings provide greater satisfaction in terms of health and comfort for those aged 18–30 and 31–50, compared to those over 51 years old. Additionally, satisfaction levels are higher for those aged 18–30 compared to those over 51 years old, and for those aged 31–50 compared to those under 18 and over 51 years old. Public buildings primarily serve middle-aged and young people, who value convenience, health, and comfort. The current public buildings have successfully prioritized these aspects.

Table 11. Difference analysis of satisfaction with age in each dimension (one-way analysis of variance).

Variables	Options	Number of Cases		t		F		Multiple Comparisons	
Types of Satisfaction	Age Group	Residential Building	Public Building	Residential Building	Public Building	Residential Building	Public Building	Residential Building	Public Building
Safety and durability satisfaction	Under 18 18–30 31–50 Age 51+	32 701 346 29	35 725 356 27	0.355	1.197	5.819	2.452	3 > 2	/
Health comfort satisfaction	Under 18 18–30 31–50 Age 51+	32 701 346 29	35 725 356 27	0.210	1.338	3.564	3.120	3 > 2	2 > 4, 3 > 4
Satisfaction with convenience of life	Under 18 18–30 31–50 Age 51+	32 701 346 29	35 725 356 27	0.151	1.156	4.329	4.469	3 > 2	2 > 4, 3 > 1, 3 > 4
Environmental habitability satisfaction	Under 18 18–30 31–50 Age 51+	32 701 346 29	35 725 356 27	0.166	0.815	5.318	2.154	3 > 2	/

In addition to the four categories of safety and durability, health and comfort, convenience of life, and livable environment, the questionnaire adds specific evaluations of these four categories in terms of their greater impact on life, better functional realization, and need to be improved.

Table 12 displays the distribution of subjective feelings across the various categories. Based on the results, 52.3% of residents believe that health and comfort have a greater impact on their lives, while 39.1% believe that their current living environment provides adequate health and comfort. Safe and durable and convenient life were ranked second, while environmental habitability was ranked the lowest. Similarly, the demand for health and comfort is significantly higher than the other three factors in terms of function. The proportion of health and comfort in public buildings' impact on life is 48.5%, the proportion of function realization is 39.2%, and the proportion of improvement is 42.3%, all of which exceed the other four factors. This indicates that public buildings play a crucial role in people's lives. Meanwhile, with regard to convenient life, 32.9% of respondents believe that the function is superior.

Table 12. The distribution of people's subjective feelings in each category.

Variables	Categories	Frequency		Percentage	
		Residential Buildings		Public Buildings	
Life impact situation	Safe and durable Healthy and comfortable Ease of living Environmental habitability	208 579 229 92	18.80% 52.30% 20.70% 8.30%	264 554 247 78	23.10% 48.50% 21.60% 6.80%
Function implementation	Safe and durable Healthy and comfortable Convenience of living Environmental habitability	254 433 329 92	22.90% 39.10% 29.70% 8.30%	258 425 376 84	22.60% 37.20% 32.90% 7.30%
Need to improve	Safe and durable Healthy and comfortable Convenient living Environmental habitability	266 437 257 148	24.00% 39.40% 23.20% 13.40%	263 483 241 156	23.00% 42.30% 21.10% 13.60%

A chi-square test was conducted on the data from the aforementioned questions to analyze the correlation between different needs, gender, and age. The results of the calculations are presented in Table 13.

Table 13. Difference analysis of correlation degree between different demands and variables (chi-square test).

Variables	Factors	χ^2	
		Residential Building	**Public Building**
Gender	An analysis of gender differences in the impact of different dimensions on life	0.001	0.047
	Age difference analysis of functional realization perception in each dimension	0.070	0.075
	Gender difference analysis of functional improvement in each dimension	0.584	0.307
Age	An analysis of the differences in the impact of various dimensions on life in age	0.435	0.760
	Analysis of the difference in functional realization experience in each dimension in age	0.582	0.002
	Age difference analysis of functional improvement in each dimension	0.178	0.895

Residential and public buildings exhibit significant differences in the importance of four dimensions of life impact based on gender. Men place a higher demand for safety and durability on both types of buildings compared to women. This suggests that men prioritize practicality and pay more attention to the safety and durability of buildings.

In residential buildings, there is no obvious difference in the demand and satisfaction of different age groups for the realization of green building functions. There are significant differences in the functional realization of all dimensions of public buildings by age. People between 31 and 50 years old have the lowest perception of the realization of safety and durable functions, while people in this age group feel better in terms of health and comfort. People in the age group of 31–50 have more frequent contact with public buildings in daily life. Compared with the safety of public buildings, they have more obvious feelings about the use of buildings. Whether there are problems in the safety and durability of public buildings is the next step to discuss.

3.2.2. Level 2 Index Satisfaction

The following table (Table 14) shows the average satisfaction scores of residential buildings and public buildings in the survey.

To visually display the distribution of satisfaction levels in green building use, a bar chart has been chosen to represent the frequency distribution across different satisfaction intervals. The bar chart allows for a quick comparison of frequencies between intervals, identifying which intervals have higher or lower satisfaction levels. This analysis examines the distribution of satisfaction by analyzing the satisfaction intervals and the frequency of residential and public buildings within these intervals.

In Figure 4, the first picture shows the satisfaction level of residential buildings, and the second picture shows the satisfaction level of public buildings. The satisfaction distribution of residential buildings is concentrated in the range of 3.65–3.8, with the highest frequency being 3.65–3.7 and 3.75–3.8. In contrast, the satisfaction distribution of public buildings in the 3.8–3.85 interval is significantly higher than that in other intervals, indicating that users in this interval are more satisfied.

Table 14. Average score of satisfaction with secondary indicators.

Residential Building Index	Average Score	Public Building Index	Average Score
SD-1-1	3.867	SD-1-2	3.893
SD-2-1	3.731	SD-2-2	3.720
SD-3-1	3.776	SD-3-2	3.815
SD-4-1	3.776	SD-4-2	3.824
SD-5-1	3.656	SD-5-2	3.808
SD-6-1	3.670	SD-6-2	3.833
HC-1-1	3.887	HC-1-2	3.917
HC-2-1	3.760	HC-2-2	3.748
HC-3-1	3.874	HC-3-2	3.884
HC-4-1	3.648	HC-4-2	3.647
HC-5-1	3.681	HC-5-2	3.733
HC-6-1	3.860	HC-6-2	3.841
LC-1-1	3.910	LC-1-2	3.927
LC-2-1	3.714	LC-2-2	3.696
LC-3-1	3.789	LC-3-2	3.780
LC-4-1	3.853	LC-4-2	3.817
LC-5-1	3.732	LC-5-2	3.830
LC-6-1	3.727	LC-6-2	3.837
EH-1-1	3.782	EH-1-2	3.758
EH-2-1	3.822	EH-2-2	3.768
EH-3-1	3.662	EH-3-2	3.706
EH-4-1	3.672	EH-4-2	3.706
EH-5-1	3.812	EH-5-2	3.777
EH-6-1	3.830	EH-6-2	3.815
EH-7-1	3.823	EH-7-2	3.820

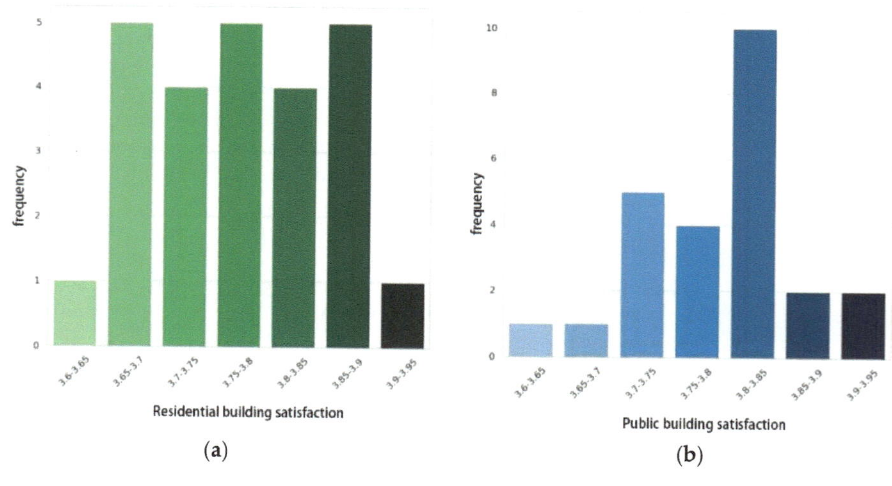

Figure 4. Distribution interval of satisfaction.

In Figure 4, subfigure a represents the distribution of satisfaction with residential buildings, and subfigure b represents the distribution of satisfaction with public buildings. When comparing the satisfaction distribution of residential and public buildings, it is found that residential buildings have six indicators of low satisfaction, with an average satisfaction score lower than 3.7, while public buildings only have two. Among the indicators of high satisfaction above 3.9, residential buildings have six indicators and public buildings have four indicators. The user satisfaction of residential buildings is evenly distributed between 3.65 and 3.8, with scattered satisfaction scores. However, the user satisfaction of public buildings shows a clear central trend in the range of 3.8–3.85. Public buildings, as a whole, have a slightly higher satisfaction rating, indicating that most users tend to rate them highly. However, there is a need to enhance the architectural design of residential buildings and improve indicators with low satisfaction. This can be achieved by increasing the use of sound-absorbing materials to strengthen indoor sound insulation and enhancing the overall satisfaction of green buildings.

3.3. Weight Analysis of Evaluation Index Based on Personnel Satisfaction

The score of the questionnaire was scored in turn, the number of people was proportional, and the quantified value was converted according to the ki value. See Table 15 for an example of constructing a judgment matrix.

Table 15. Judgment Matrix (example).

Residential Building Satisfaction	Safe and Durable	Healthy and Comfortable	Convenient Life	Environmental Habitability	ω
Safe and durable	1	0.3333	0.5	2	0.1601
Healthy and comfortable	3	1	2	4	0.4673
Convenient life	2	0.5	1	3	0.2772
Environmental habitability	0.5	0.25	0.3333	1	0.0954

The consistency ratio of satisfaction of residential buildings is 0.0116; the weight of "residential building satisfaction" is 1.0000; $\lambda_{\max}$ is 4.0310. The satisfaction consistency ratio 0.0116 < 0.1, and the judgment matrix satisfies the consistency test.

Similarly, the two-level index judgment matrix was constructed, and a consistency test was carried out. The weight results of each classification of residential buildings are shown in Table 14.

The calculation formula of standard weight P_i is as follows:

$$P_i = \frac{P_n}{P_N} \tag{17}$$

where P_n is the single question score; P_N is the sum of question scores (the standard item score is the corresponding item score in ASGB 2019).

An evaluation function P is introduced as the weight ratio index.

$$P = \frac{P_i}{P_t} \tag{18}$$

where P_t is the Level 2 index weight.

The evaluation level P can be set as four evaluation levels: poor function realization ($P \leq 50\%$), functional implementation is mediocre ($50\% < P \leq 80\%$), the function is better ($80\% < P \leq 150\%$), the function is very good ($P < 150\%$). The calculation results are shown in the table below.

3.3.1. Residential Building

Figure 5 shows the weight of residential first-level indicators. In the subjective evaluation of owners, the subjective satisfaction of health and comfort is lower than the weight value calculated by the comprehensive calculation of each item, indicating that when the users of the building users use the building, the advantages of the building in health and comfort are not obvious, and the evaluation of the building in this aspect cannot be intuitively given. However, the subjective weight of safety durability is higher than the calculated weight, which shows that the current building performs relatively well in the stage from handover to operation and management.

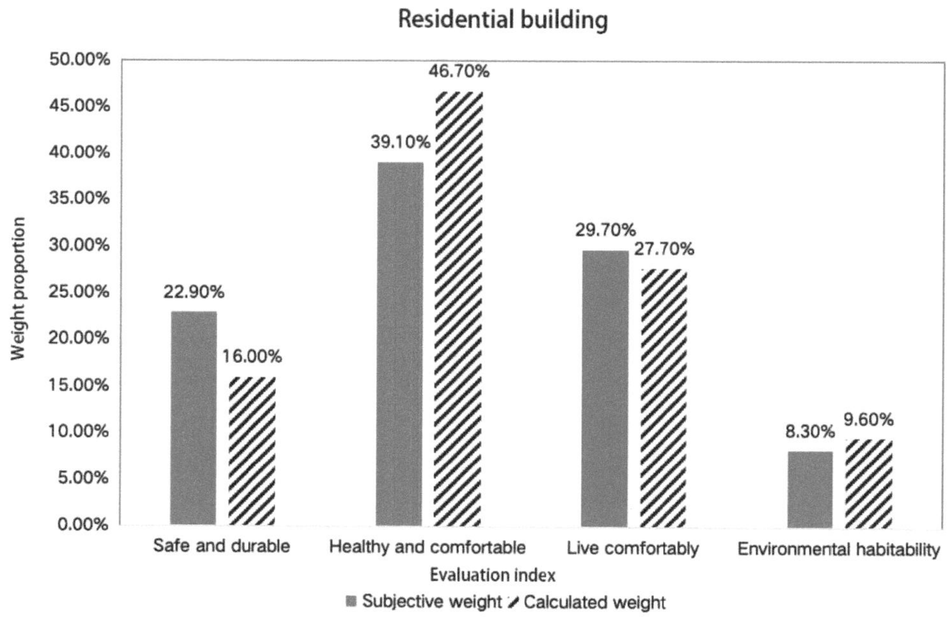

Figure 5. Level 1 index weight of residential buildings.

The proportion of the weight function of the secondary index is shown in Figure 6. The figure above presents the *p*-value for each factor. Based on the weight data, it is evident that personnel involved in residential construction are highly satisfied with indoor air quality, water safety, indoor natural ventilation comfort, convenience of public transportation, and open spaces such as cities and public venues within walking distance. These factors are represented as 'function realization is very good'. However, the level of satisfaction with indoor wall cracking, building exterior paint discoloration, water seepage, outdoor water accumulation, green comfort, outdoor smoking area provision, and noise and light pollution control is low, indicating poor functional realization. The satisfaction of outdoor ground non-slip surfaces, indoor sound insulation, lighting, outdoor fitness activity area setting, and surrounding public service facilities falls under the category of 'function realization is general'.

In terms of residential buildings, less than 50% of the standard of 85.3% of environmental livability has been achieved, indicating poor implementation in this field. In addition, 66.7% of the indicators exceeded 150% for health and comfort, which was the best performance among the four evaluation indicators. Convenience of life followed closely behind. Overall, it can be concluded that the evaluation system meets the objective requirements.

3.3.2. Public Building

The two-level index judgment matrix was constructed, and a consistency test was carried out. The weight results of each classification index of public buildings are shown in Table 16.

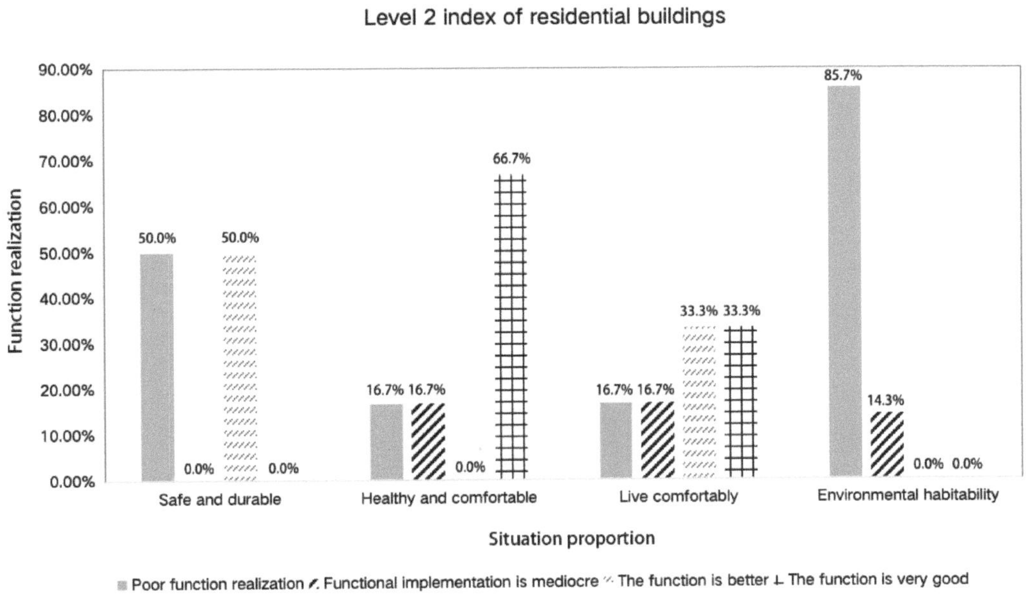

Figure 6. Level 2 index weight of residential buildings.

Table 16. Weights of evaluation indicators.

Level 1 Index	Level 1 Index Weight	Residential Building				Level 1 Index Weight	Public Building			
		Topic Number	Gauge Weight P_i	Level 2 Index Weight P_t	P		Topic Number	Gauge Weight P_i	Level 2 Index Weight P_t	P
Safe and durable	0.1601	SD-1-1	0.0615	0.0548	89.14%	0.1601	SD-1-2	0.0591	0.0603	102.11%
		SD-2-1	0.0410	0.0195	47.58%		SD-2-2	0.0394	0.0087	22.10%
		SD-3-1	0.0328	0.0329	100.35%		SD-3-2	0.0315	0.0141	44.77%
		SD-4-1	0.0410	0.0329	80.28%		SD-4-2	0.0709	0.0240	33.87%
		SD-5-1	0.0410	0.0079	19.28%		SD-5-2	0.0394	0.0141	35.81%
		SD-6-1	0.0369	0.0121	32.80%		SD-6-2	0.0354	0.0389	109.78%
Healthy and comfortable	0.4673	HC-1-1	0.0492	0.1751	356.04%	0.4673	HC-1-2	0.0472	0.1787	378.25%
		HC-2-1	0.0328	0.0648	197.64%		HC-2-2	0.0315	0.0470	149.23%
		HC-3-1	0.0369	0.0975	264.33%		HC-3-2	0.0354	0.1170	330.20%
		HC-4-1	0.0410	0.0192	46.85%		HC-4-2	0.0394	0.0200	50.80%
		HC-5-1	0.0492	0.0288	58.56%		HC-5-2	0.0472	0.0299	63.29%
		HC-6-1	0.0328	0.0819	249.80%		HC-6-2	0.0315	0.0746	236.86%
Live comfortably	0.2772	LC-1-1	0.0328	0.1044	318.42%	0.2772	LC-1-2	0.0315	0.1060	336.55%
		LC-2-1	0.0328	0.0150	45.75%		LC-2-2	0.0315	0.0119	37.78%
		LC-3-1	0.0410	0.0416	101.50%		LC-3-2	0.0394	0.0178	45.21%
		LC-4-1	0.0205	0.0674	328.91%		LC-4-2	0.0197	0.0279	141.73%
		LC-5-1	0.0410	0.0244	59.54%		LC-5-2	0.0394	0.0442	112.27%
		LC-6-1	0.0287	0.0244	85.05%		LC-6-2	0.0354	0.0694	195.86%
Environmental habitability	0.0954	EL-1-1	0.0410	0.0075	18.30%	0.0954	EL-1-2	0.0394	0.0062	15.75%
		EL-2-1	0.0656	0.0189	28.82%		EL-2-2	0.0630	0.0097	15.40%
		EL-3-1	0.0369	0.0033	8.95%		EL-3-2	0.0354	0.0039	11.01%
		EL-4-1	0.0410	0.0049	11.96%		EL-4-2	0.0394	0.0039	9.91%
		EL-5-1	0.0410	0.0117	28.55%		EL-5-2	0.0394	0.0150	38.10%
		EL-6-1	0.0410	0.0302	73.69%		EL-6-2	0.0394	0.0228	57.91%
		EL-7-1	0.0410	0.0189	46.12%		EL-7-2	0.0394	0.0338	85.85%

The Figure 7 indicates that the subjective satisfaction of health and comfort is lower than the weight value calculated by the comprehensive calculation of each item. This suggests that people's intuitive feeling of health and comfort is lower than the objective situation in the use of public buildings. It is evident that the current function of public buildings in health and comfort is relatively complete, exceeding the users' expectations. At the same time, the importance of safety, durability, and convenience in daily life outweighs that of mere calculation. It is evident that current public buildings have some shortcomings in these areas, which are noticeable and affect the users' experience.

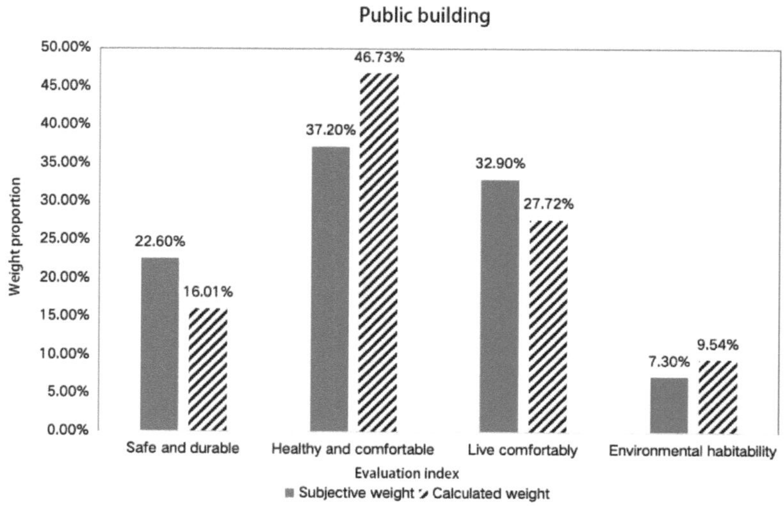

Figure 7. Level 1 index weight of public buildings.

Figure 8 shows the proportion of the implementation of the weight function of public secondary indicators. The figure shows the p-value for each factor. Based on the weight data, it is evident that public construction personnel are highly satisfied with indoor air quality, water safety, natural ventilation comfort, and the convenience of public transportation, with a 'good function realization' performance. However, they are less satisfied with outdoor water, green comfort, outdoor smoking area setting, and noise control, which fall under 'poor function realization'. Many other aspects of satisfaction fall somewhere in between.

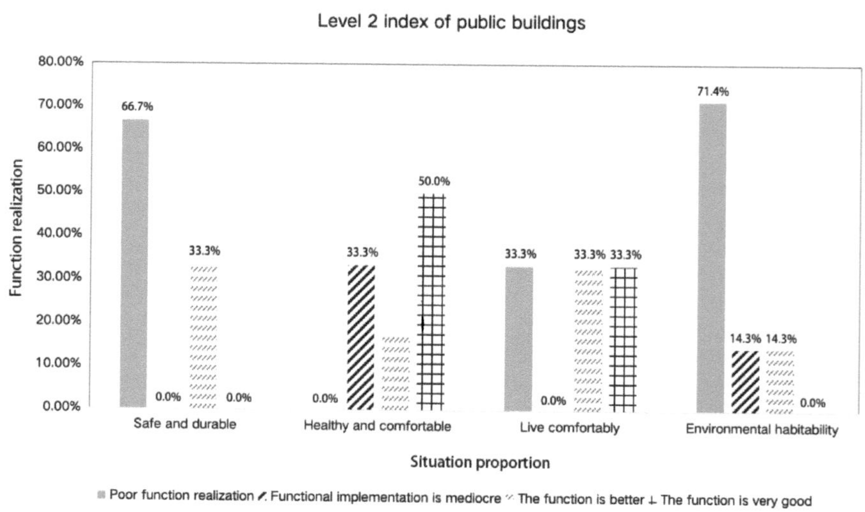

Figure 8. Level 2 index weight of public buildings.

Regarding safety durability and environmental livability of public buildings, 66.7% and 71.4% of the functional implementation standards are less than 50%, indicating poor implementation that needs improvement. As for health and comfort, 50% of the functional realization standards exceed 150%, indicating excellent performance. Meanwhile, in terms of living convenience, 33.3% of the indicators are better, and another 33.3% are rated as very good. This places it second only to health and comfort, indicating an overall positive effect of this factor.

4. Discussion

This study shows progress in the field of green building in China, particularly in terms of health and comfort, which are highly rated by building users. However, we have also observed that green buildings have some imperfections in certain functions, mainly in the areas of environmental livability and safety and durability, and user satisfaction is low in these two aspects. The future development of green buildings requires a better balance between various performance indicators. The study identified deficiencies in current green building technologies, including outdoor drainage after rainfall, the comfort of greenery around buildings, the layout of outdoor smoking areas, and outdoor noise control. Residential and public buildings exhibit a consistent trend in these key indicators, as evidenced by the weight of the indicators. It is apparent that certain indicators with low satisfaction require improvement.

Compared to the quantitative indicators in ASGB2019, this study offers a new perspective for understanding the needs of building users. It emphasizes that future green building designs should address the practical concerns of users more effectively. Hubei Province is located in the central region of China, making it both geographically representative and suitable for extending research results to a national scale. Additionally, this study presents a novel evaluation method that compares satisfaction weight with index weight in ASGB2019 to analyze the pros and cons of green buildings in a particular region. This provides a robust framework for differentiated green building development in the region. As a whole, this study enhances our comprehension of the user experience of green buildings and offers valuable insights to drive the continuous improvement of green buildings in China. This will help to further optimize the advantages of green buildings and address their shortcomings.

This research focuses on the content of building comfort in the current ASGB 2019 in China. However, there are still some limitations in this study, which did not investigate environmental performance, energy conservation, and carbon footprint. Therefore, it can be improved with follow-up research of this kind.

5. Conclusions

In summary, through an in-depth analysis of the satisfaction of existing green building users in Hubei Province, the following conclusions can be drawn:

(1) Health and comfort are the most important factors for users of residential buildings and public buildings, followed by convenience of life.

(2) Residential buildings and public buildings performed poorly in terms of livability, while health and comfort performed best.

(3) According to the average satisfaction, public buildings have a slightly higher overall satisfaction rating than residential buildings. However, some satisfaction indicators show that residential buildings have higher scores than public buildings. The sound insulation effect was rated lowest in both public and residential buildings.

Through this survey, we can more intuitively understand some problems existing in the current development of green buildings. In future green building design, it is important to prioritize the actual needs and satisfaction of users. Simultaneously, it is recommended that architectural designers focus on aspects with low user satisfaction while enhancing the overall performance of buildings. This will promote the development of green buildings in a more humanized and comfortable direction. Continuous optimization and adjustment are expected to lead to a higher level of green building development, achieving energy savings and emission reductions while providing residents with an improved living experience.

Author Contributions: Conceptualization, Q.D. and S.O.; methodology, X.S. and W.W.; resources, Q.D. and Z.R.; data curation, T.M. and Y.W.; writing—original draft preparation, S.O.; writing—review and editing, S.O.; visualization, Q.D. All authors have read and agreed to the published version of the manuscript.

Funding: This project was funded by Central-Southern China Engineering Consulting and Design Group Co., Ltd. (Grant No. 2019-77-216).

Data Availability Statement: The raw data supporting the conclusions of this article will be made available by the authors on request.

Conflicts of Interest: Author Yinguang Wu was employed by the company Central-South Architectural Design Institute Co., Ltd. The remaining authors declare that the research was conducted in the absence of any commercial or financial relationships that could be construed as a potential conflict of interest.

Appendix A

Residential building questionnaire

Dear owner! In order to understand your actual needs, understand your service experience, in order to provide you with better service and experience later, I hope you fill in the following questionnaire according to your real ideas, thank you for your cooperation! Satisfaction scores range from low to high on a scale of 1–5.

Your gender: Male female

Your age: Under 18 18–30 31–50 51+

1. Balconies, outside Windows, windowsills, protective railings, etc. have strengthened anti-fall design to reduce the risk of falling objects hurting people. ( )
2. Put anti-slip measures on the ground or pavement outside. ( )
3. Human-vehicle diversion, outdoor lighting effects. ( )
4. Original hardware fittings, pipe valves, switch taps, etc., good quality and easy to replace. ( )
5. Interior wall cracking. ( )
6. Building exterior paint fading, water seepage condition. ( )
7. The indoor public space has good ventilation effect and no odor. ( )
8. Outdoor landscape water is clean and pollution-free. ( )
9. Domestic water is clean and pollution-free. ( )
10. Indoor sound insulation effects. ( )
11. Indoor and basement lighting conditions. ( )
12. Indoor natural ventilation comfort. ( )
13. Accessibility to public transportation. ( )
14. The indoor and outdoor public areas of the building meet the requirements of all-age design. ( )
15. The surrounding area is equipped with corresponding public service facilities, such as hospitals, cultural centers, etc. ( )
16. Open Spaces such as urban green Spaces, squares and public sports fields within walking distance. ( )
17. Set up a centralized outdoor fitness activity area. ( )
18. Real-time information on water use available online. ( )
19. Conditions for standing water outside during rainy weather. ( )
20. Comfort of greenery around this building. ( )
21. Reasonable setting of outdoor smoking areas. ( )
22. The noise level of the outdoor site. ( )
23. Outdoor light pollution levels. ( )
24. Comfort with natural ventilation outside. ( )

25. Outdoor shade comfort. ( )

Public building questionnaire

Dear Sir/Madam! In order to understand your actual needs, understand your actual experience, in order to provide you with better projects and services later, I hope you fill in the following questionnaire according to your real ideas, thank you for your cooperation! Satisfaction scores range from low to high on a scale of 1–5.

Your gender: Male female

Your age: Under 18 18–30 31–50 51+

1. Balconies, exterior Windows, windowsills, protective railings, etc., have strengthened anti-fall design to reduce the risk of falling objects hurting people. ( )
2. Put anti-slip measures on the ground or pavement outside. ( )
3. Human-vehicle diversion, outdoor lighting effects. ( )
4. Durability of equipment such as lighting, faucets, plumbing, etc. in public areas. ( )
5. Use durable steel, anticorrosive wood and other building materials. ( )
6. Indoor floor tiles have good wear resistance and low breakage rate. ( )
7. Indoor public Spaces are well ventilated and odor-free. ( )
8. Outdoor landscape water is clean and pollution-free. ( )
9. Domestic water is clean and pollution-free. ( )
10. Indoor sound insulation effects. ( )
11. Indoor and basement lighting. ( )
12. Natural ventilation comfort in the building during the transition season. ( )
13. Accessibility to public transportation. ( )
14. The indoor and outdoor activity areas of the building meet the requirements of all-age design. ( )
15. The building has shared meeting facilities, fitness facilities, dining facilities, etc. ( )
16. Open Spaces such as urban green Spaces, squares and public sports fields within walking distance. ( )
17. Set up outdoor walking slow lanes. ( )
18. Mobile data network connection effect. ( )
19. Conditions for standing water outside during rainy weather. ( )
20. Comfort of greenery around this building. ( )
21. Reasonable setting of outdoor smoking areas. ( )
22. Noise levels in outdoor activity areas. ( )
23. Outdoor light pollution levels. ( )
24. Comfort in natural ventilation outside. ( )
25. Outdoor shade comfort. ( )

References

1. Hassanein, H.; Zorba, N.; Han, S.; Kanhere, S.S.; Shukair, M. Crowd Management. *IEEE Commun. Mag.* **2019**, *57*, 18–19. [CrossRef]
2. Ma, G.; Liu, T.; Shang, S. Improving the climate adaptability of building green retrofitting in different regions: A weight correction system for Chinese national standard—ScienceDirect. *Sustain. Cities Soc.* **2021**, *69*, 102843. [CrossRef]
3. Chan, A.P.C.; Darko, A.; Ameyaw, E.E.; Owusu-Manu, D. Barriers Affecting the Adoption of Green Building Technologies. *J. Manag. Eng.* **2017**, *33*, 04016057. [CrossRef]
4. Han, F.; Wang, Y.; Feist, W.; Cao, X.; Yu, Z.; Song, B.; Benli, H.; Dermentzis, G. Exploring solutions to achieve carbon neutrality in China: A comparative study of a large-scale passive House district and a Green building district in Qingdao. *Energy Build.* **2022**, *268*, 112224. [CrossRef]
5. Su, S.; Zhu, C.; Li, X.; Wang, Q. Dynamic global warming impact assessment integrating temporal variables: Application to a residential building in China. *Environ. Impact Assess. Rev.* **2021**, *88*, 106568. [CrossRef]
6. Yeo, Z.; Ng, R.; Song, B. Technique for quantification of embodied carbon footprint of construction projects using probabilistic emission factor estimators. *J. Clean. Prod.* **2016**, *119*, 135–151. [CrossRef]
7. Zhao, D.-X.; He, B.-J.; Johnson, C.; Mou, B. Social problems of green buildings: From the humanistic needs to social acceptance. *Renew. Sustain. Energy Rev.* **2015**, *51*, 1594–1609. [CrossRef]

8. Cohen, C.; Pearlmutter, D.; Schwartz, M. A game theory-based assessment of the implementation of green building in Israel. *Build. Environ.* **2017**, *125*, 122–128. [CrossRef]
9. Ge, J.; Zhao, Y.; Luo, X.; Lin, M. Study on the Suitability of Green Building Technology for Affordable Housing: A Case Study on Zhejiang Province, China. *J. Clean. Prod.* **2020**, *275*, 122685. [CrossRef]
10. Abdelaal, F.; Guo, B.H. Knowledge, attitude, and practice of green building design and assessment: New Zealand case. *Build. Environ.* **2021**, *201*, 107960. [CrossRef]
11. Chen, L.; Chan, A.P.; Owusu, E.K.; Darko, A.; Gao, X. Critical success factors for green building promotion: A systematic review and meta-analysis. *Build. Environ.* **2022**, *207*, 108452. [CrossRef]
12. Sievanen, L.; Leslie, H.M.; Wondolleck, J.M.; Yaffee, S.L.; McLeod, K.L.; Campbell, L.M. Linking top-down and bottom-up processes through the new U.S. National Ocean Policy. *Conserv. Lett.* **2011**, *4*, 298–303. [CrossRef]
13. Chen, L.; Gao, X.; Hua, C.; Gong, S.; Yue, A. Evolutionary process of promoting green building technologies adoption in China: A perspective of government. *J. Clean. Prod.* **2020**, *279*, 123607. [CrossRef]
14. Assefa, S.; Lee, H.-Y.; Shiue, F.-J. Sustainability Performance of Green Building Rating Systems (GBRSs) in an Integration Model. *Buildings* **2022**, *12*, 208. [CrossRef]
15. Bohari, A.A.M.; Skitmore, M.; Xia, B.; Teo, M. Green oriented procurement for building projects: Preliminary findings from Malaysia. *J. Clean. Prod.* **2017**, *148*, 690–700. [CrossRef]
16. Fan, K.; Chan, E.H.W.; Chau, C.K. Costs and Benefits of Implementing Green Building Economic Incentives: Case Study of a Gross Floor Area Concession Scheme in Hong Kong. *Sustainability* **2018**, *10*, 2814. [CrossRef]
17. Yamany, S.E.; Afifi, M.; Hassan, A. Applicability and Implementation of U.S. Green Building Council Rating System (LEED) in Egypt (A Longitudinal study for Egyptian LEED Certified Buildings). In Proceedings of the International Conference on Improving Sustainability Concepts in Developing Countries, Cairo, Egypt, 15–17 January 2017.
18. *GB/T 50378-2019*; Assessment Standard for Green Building. China Architecture & Building Press: Beijing, China, 2019.
19. Wuni, I.Y.; Shen, G.Q.; Osei-Kyei, R. Scientometric review of global research trends on green buildings in construction journals from 1992 to 2018. *Energy Build.* **2019**, *190*, 69–85. [CrossRef]
20. Diamond, R.C.; Ye, Q.; Feng, W.; Yan, T.; Mao, H.; Li, Y.; Guo, Y.; Wang, J. Sustainable Building in China—A Green Leap Forward? *Buildings* **2013**, *3*, 639–658. [CrossRef]
21. Tan, S.-K.; Lim, H.-H.; Tan, S.-H.; Kok, Y.-S. A Cultural Creativity Framework for the Sustainability of Intangible Cultural Heritage. *J. Hosp. Tour. Res.* **2020**, *44*, 439–471. [CrossRef]
22. *GB/T 50378-2006*; Assessment Standard for Green Building. China Architecture & Building Press: Beijing, China, 2006.
23. Han, J.; Liao, L.; Zhang, G. Review and Trend Outlook on the Development History of China's Green Building Evaluation Standard System. *Constr. Sci. Technol.* **2017**, *8*, 10–13.
24. Zhang, L.; Wu, J.; Liu, H. Policies to enhance the drivers of green housing development in China. *Energy Policy* **2018**, *121*, 225–235. [CrossRef]
25. Li, Y.; Yang, L.; He, B.; Zhao, D. Green building in China: Needs great promotion. *Sustain. Cities Soc.* **2014**, *11*, 1–6. [CrossRef]
26. Wang, C. Comparison of the Development for Green Building and Low-carbon Building Evaluation Systems in China and Optimization Suggestions. *Build. Sci.* **2023**, *39*, 235–244.
27. Li, J.; Deng, Y.; Meng, C.; Zhao, N.; Li, G.; Wang, X.; Ma, Q. Research on the Internationalization Upgrade and Promotion Path of Green Building Standard System. *Constr. Econ.* **2023**, *44*, 69–76.
28. Yang, Z.; Chen, H.; Mi, L.; Li, P.; Qi, K. Green building technologies adoption process in China: How environmental policies are reshaping the decision-making among alliance-based construction enterprises. *Sustain. Cities Soc.* **2021**, *73*, 103122. [CrossRef]
29. Liu, W.; Tian, Z.; Xun, T. Analysis of High Quality Development of Green Buildings in China under the Background of Dual Carbon Goals. *China Hous. Facil.* **2022**, *10*, 115–117.
30. Wang, Q. Green buildings and green building standards. *Stand. Eng. Constr.* **2022**, *9*, 15–20.
31. Duan, H.; Li, G. Application of Architectural Design Strategy Based on Green Building Evaluation Standard. *Build. Tech. Dev.* **2021**, *48*, 14–16.
32. Li, Y.; Fan, L.; Zhang, Z.; Wei, Z.; Qin, Z. Exploring the design risks affecting operation performance of green commercial buildings in China. *J. Build. Eng.* **2023**, *64*, 105711. [CrossRef]
33. Sheng, Y. Research on the Characteristics and Evaluation System of Green Buildings. *Urban Archit. Space* **2022**, *29*, 122–123.
34. Li, X.; Feng, W.; Liu, X.; Yang, Y. A comparative analysis of green building rating systems in China and the United States. *Sustain. Cities Soc.* **2023**, *93*, 104520. [CrossRef]
35. Li, F.; Zhang, Y.; Zheng, B.; Wang, X. Evaluation of harmonious development for economy-environment system in coastal cities: A case study of Tianjin Municipality. *Ecol. Econ.* **2006**, 152–160. [CrossRef]
36. Gou, Z.; Prasad, D.; Lau, S.S.Y. Are green buildings more satisfactory and comfortable? *Habitat Int.* **2013**, *39*, 156–161. [CrossRef]
37. Xiong, C.; Liu, J. Improving Expert Scoring Method for Complex Environmental Risk Assessment. *Constr. Saf.* **2022**, *37*, 73–75.
38. Ding, D.; Wu, J.; Zhu, S.; Mu, Y.; Li, Y. Research on AHP-based fuzzy evaluation of urban green building planning. *Environ. Chall.* **2021**, *5*, 100305. [CrossRef]
39. Sakhardande, M.J.; Gaonkar, R.S.P. On solving large data matrix problems in Fuzzy AHP. *Expert Syst. Appl.* **2022**, *194*, 116488. [CrossRef]

40. Han, L.; Mei, Q.; Lu, Y.; Ji, M. Analysis and Study on AHP-Fuzzy Comprehensive Evaluation. *China Saf. Sci. J.* **2004**, *7*, 89–92+3.
41. Pan, X. Investigating aging Americans' transportation options in the era of crowdsourcing: Questionnaire design and survey implementation through Amazon Mechanical Turk. *Res. Transp. Bus. Manag.* **2019**, *30*, 100372. [CrossRef]

buildings

MDPI

Article

Study on the Impact of Design Factors of Piloti Forms on the Thermal Environment in Residential Quarters

Jinhan Li [1], Xiaofang Shan [1,2] and Qinli Deng [1,2,*]

[1] School of Civil Engineering and Architecture, Wuhan University of Technology, No. 122 Luoshi Road, Wuhan 430070, China; l1049732104702@whut.edu.cn (J.L.); xfshan@whut.edu.cn (X.S.)
[2] Hainan Institute, Wuhan University of Technology, No. 5 Chuangxin Road, Sanya 572024, China
* Correspondence: deng4213@whut.edu.cn

Abstract: According to piloti design, the outdoor thermal environment can be improved in cities with hot summer conditions. Taking Chinese cities with a hot summer and cold winter as the research object, this paper discusses the improvement of the outdoor thermal environment of residential districts in summer by considering piloti design factors. In this article, according to our investigation of piloti design in Wuhan, a basic model of the overhead layer in the Wuhan residential area is presented, along with the effects of different piloti ratios (0–80%), piloti heights (2–6 m), and greening rates (30–35%) on the outdoor thermal environment of buildings. The average air temperature and average wind speed at the pedestrian level are used as outdoor thermal environment indicators, the average PET is used as the outdoor thermal comfort indicator, and the comfort wind ratio is used as the outdoor wind comfort indicator. The results show that increasing the ratio of corridor columns has the greatest thermal comfort enhancement effect in the corridor area, and when the piloti ratio increases from 20% to 80%, the PET in piloti areas reduces by 2.926 °C. Improving the greening rate has the greatest thermal comfort enhancement effect in the passageway area, and when the greening rate increases from 20% to 80%, the PET in piloti areas reduces by 0.9 °C. Furthermore, the increases in both the piloti ratio and piloti height have an enhancement effect on the outdoor wind environment and wind comfort, with thresholds of a piloti ratio over 60% and a piloti height over 5 m. In contrast, the increase in the greening rate will deteriorate the outdoor wind environment and wind comfort. The conclusions of this study are of great significance for the planning and design of overhead layers in residential areas in hot and humid areas in summer.

Keywords: outdoor thermal environment; piloti design; greening rate; residential quarters; human thermal comfort

Citation: Li, J.; Shan, X.; Deng, Q. Study on the Impact of Design Factors of Piloti Forms on the Thermal Environment in Residential Quarters. *Buildings* **2024**, *14*, 1303. https://doi.org/10.3390/buildings14051303

Academic Editor: Eusébio Z.E. Conceição

Received: 25 March 2024
Revised: 28 April 2024
Accepted: 1 May 2024
Published: 5 May 2024

1. Introduction

A comfortable outdoor environment leads to a greater willingness of city dwellers to engage in outdoor activities, including walks, cycling, and outdoor ball games, which offers people a greater sense of well-being. At the same time, outdoor activities can also reduce energy consumption in built-up areas [1]. However, in most parts of China, where summers are always hot, outdoor thermal comfort is often affected by unfavorable factors, such as strong solar radiation and wind conditions, as well as high temperatures from the urban heat island effect [2,3]. At the same time, land use has changed significantly [4,5]. Early studies have shown that the loss of green spaces and the application of paving, together with a significant increase in anthropogenic heat released, have affected the urban climate, leading to severe environmental degradation and significantly increasing the ecological footprint of the city [6–8]. Because of their growing populations, cities require large amounts of energy to function properly. In fact, urban dwellers consume more than 75% of the total energy due to activities carried out in the urban environment [9,10]. Relevant studies have shown that the heat island effect in Wuhan is significant and that there is a strong positive correlation between the number of people in the urban area of Wuhan and the heat island effect [11]. At the same time, the population of Wuhan has increased by 2.54 million people from 2011 to 2021, that is, by nearly 20%. The rising population trend will continue in China's urbanization process [12], which will undoubtedly have a negative impact on the thermal environment and thermal comfort in Wuhan, and the duration of extreme high temperatures will also increase significantly in the future.

In recent years, a variety of strategies have been used to improve the outdoor thermal environment and thermal comfort in cities. One of the strategies is to use pilotis to provide shading for building areas and affect natural ventilation, which can facilitate air circulation within the building area, thereby improving wind comfort and average wind speeds [13–19]. Piloti design has also been considered as an effective means of cooling. Studies on the outdoor thermal environment of buildings with different piloti design variables have mainly focused on the piloti rate, the height of the piloti core, and the position of the piloti setting relative to the buildings [20]. Xi et al. [21,22] used questionnaires and numerical simulations to calculate the thermal sensation and thermal comfort in the campus area of Guangzhou on 14 July and 15 July in summer, recording data changes in building pedestrian height SET* at piloti ratios of 0–100%. The results showed that an increase in the piloti ratio could improve the thermal comfort in spaces sheltered by the pilotis, and that when the piloti ratio was 100%, the SET* in this space could satisfy the human body limit. Zhou [23] investigated the outdoor thermal environment of Wuhan in summer (July) and winter (January). Based on the optimal location of the piloti arrangement (both ends), the relationship between the piloti ratio and the outdoor wind environment was investigated. Their results showed that the wind speed increased with the growing piloti ratio in summer, and that it slightly changed in winter. Thus, it was surmised that the piloti ratio should be between 12% and 38% to avoid discomfort caused by wind in winter. Chen et al. [24] simulated the influence of indoor ventilation on the outdoor thermal environment of a campus building in Guangzhou, and they researched the improvement of the outdoor thermal environment in the campus area with different piloti heights and piloti ratios.

Apart from buildings, trees are another major determinant of the thermal environment in residential areas, with effects through shading, evapotranspiration, and increased albedo [25–27]. Canopies of trees can act as shading like buildings to block solar radiation and thus have a reducing effect on the surface temperature as well as the air temperature in residential areas [28]. At the same time, evapotranspiration from trees can reduce sensible heat and increase the latent heat flux into the air [29,30]. In addition, since the albedo of trees is higher than that of built-up areas, more trees can reflect more solar radiation, thus lowering the surface temperature [31]. In residential areas, where buildings are usually planned or constructed first, trees are considered a great method to mitigate the effects of the thermal environment. Therefore, it is necessary to evaluate the ability of trees to influence the thermal environment with appropriate methods. Several previous studies have been conducted on the effectiveness of trees in improving the thermal environment in residential areas during the summer and have confirmed that trees can improve the thermal environment of buildings to a certain extent [32–34]. However, trees also have a negative effect on the wind speed in the area to a certain extent, which is contrary to the design concept of increasing wind speed for piloti design. The effect of the tree greening rate on the thermal environment with respect to the piloti building area is yet to be investigated.

Most of the existing studies on the thermal environment of piloti buildings have failed to consider the influence of complex urban blocks on the thermal environment of buildings. Most of them have focused on a single block scale, where the buildings are replaced by ideal blocks, or purely on a single building, without considering the role of vegetation and the influence of the subsurface on the thermal environment. In addition, there is a lack of research on thermal comfort at the pedestrian level in residential quarters based on piloti design variables (Table 1). Furthermore, there is a lack of qualitative and quantitative studies on greening that study the effect of the greening rate on outdoor thermal comfort in piloti residential quarters.

Therefore, in order to address the research gaps in regard to the thermal environmental aspects of elevated buildings and the lack of studies on greenery settings, efforts have been made to diversify the scope of research in this area. This article presents a basic model of piloti residential quarters in Wuhan. Three design variables are considered: two piloti design variables, i.e., piloti ratio and piloti height, to analyze the impact of piloti installations on the outdoor thermal environment residentially in terms of changes caused by changes in these two design variables; and one overall architectural design variable, i.e., the greening rate of the residential area, to qualitatively analyze outdoor thermal comfort by increasing greenness in piloti residential quarters and to explore whether a threshold exists. A total of 15 models were established for comparison, to derive the effects of each design variable on the thermal environment of piloti buildings. This study offers a reference for the design of piloti buildings in hot and humid areas, along with assistance for urban design planners seeking to mitigate the increasingly severe heat island effect.

2. Methods

2.1. Study Area

Located in central China, Wuhan is at 30°35′ north latitude, 114°17′ east longitude, standing at 23 m above sea level. It is the fifth largest city in China, with about 13.74 million urban residents. Wuhan has a typical subtropical monsoon climate with hot summers and cold winters. Summer extends from June to September, with July and August being the two hottest months. Wuhan's urban area is sweltering in summer, with daytime temperatures often around 37 °C and nighttime temperatures around 30 °C, with an extreme maximum of 41.3 °C. The wind speed is low in summer, with an average wind speed of about 1 m/s from the southwest. Such meteorological conditions may exacerbate people's feeling of high temperatures and make the heat more unbearable. In contrast, winter is a cold and dry season from December to February, with an average temperature of about 3 °C. Wind from the north is more common, with an average wind speed of 4–5 m per second. Wuhan is in the eastern part of the Jianghan Plain, with 166 lakes of various sizes scattered throughout the city, and the average relative humidity of the air is high at about 73–77% throughout the year (Figure 1).

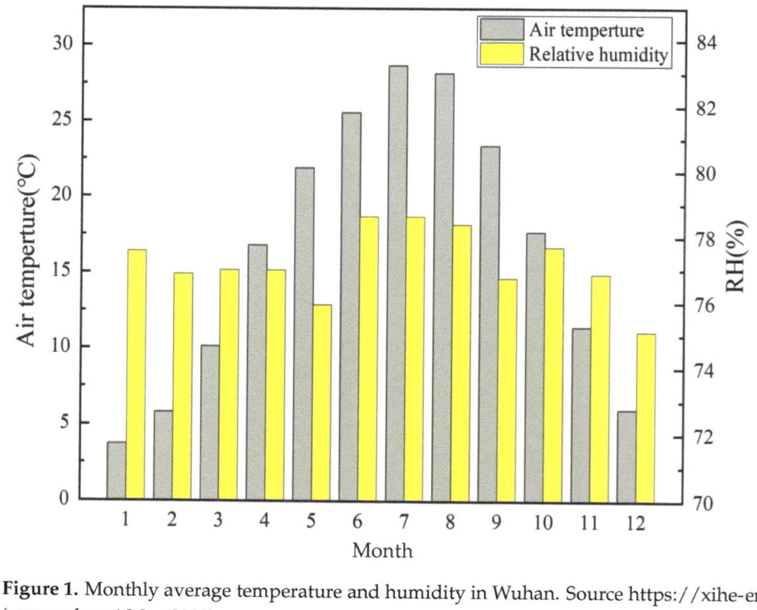

Figure 1. Monthly average temperature and humidity in Wuhan. Source https://xihe-energy.com/ (assessed on 4 May 2022).

2.2. Research on the Status of Piloti Residential Quarters in Wuhan City

The purpose of this study was to facilitate an understanding of the status quo of the construction of pilotis in Wuhan's residential areas, and to statistically analyze the form, layout, and use of the pilotis. This study adopted an offline research method to gather statistics on the piloti residential districts in Wuhan's various central urban areas, and the specific data of the research are shown in Table 2. Between the 19 residential quarters with pilotis (Figure 2), the following common features were found in Wuhan:

1. The piloti ratio is below 50%.
2. Most pilotis are located on both sides of the buildings; only individual buildings in individual districts use middle pilotis due to the topography and orientation of these buildings. In those districts where there is a single middle-piloti building, most of the buildings are elevated at both ends.
3. The greening rate is less than 50% but higher than 25%.
4. The piloti height is between 2 and 6.5 m: when it is less than 2.5 m, piloti areas act as underground car parks; and only a small number are more than 5.5 m. Based on the status quo of the pilotis in Wuhan residential districts, the ideal district model was extracted and can be used as a substitute: a 40% piloti ratio, 30% greening rate, 4 m piloti height, and parallel rows and columns as the building layout.

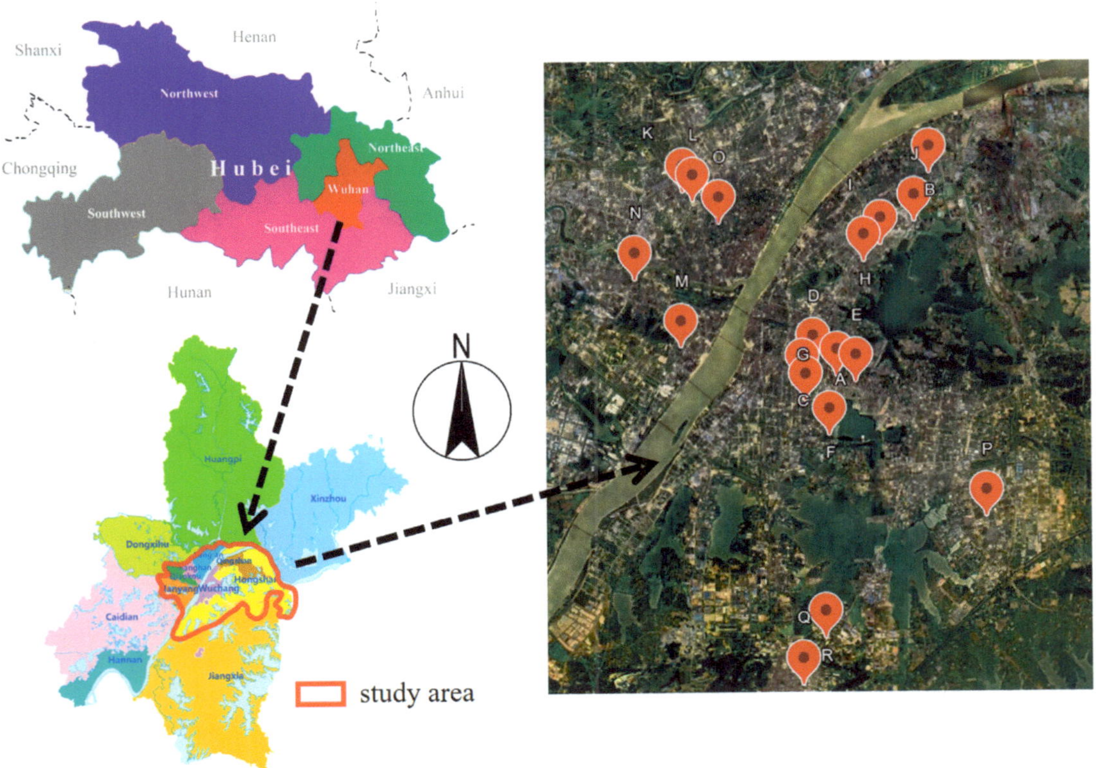

Figure 2. Distribution map of the research community.

Table 1. Difference between selected piloti building research papers and this study.

References	Content of Research	Differences from this Research
[13]	Study of the impact of piloti ratio, piloti height, and geometry on the wind environment	No consideration of the thermal environment, underpinning, or greening
[15]	Impact of piloti ratio and wind direction on the wind environment	No consideration of the thermal environment, underpinning, or greening
[18]	Study on the impact of piloti buildings with different piloti heights and widths on the wind environment	No consideration of the impact on the thermal environment
[21]	Study on the impact of buildings with different piloti ratios on the outdoor thermal environment	No categorization of the underlay and no consideration of the impact of greening
[23]	Analysis of the cooling and thermal environment improvement effects of piloti buildings through measured methods	Using measured methods while not considering underpinning or greening

2.3. Assessment Methodology

In this study, the physiological equivalent temperature (PET) [35] was used as the thermal environment assessment index. The PET is calculated based on environmental parameters such as air temperature, wind speed, mean radiant temperature, relative humidity, clothing thermal resistance, and activity level. The personal parameters of the PET are set as male, 35 years old, height 1.75 m, weight 75 kg, physiological metabolism 80 W, labor metabolism 85 W, thermal resistance value 0.31, and height 1.5 m above the ground.

A comfortable wind ratio is used as the evaluation standard of outdoor wind comfort. The comfortable wind ratio is the area occupied by comfortable wind in the target area divided by the area of the target area, and the evaluation criteria of comfort wind are shown in Table 3 [36].

Table 2. Basic information table of piloti in residential quarters of Wuhan.

Residential Quarter	Greening Rate	Piloti Height	Piloti Arrangement
A	35%	5 m	at the two ends of the building
B	15%	4.2 m	at the two ends of the building
C	28%	6.5 m	at the two ends of the building
D	30%	3.9 m	at the two ends of the building
E	28%	4.4 m	in the middle of the building
F	30%	4.5 m	at the two ends of the building
G	35%	5.4 m	at the two ends of the building
H	36%	2 m	at the two ends of the building
I	35%	4.6 m	at the two ends of the building
J	35%	2.6 m	at one end of the building
K	35%	3.9 m	at the two ends of the building
L	40%	2.6 m	at the two ends of the building
M	30%	4.2 m	in the middle of the building
N	30%	2.6 m	at the two ends of the building
O	35%	3.9 m	at the two ends of the building
P	35%	5.2 m	at the two ends of the building
Q	35%	2.8 m	at the two ends of the building
R	30%	3.4 m	at the two ends of the building
S	38%	3.2 m	at the two ends of the building

2.4. Envi-Met Software Thermal Environment Validation

2.4.1. Introduction of the Validation Case

Envi-met validation is carried out using Li's measurements in the Xixiu Village neighborhood of South China University of Technology [37]. This is located in the western district of the university and forms a faculty and staff residential neighborhood. The whole community covers an area of nearly 24,000 square meters, with a building density of 32%, a greening rate of 33%, and a row-type building layout. Between the 5th and 7th faculty flats is the central garden, covering an area of about 1086 m^2, equipped with slides and other activity facilities. Between the 12th faculty flat that is near the entrance to the community and adjacent to the western student dormitory and 13th faculty flats is a space raised at about 3.2 m, of which the lower part is a garage and the upper part is the pavement. Measurement point A1 is between the 12th and 13th buildings in Xixiu Village, with no shade; A5 is between the 2nd and 3rd buildings under mango trees, with relatively dense shade; and A7 is between the 3rd and 4th buildings under street trees, with sparse shade.The distribution of measurement points in the model is shown in Figure 3.

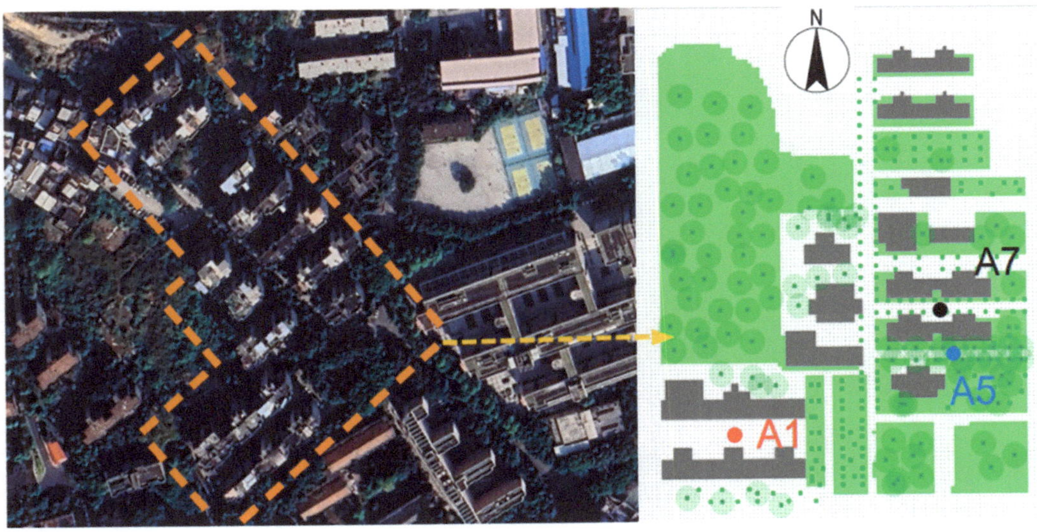

Figure 3. Distribution of measurement points in the Xixiu District cluster.

2.4.2. Validation Results

We compared the measured air temperature and relative humidity data of three measurement points [38], A1, A5, and A7, in the Xixiu Village community at a pedestrian-level height of 1.5 m with the simulated data from the Envi-met 5.5.1. The evaluation indexes commonly used in the validation of thermal environmental simulation are root mean square error (RMSE), mean absolute error (MAE), mean absolute percentage error (MAPE), and Willmott's index of agreement (d) [39], and the specific calculation methods of each index are as follows:

$$\mathrm{RMSE} = \sqrt{\frac{1}{\mathrm{n}}\sum_{t=1}^{n}(\mathrm{P_i} - \mathrm{O_i})^2} \quad (1)$$

$$\mathrm{MAE} = \frac{1}{\mathrm{n}}\sum_{i=1}^{n}|\mathrm{P_i} - \mathrm{O_i}| \quad (2)$$

$$\mathrm{MBE} = \frac{1}{\mathrm{n}}\sum_{i=1}^{n}\frac{\mathrm{P_i} - \mathrm{O_i}}{\mathrm{O_i}} \quad (3)$$

$$d = 1 - \frac{\sum_{i=1}^{n} |O_i - P_i|}{\sum_{i=1}^{n} (|P_i - \widehat{O}\,| - |O_i - \widehat{O}|)} \quad (4)$$

In the formulas above, "P_i" represents simulation data, "O_i" represents measured data, "$\widehat{O}$" is the average value, and "n" is the number of measurement points. The comparison results are shown in Figure 4. Envi-met shows a better simulation of air temperature than relative humidity and is also better during the day than at night. Meanwhile, when comparing the data agreement of the three different measurement points, it can be found that the simulation effect of point A1 is the best, which is due to the influence of vegetation at points A5 and A7 (Table 4). Therefore, the consistency for temperature and relative humidity is not as good as that of point A1. The consistency between the simulated and measured values of temperature and relative humidity at 1.5 m for the three groups of points is acceptable (D > 0.6). Based on validation and comparison results above, it can be concluded that Envi-met version 5.5.1 offers a reasonable prediction of the thermal environment in a complex residential area in hot and humid environments.

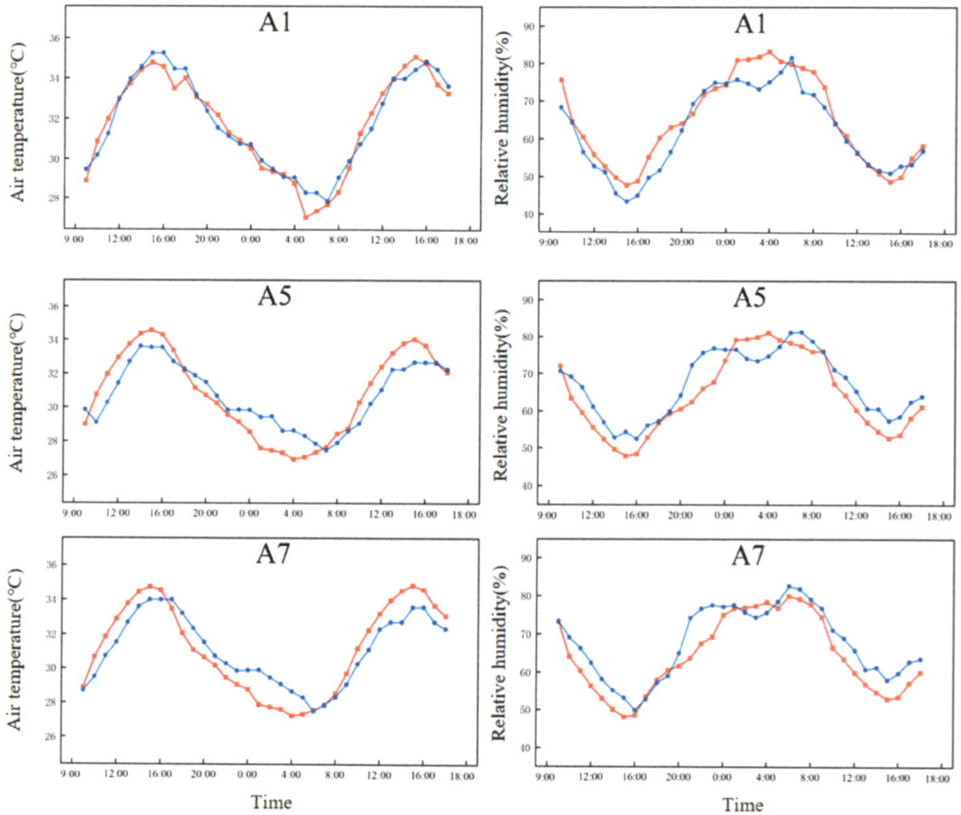

Figure 4. Comparison of the simulated and measured results for air temperature and relative humidity. (The blue curve represents simulated values, and the red curve represents measured values).

2.5. Analysis Model and Cases

The analytical model of this study was formed with reference to the research on the current situation of piloti buildings in Wuhan, as shown in Figure 5. The overall model is a row-type building cluster consisting of 7 rows and 6 columns with a total of 42 single buildings, and the size of the whole computational domain is 456 m (X) × 484 m (Y) × 220 m (Z). The buildings are separated by 36 m from left to right and 16 m from front to back, and the building height is 65 m. The target area is the six buildings in the center of the cluster, which are used to remove the uncertainty brought about by the influence of the complex urban block on the wind and thermal environments in the study area [33]. The dimensions of the target area are 136 m (X) × 168 m (Y) and the dimensions of the aisle area are 8 m (X) × 168 m (Y). The boundary condition settings of the model are shown in Table 5. The overall greening rate of the building area varies in the range of 30–50%, in which the ratio of the area of trees, shrubs, and lawns remains unchanged. The variation range of piloti height is 2–6 m. A total of 15 calculation cases were set up in this study with differences in greening rate, piloti ratio, and height (Table 6).

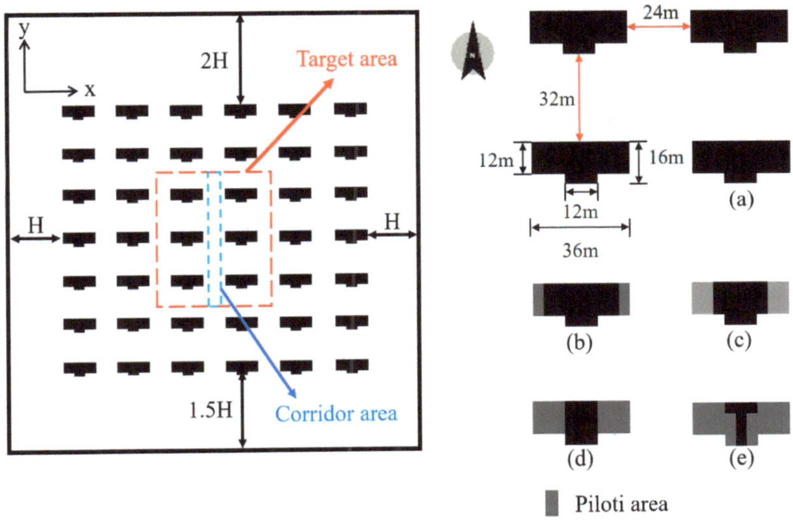

Figure 5. Architectural layout and building model piloti diagram. (**a**–**e**) are schematic diagrams of piloti ratios ranging from 0% to 80%. The gray areas represent piloti spaces.

Table 3. Evaluation principle of velocity considering temperature influence.

Evaluation Range	Temperature Range (°C)		
	<10	**10~25**	**>25**
Range of breezes that cause thermal discomfort in the human body (m/s)	——	——	<0.7
Human comfortable wind speed range (m/s)	<1.3	<1.5	0.7~1.7
Excessive range between comfortable and uncomfortable winds (m/s)	1.3~2.0	1.5~2.3	1.7~2.9
Range of strong winds that cause human discomfort (m/s)	>2.0	>2.3	>2.9

Table 4. Index values of air temperature and relative humidity at the measuring points.

Measuring point	RMSE	MAE	MBE	D
A1 (T_a)	0.538	0.442	−0.100	0.89
A5 (T_a)	1.1	0.962	0.14	0.64
A7 (T_a)	1.064	0.959	0.087	0.712
A1 (RH)	4.41	3.54	0.058	0.823
A5 (RH)	5.002	4.416	−2.986	0.669
A7 (RH)	4.7	3.985	−3.398	0.74

Table 5. Envi-met parameter setting.

	Parameter Setting	
Location	Wuhan Roughness	(114.30, 30.5) 0.1
Time	Start time Total simulation time (h) Pre-run time	7.31 00:00 21 h 7 h
Building Parameter	Wall heat transfer coefficient (w/m^2·k) Roof heat transfer coefficient (w/m^2·k) Wall reflectance (%) Roof reflectance (%)	1.7 2.2 0.3 0.15
Vegetation	Lawn Hedge Tree	XY (glass 50 cm aver. dens.) H2 (Hedge 2 m dens.) T1 (Tree 10 m dens.)
Underlying surface	Brick road (red stones) Brick road (yellow stones) Dark concrete pavement Asphalt road Sandy loam	reflectance: 0.3; emissivity: 0.9 reflectance: 0.4; emissivity: 0.9 reflectance: 0.2; emissivity: 0.9 reflectance: 0.12; emissivity: 0.9 reflectance: 0.2; emissivity: 0.9
Grid setting	Grid size Number of grids Number of nested grids Size of the computational domain	4 m(X) × 4 m(Y) × 3 m(Z) 114(X) × 121(Y) × 25(Z) 6 456 m(X) × 484 m(Y) × 220 m(Z)

Table 6. Analysis cases.

Case	Floor Space (m^2)	Piloti Area Space (m^2)	Piloti Ratio	Greening Rate	Piloti Area Height (m)
1	480	0	0%	30%	4 m
2	480	96	20%	30%	4 m
3	480	192	40%	30%	4 m
4	480	288	60%	30%	4 m
5	480	384	80%	30%	4 m
6	480	192	40%	30%	2 m
7	480	192	40%	30%	3 m
8	480	192	40%	30%	4 m
9	480	192	40%	30%	5 m
10	480	192	40%	30%	6 m
11	480	192	40%	30%	4 m
12	480	192	40%	35%	4 m
13	480	192	40%	40%	4 m
14	480	192	40%	45%	4 m
15	480	192	40%	50%	4 m

3. Results and Analysis

This study examined the effects of piloti ratio, piloti height, and the overall greenness of piloti residential quarters on the thermal environment and thermal comfort of the piloti residential area. The results related to changes in mean air temperature (Ta), mean radiant temperature (MRT), mean physiological equivalent temperature (PET), mean wind speed, and comfort wind ratio at the pedestrian level. All of those are discussed herein for a comprehensive and accurate analysis and evaluation of the thermal environment. During the hot summer months, the main activity areas of pedestrians and occupants walking in residential neighborhoods are mainly in the aisles between two buildings as well as in the piloti areas (Figure 5). Therefore, this study analyzed the aisle and piloti areas along with the overall thermal environment of the target area. Three time points, 8:00, 14:00, and 18:00, were selected for analysis in order to understand the changes in the thermal environment during the day. Terms and abbreviations used in this paper can be found in Appendix A and Vector wind speed cloud chart can be found in Appendix B.

3.1. Effect of Piloti Ratio on the Outdoor Thermal Environment and Thermal Comfort in Piloti Residential Quarters

3.1.1. Air Temperature

The air temperature (Ta) at the pedestrian level for buildings with different piloti rates at the 8:00, 14:00, and 18:00 points are shown in Figure 6. It can be concluded that the 14:00 period has the greatest cooling. Meanwhile, the thermal environment is also in the worst condition at this time, because it is at the hottest time of the day in summer in Wuhan. The average air temperature of the target area decreases with the rise in piloti ratio. When the piloti rate is from 0% to 20%,it decreases the most; while from 60% to 80%, it decreases the least. The average air temperature decreases by 0.327 °C at an 80% piloti ratio compared to the temperature at a 0% piloti ratio, and the box plot shows that the upper limit of the air temperature in the area also decreases as the piloti ratio increases.

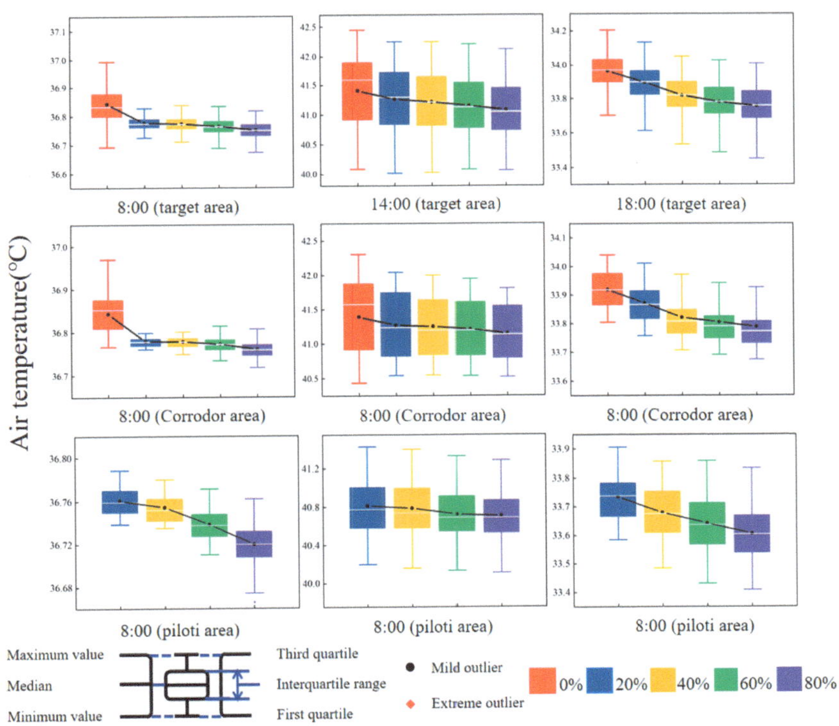

Figure 6. Average air temperature at 1.5 m height in the study area (target area, aisle area, piloti area) for buildings with different piloti ratios (0–80%) at different times of the day: morning (8:00), midday (14:00), nightfall (18:00).

In the aisle area, the increase in piloti ratio also has an effect of decreasing the average air temperature at the pedestrian level. The most significant decrease in temperature is observed from a 0% to 20% piloti ratio, which indicates that the increase in piloti ratio leads to a decrease in the total amount of solar radiation reflected by the buildings into the aisle area, thus causing a certain decrease in the air temperature in the aisle area. The maximum temperature drop still occurs at 14:00. The average air temperatures in the aisle area decrease by 0.126 °C, 0.14 °C, 0.181 °C, and 0.244 °C for the 20% to 80% piloti ratios compared to the no piloti case. The smallest temperature drop is at 18:00.

In the piloti area, the increase in the piloti ratio has a limited effect on the air temperature reduction in the piloti region. Different from the other two regions, the cooling in the piloti area is greater at 8:00 than at 14:00. Because the sun's angle is lower in the morning, and buildings with a low piloti ratio are exposed to direct sunlight, while buildings with a high piloti ratio are not exposed to direct sunlight in some areas. At 14:00, since the sun's angle is higher, the amount of radiation received by piloti areas with different overhead ratios is the same. At 14:00, the average air temperature drops in the piloti area of the 20% piloti ratio case are only 0.021 °C, 0.086 °C, and 0.104 °C when compared with the 40–80% piloti ratio cases.

An increase in piloti ratio has a cooling effect on the overall air temperature at the pedestrian level in the target area. This cooling effect is mainly achieved through the increase in the shading area caused by the increase in the piloti ratio, thus lowering the overall average air temperature. In the aisle area, although a piloti does not provide shading, it can still reduce the average air temperature to a certain extent. For the piloti area, the increase in the piloti ratio does not have a significant effect on the average air temperature, which only decreases slightly.

3.1.2. Wind Speed and Comfort Wind Ratio

The average wind speeds in the target area of buildings with different piloti ratios are shown in Figure 7, where the average wind speeds at 8:00 and 18:00 are increasing with the increase in piloti ratio, while the average wind speed at 14:00 is decreasing with the increase in piloti ratio. The average wind speed in the target area is very low owing to the obstruction provided by external buildings, so wind speed variation caused by the change in the piloti ratio is not observed. At 14:00, when the variation in wind speed is at its maximum, the wind speeds of the buildings with 20–80% piloti ratios increase by 0.129 m/s, 0.147 m/s, 0.180 m/s, and 0.220 m/s, respectively, compared to buildings without pilotis, which has a very limited effect on the comfort of the outdoor thermal environment.

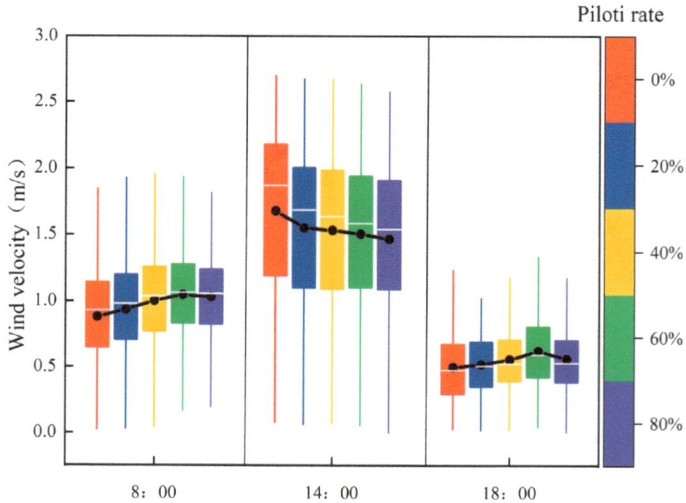

Figure 7. Wind speeds in the target area at 8:00, 14:00, and 18:00.(Different piloti ratio model)

A histogram of the proportion of comfortable wind points at the pedestrian level for different overhead rates in the target area is shown in Figure 8. In the target area, the proportion of comfortable wind increases with the increase in piloti ratio. This indicates that an increase in piloti ratio can increase the proportion of comfortable wind in the target area. When the piloti ratio is from 0% to 60%, the proportion of comfortable wind increases faster, but when the piloti ratio is higher than 60%, it increases slowly, and it starts to decrease at 18:00.

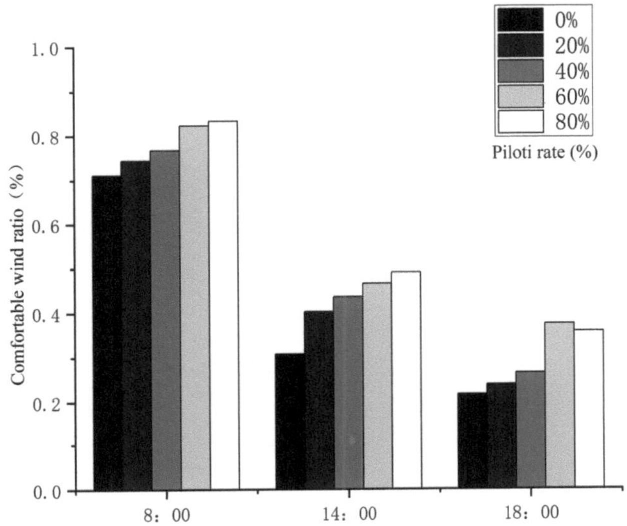

Figure 8. Schematic representation of the proportions of comfortable winds at 8:00, 14:00, and 18:00 for buildings with different piloti ratios in the target area.

The increase in piloti ratio does not have a significant effect on the variation in pedestrian average wind speed in the target area, due to changes in wind direction. Moreover, the increase in the piloti ratio improves the proportion of comfortable wind in the target area, and its influence is reflected most obviously when piloti ratio is from 0% to 60%. When the piloti ratio is between 60% and 80%, an increase in the piloti ratio can bring a decrease in the proportion of comfortable winds. Therefore, the wind environment and wind comfort in the target area can be improved when the piloti ratio does not exceed 60%.

3.1.3. PET

Box plots of PET at three time points for buildings with different greenness rates are shown in Figure 9. The decrease in PET is greater compared to the air temperature because the PET is significantly correlated with the mean radiant temperature, which decreases significantly. The average radiant temperatures for the three regions at 14:00 are shown in Figure 10. At 8:00, the PET decreases by 0.424 °C, 0.039 °C, and 0.357 °C per a 20% increase in piloti ratio for the target, aisle, and piloti area, respectively. The aisle area between buildings is less affected by the piloti setting, so the PET in the aisle area decreases less in comparison.

Like air temperature, the largest PET drop also occurs at 14:00, when for every 20% increase in piloti ratio, the PET decreases by 0.546 °C, 0.029 °C, and 0.946 °C in the target, aisle, and piloti areas, respectively. At 18:00, the average PET of the target region as well as the aisle region change very little. Moreover, the average PET in the aisle area increases as the piloti ratio increases from 0% to 60%. In the piloti area, the average PET decreases by about 0.431 °C for each 20% increase in the piloti ratio.

It can be concluded by analyzing the three study areas of the building model with different piloti ratios that an increase in the piloti ratio reduces the average PET in the three regions, resulting in better pedestrian thermal comfort. In the target and piloti areas, a decrease in piloti ratio can reduce the lower limit of PET, while in the aisle area, it can raise that lower limit.

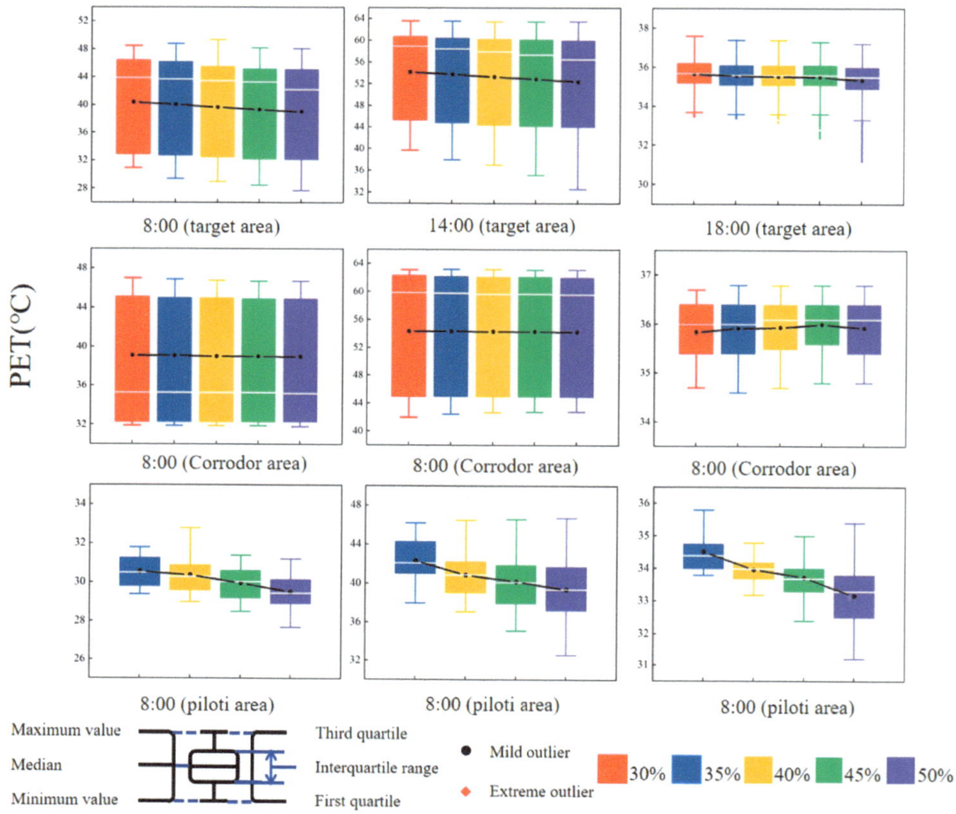

Figure 9. The 1.5 m height PET of the study area (target area, aisle area, piloti area) at different times of the day for buildings with different piloti ratios (0–80%): morning (8:00), midday (14:00), nightfall (18:00).

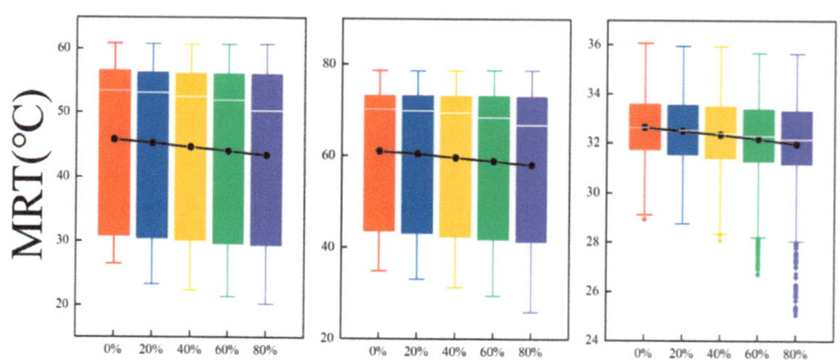

Figure 10. MRTs of target areas for buildings with different piloti ratios (8:00, 14:00, 18:00).

3.2. Analysis of the Impact of Piloti Height on the Outdoor Thermal Environment of Piloti Residential Areas

3.2.1. Air Temperature

The air temperatures (Ta) in the target areas of buildings with different piloti heights at 8:00, 14:00k, and 18:00 are shown in Figure 11. At 8:00, the overall temperature of the target area has a decreasing trend to an extent as the piloti height increases from 2 m to

4 m. At 14:00 and 18:00, the average temperature of the target area increases gradually as the piloti height increases. At 14:00, when the height of the piloti rises to 6 m, the average temperature of the pedestrian level rises by 0.231 °C when compared with that when the piloti height is 2 m. The main reason for the increasing temperature is that the elevation of the piloti height increases solar radiation. Accordingly, the average MRTs of the target areas of the buildings with piloti heights of 3–6 m rise by 0.062, 0.373, 0.546, and 0.677 °C at 14:00 compared to that for buildings of 2 m, respectively.

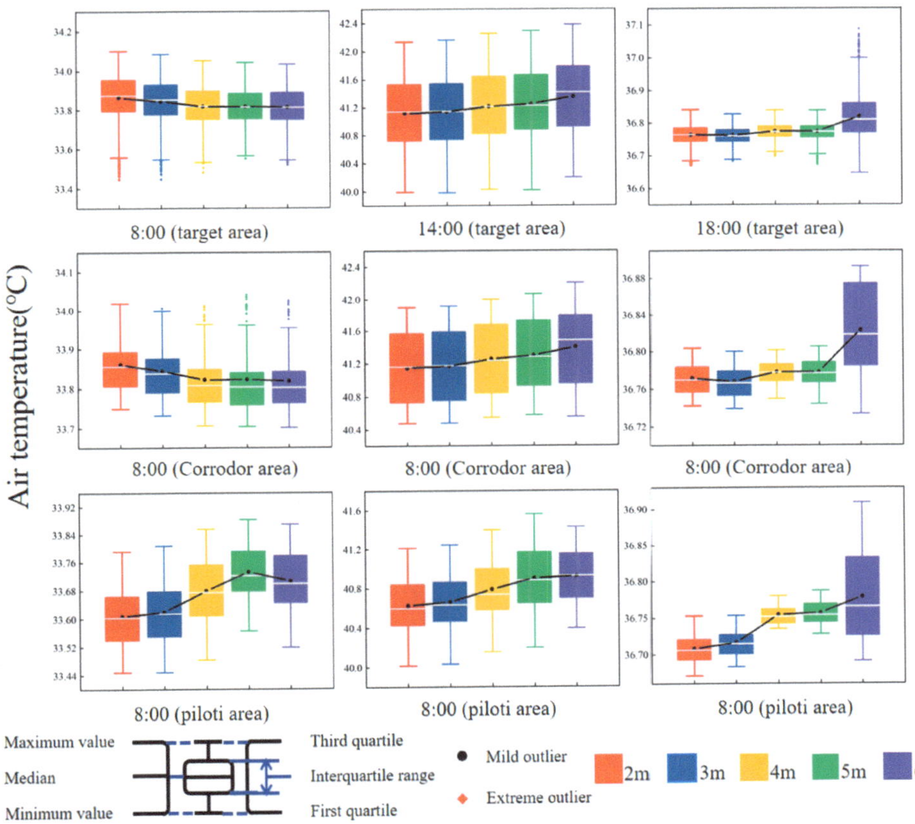

Figure 11. Average air temperature at 1.5 m height in the study area (target area, aisle area, piloti area) for buildings with different piloti heights (2–6 m) at different times of the day: morning (8:00), midday (14:00), nightfall (18:00).

In the aisle area, the trend and magnitude of change in the average pedestrian-level temperature are consistent with those in the target area. The average T_{mrt} in the aisle region is essentially unchanged at all three time points, indicating that the increase in piloti height does not affect the average radiant temperature in the aisle region.

In the piloti area, the average air temperature at all three time points increases with the increase in piloti height, and the temperature has the largest rise. Furthermore, the increasing piloti height has a greater effect on the average radiant temperature in the piloti area. At 14:00, the average radiant temperatures in the piloti areas of buildings with piloti heights of 2–6 m are 30.772 °C, 32.336 °C, 37.29 °C, 41.278 °C, and 42.761 °C, respectively. The increase in the average radiant temperature is fairly significant: compared to a piloti height of 2 m, the average radiant temperature of buildings with a height of 6 m increases by 4 °C.

3.2.2. Wind Speed and Comfort Air Ratio

The wind speed simulation results in Figure 12 show that the average wind speed at the pedestrian level at 8:00 and 18:00 becomes larger as the piloti height increases. The change in wind speed at 8:00 is the largest, and the average wind speed essentially shows a strong linear relationship with the height of the piloti area. For every 1 m increase in the piloti area, the wind speed changes by about 0.03 m/s. At 14:00 and 18:00, the average wind speeds in the target area of buildings with a piloti height of 6 m are lower than those with buildings at 5 m. The average wind speed in the target area of the buildings with a height of 6 m is higher than that with buildings at 5 m. As shown in Figure 13, the presence of a reverse wind region in the target area leads to a strange phenomenon where the average wind speed at a piloti height of 6 m is lower than that at a piloti height of 5 m, because the reverse wind return reduces the average wind speed in the target area. The proportions of average comfortable wind for different piloti heights in the target area are shown in Figure 14. When the piloti height changes from 2 m to 5 m, the proportions of comfort wind at all three time points increase to an extent. Compared to the change in the proportion of comfortable wind brought about by changes in the piloti ratio, the change in the proportion of comfortable wind resulting from increasing the piloti height is smaller. For instance, at 8:00 in the morning, although there is a slight increase in the average wind speed, the proportion of comfortable breeze remains largely unchanged. This is because the wind speeds at 8:00 in the morning fall within the comfortable breeze range (0.7–1.7 m/s). Similar to the change in average wind speed, the proportion of comfortable wind speed decreases when the piloti height increases to 6 m from 5 m.

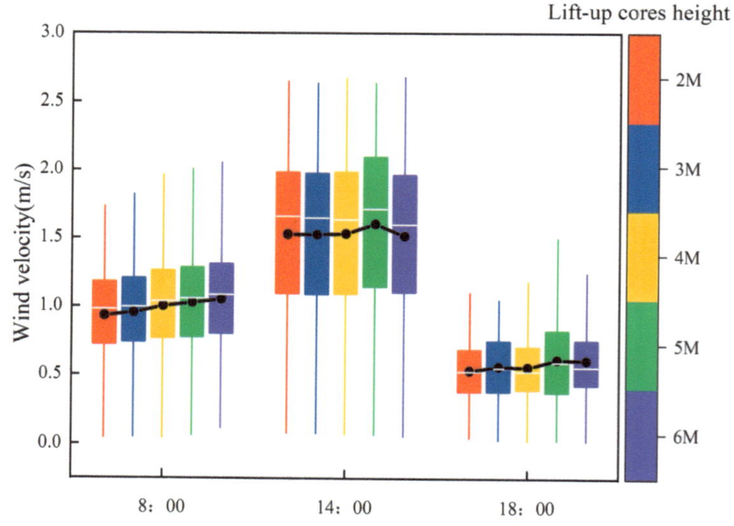

Figure 12. Wind speeds in the target area at 8:00, 14:00, and 18:00 (Different piloti height model).

From the changes in average wind speed and comfortable wind ratio, it can be found that increasing the piloti height has a certain optimization effect on the overall wind environment of the target area, because the elevation reduces the obstruction of the air in the process of circulation. However, there is a certain threshold. When the piloti height is increased to 6 m, the average wind speed and the proportion of comfortable wind in the target area can decrease to a certain extent, worsening the wind environment.

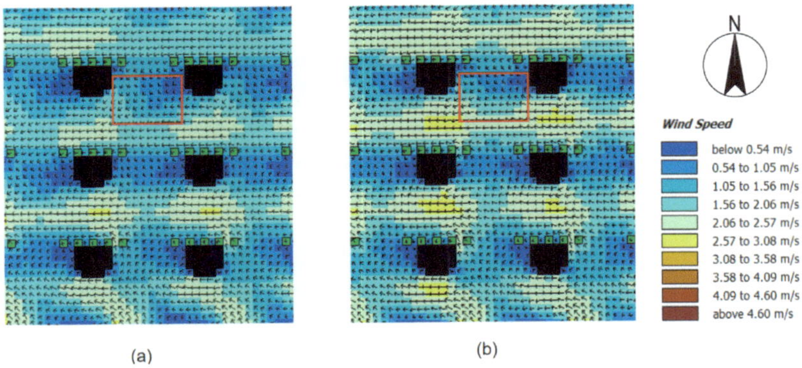

Figure 13. Vector diagram of wind speed, where (**a**,**b**) are building models with piloti heights of 6 m and 5 m, respectively.

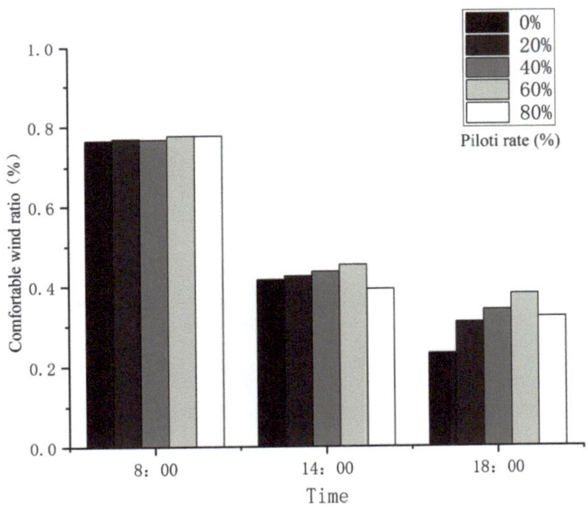

Figure 14. Schematic representation of the proportion of comfortable winds at 8:00, 14:00, and 18:00 for buildings with different piloti heights in the target area.

3.2.3. PET

The simulation results of PET for buildings with different piloti heights are shown in Figure 15. In the target and aisle areas, the changes in PET with changes in the piloti height are small, both within 0.3 °C. At 8:00, the increase in piloti height decreases the average PET in the target and aisle areas, while the PET increases at 14:00 and 18:00. Meanwhile, an increase in the piloti height significantly increases the lower limit of PET in the target area. At 14:00, this lower limit in the target area of buildings with a piloti height of 6 m rises by almost 5 °C compared to the buildings with a piloti height of 2 m.

In the piloti area, the average PET increases with the rising piloti height, and the difference between the maximum and minimum values of PET increases significantly. At 8:00, 14:00, and 18:00, the average PETs rise by 0.46 °C, 0.97 °C, and 0.54 °C for every 1 m increase in the piloti height, respectively. Meanwhile, due to the rising piloti height, the outermost PETs of the overhead region become high, which are shown as abnormal values in the box plot.

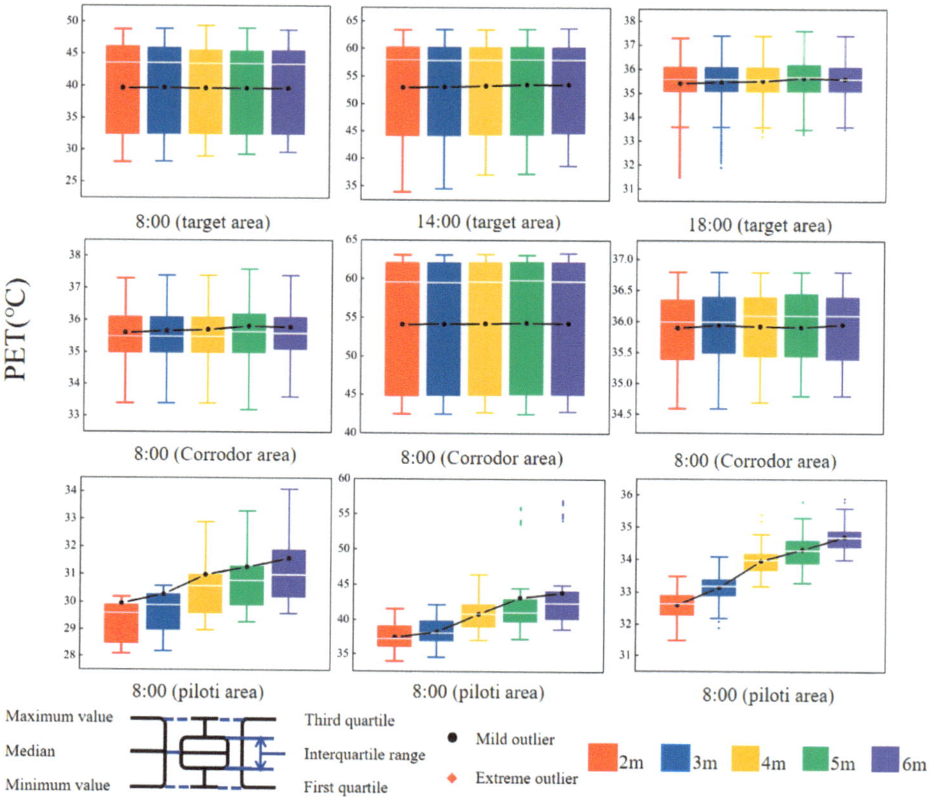

Figure 15. The 1.5 m height PET of the study area (target area, aisle area, piloti area) for buildings with different piloti heights (2 m–6 m) at different times of the day: morning (8:00), midday (14:00), nightfall (18:00).

Therefore, it can be concluded that increasing the piloti height not only increases the average PET value in the target area but also increases the lower PET limit more significantly. In summer, pedestrians tend to walk where there is greater thermal comfort, so a significant increase in the lower PET limit can lead pedestrians to feel hotter while walking. In the aisle area, the effect of piloti height on the average PET is not huge. In the piloti area, the increase in piloti height leads to a significant increase in PET at all three time points, creating an unfavorable thermal environment.

3.3. *Analysis of the Impact of Different Greening Rates on the Thermal Environment in Piloti Residential Areas*

3.3.1. Air Temperature

The effects of different greening rates on the air temperature are shown in Figure 16. The physical properties of vegetation such as shading and reflection lower the reception of solar radiation by the surrounding environment. At the same time, the transpiration of vegetation consumes heat and reduces the air temperature in residential areas, so the average air temperature at the pedestrian level in the target area decreases as the greening rate increases. In the target area, every 5% increase in the greening rate reduces the air temperature by 0.032 °C, 0.055 °C, and 0.04 °C at 8:00, 14:00, and 18:00, respectively.

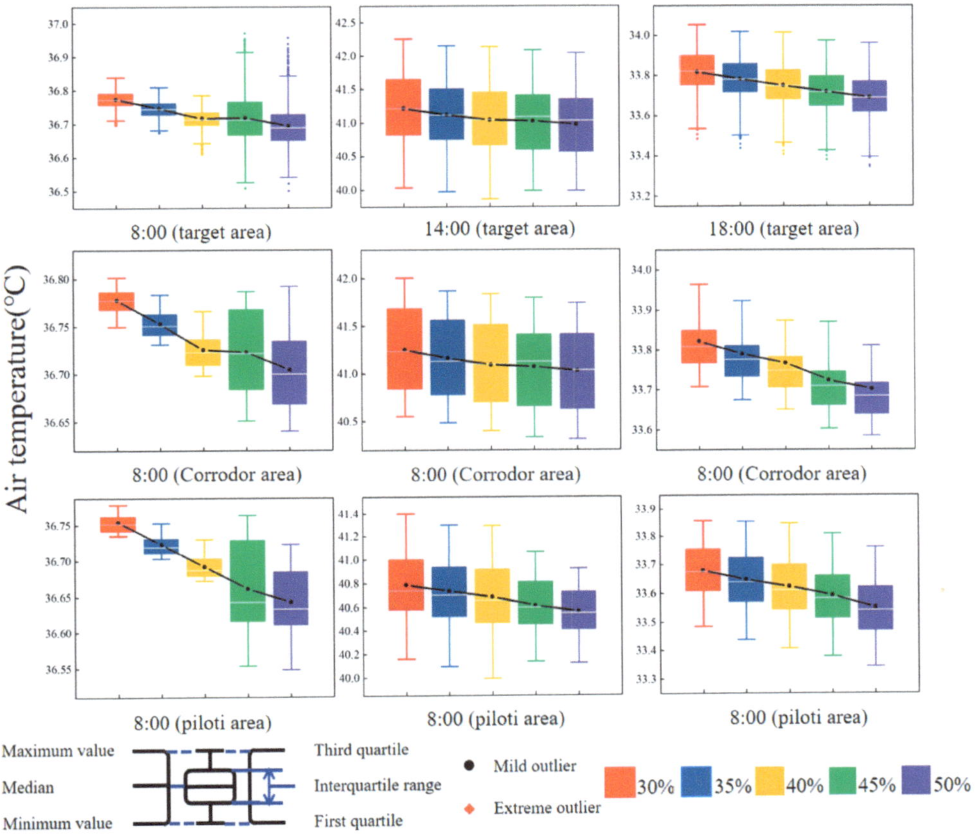

Figure 16. Average air temperature at 1.5 m height in the study area (target area, aisle area, piloti area) at different times of the day for buildings with different greening rates (30–50%): morning (8:00), midday (14:00), and nightfall (18:00).

Like the target area, the average air temperatures at the pedestrian level at 8:00, 14:00, and 18:00 decrease by 0.03 °C, 0.054 °C, and 0.016 °C in the aisle area and by 0.031 °C, 0.057 °C, and 0.0187 °C in the piloti area, respectively, as the greening rate continues to increase. Unlike the temperature changes brought about by the piloti ratio, the temperature changes brought about by the increase in greening rate are essentially the same for the target area and the aisle area. We suggest that the temperature changes from the increasing greening rate are distributed more uniformly, i.e., the average air temperature at the pedestrian level within the target area is reduced on the whole.

3.3.2. Wind Speed and Comfort Wind Ratio

The simulation results for wind speed are shown in Figure 17, from which it can be seen that the greening rate has a mitigating effect on the overall average wind speed at the pedestrian level in the target area. At 8:00, the wind speed and greening rate have a linear relationship, and for every 5% increase in greening rate, the wind speed decreases by 0.0165 m/s. At 14:00, when the greening rate varies from 30% to 40%, the wind speed decreases by 0.082 m/s for every 5% increase in greening rate; meanwhile, when the greening rate varies from 40% to 50%, the wind speed varies within a small range, decreasing by about 0.0236 m/s for every 5% increase in greening rate. At 18:00, the change in wind speed is slight because the wind speed in the target area is very low. The average comfortable wind ratios for the models with different greening rates in the target area are shown in Figure 18. As the greening rate increases, the comfortable wind proportion decreases, but the decline is very limited.

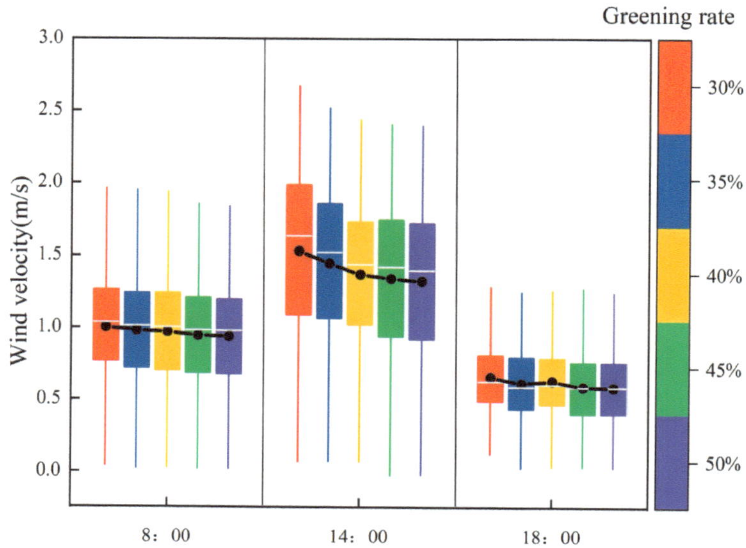

Figure 17. Wind speeds in the target area at 8:00, 14:00, and 18:00 (Different greening rate model).

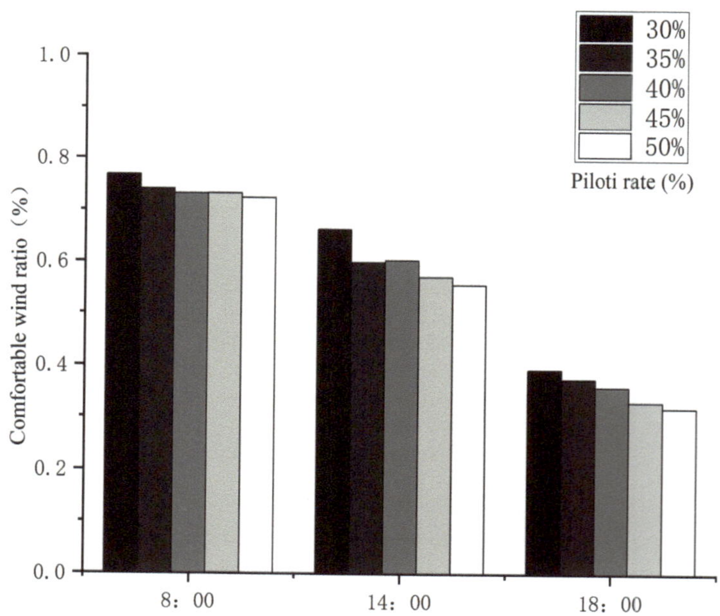

Figure 18. Histogram of the proportion of comfortable winds in the target area.

3.3.3. PET

PET box plots of buildings with different greening rates at the three time points are shown in Figure 19. An overall trend of PET decreasing with increasing greenness is shown.

At 8:00, for every 5% increase in greening rate, the PETs in the target area, aisle area, and piloti area decrease by 0.271 °C, 0.071 °C, and 0.069 °C, respectively. The average PET of the pedestrian layer in the target area demonstrates a clear change because the trees can provide shade and thus the PET of the shaded area decreases significantly. This temperature difference can reach more than 10 °C, thus lowering the average PET of the target area. Both the aisle area and the piloti area have a much slighter change in PET due to the lack of shade provided by the trees; at the same time, the upper limit of the PET decreases with the increase in the greening rate.

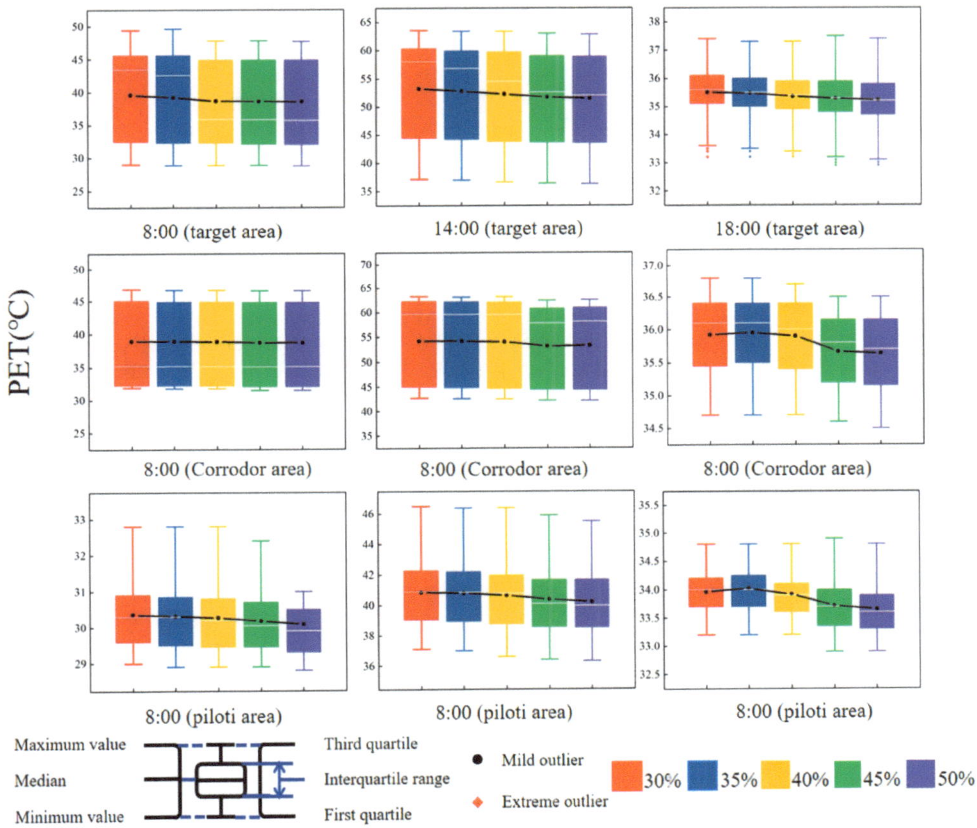

Figure 19. Average PET at 1.5 m height of the study area (target area, aisle area, piloti area) for buildings with different greening rates (30–50%) at different times of the day: morning (8:00), midday (14:00), and nightfall (18:00).

At 14:00, for every 5% increase in the greening rate, the PETs in the target and piloti areas decrease by 0.464 °C and 0.177 °C, respectively. In contrast, there is an abrupt change in the decrease in average PET in the aisle area: when the greening rate increases from 40% to 45%, this decreases by 0.9 °C; when the greening rate increases from 30% to 40%, every 5% increase in greening rate can only decrease the average PET in the aisle area by about 0.1 °C. At 18:00, there is very little change in the PET due to the decreases in the average radiant temperature and the air temperature. As the greening rate increases, for each 5%, the PETs decrease in the target, aisle, and piloti areas by 0.077 °C, 0.087 °C, and 0.094 °C, respectively.

When analyzing the PET distributions in the three study areas of the building models with different greening rates, it can be concluded that an increase in the greening rate can reduce the average value of PET at all three time points in the three study areas, and it can reduce the upper and lower bounds of the PET to improve the thermal environment. Meanwhile, the optimization of the thermal environment through the increase in the greening rate is relatively even, and the trend is largely the same in each area. This indicates that an increase in the greening rate can optimize the thermal environment in each area. Even though the aisle area is not shaded by trees and shrubs, an increase in the greening rate can still reduce the average PET.

4. Discussion

This study generated the ideal model of a piloti residential area, which is based on the piloti ratio, piloti height, and greening rate of typical residential areas in Wuhan. Numerical simulation was adopted to investigate building models with different piloti ratios, piloti heights, and greening rates. The results showed that piloti design can support passive climate adaptability in Wuhan, a city with a climate that is hot summer and cold winter in china. In this study, after taking the area's underlay and greening factors into account, the relationships between the summer outdoor thermal environment and factors including piloti design factors and the greening rate in Wuhan were investigated by applying Envi-met software. The results provide a certain reference for other hot and humid areas with similar climates to Wuhan in summer with a certain degree of adaptability.

In this study, after conducting research on buildings with different piloti heights and piloti ratios, it was found that the combined wind comfort and thermal comfort of piloti areas was great when the piloti ratio was about 60% and the piloti height was about 5 m. Since there is a lack of design codes for piloti Residential Quarters in China, this study conducted a survey on the current architectural status of piloti Residential Quarters in Wuhan. From this, we found that the piloti height of piloti residential areas is generally around 5 m, but the piloti ratio is around 40%. Therefore, the findings presented in this paper offer a certain reference for the formulation of improved norms in the piloti residential area. And the passive piloti architectural design can improve the thermal environment of the residential area, which can reduce energy consumption. However, there are some limitations to the findings presented in this article. Firstly, due to our limited computational resources, the span of the piloti ratio varies greatly among building models with different piloti ratios. In the future, by reducing the division of the piloti ratio from 20% to 10% and the piloti height from 1 m to 0.5 m, more detailed results can be obtained. Secondly, the design of the building layout was fairly homogeneous in this case, with only rows and columns of buildings considered. Thirdly, for the greening of the residential area, we only took basic vegetation types into consideration such as shrubs, lawns, and trees.

In order to further investigate different design factors' effects on the thermal environment of the piloti residential area, research in the future should incorporate more types and arrangement forms of vegetation and architectural layouts.

5. Conclusions

The purpose of this study was to investigate the thermal environmental improvement effects of various design factors in piloti residential areas. We generated a basic model of Wuhan's piloti residential quarters following research, and we considered three design variables as affecting piloti buildings in the current situation in Wuhan (piloti ratio, piloti height, and greening rate). In total, three groups with 15 sets of models were set up. The average air temperature, PET, T_{mrt}, average wind speed, and wind comfort ratios of the target area, aisle area, and piloti area were compared in a comprehensive analysis of the outdoor thermal environment of the buildings, and the following conclusions could be drawn:

1. Improving the piloti ratio and overall greening rate of residential areas can effectively reduce the air temperature, as we found in the study area. Every 5% increase in piloti ratio and 5% increase in greening rate can decrease the air temperature by 0.076 °C and 0.058 °C at the hottest time of the day, respectively. However, an increase in pilotis

reduces the air temperature in the study area only when it is the morning (8:00) and when the piloti height increases from 2 to 4 m, while it leads to an increase in air temperature at other time points.

2. During the daytime, an increase in the piloti ratio caused the most significant enhancement of the average PET value in the pedestrian activity area. When compared with buildings with a 20% piloti ratio, the average PET for an 80% piloti ratio decreased by 2.926 °C. Meanwhile, enhancement of the greening rate was a little less effective than changing the piloti ratio. Beyond this, an increase in piloti height led to the piloti area receiving more solar radiation and surrounding thermal radiation, which greatly affected its thermal comfort. The average PETs of the piloti area rose by 0.46 °C, 0.97 °C, and 0.54 °C for every 1 m increase in the piloti height at 8:00, 14:00, and 18:00, respectively.
3. An increase in the piloti ratio can improve the average wind speed, but the overall effect is small and can be reversed by a change in the wind direction. An increase in the greening rate reduces the average wind speed and the proportion of comfortable wind. Meanwhile, increasing the piloti ratio and piloti height can increase the proportion of comfortable wind within a certain limit, but when the piloti rate exceeds 60% and the piloti height exceeds 5 m, the proportion of comfortable wind will be reduced.
4. If only the piloti height is increased without providing corresponding shading measures, the average radiation temperature of the buildings in the piloti area can rise quickly. Therefore, when designing high pilotis, designers should arrange green plants to create a shading effect. These have little impact on the wind speed or the wind environment, while they significantly improve the thermal environment.

The results of this study offer a reference for the planning and design of urban residential areas with hot and wet summers. Scientific and reasonable setting of piloti design variables and the greening rate can facilitate the creation of an improved urban outdoor thermal environment and improve the thermal comfort of pedestrians.

Author Contributions: Conceptualization, Q.D. and J.L.; methodology, Q.D.; software, X.S.; resources, X.S. and J.L.; data curation, J.L.; writing—original draft preparation, J.L.; writing—review and editing, J.L. and Q.D.; visualization, Q.D. All authors have read and agreed to the published version of the manuscript.

Funding: Central-Southern China Engineering Consulting and Design Group Co., Ltd.: 2023-QT85-731.

Data Availability Statement: The data presented in this study are available on request from the corresponding author.

Conflicts of Interest: The authors declare no conflicts of interest.

Appendix A

Table A1. Terms and abbreviations used in this paper.

Abbreviations	Full Name
PET	Physiological equivalent temperature
T_{mrt}	Mean radiant temperature
Ta	Air temperature
RH	Relative humidity
RMSE	Root mean square error
MAE	Mean absolute error
MAPE	Mean absolute percentage error
d	Willmott's index of agreement

Appendix B

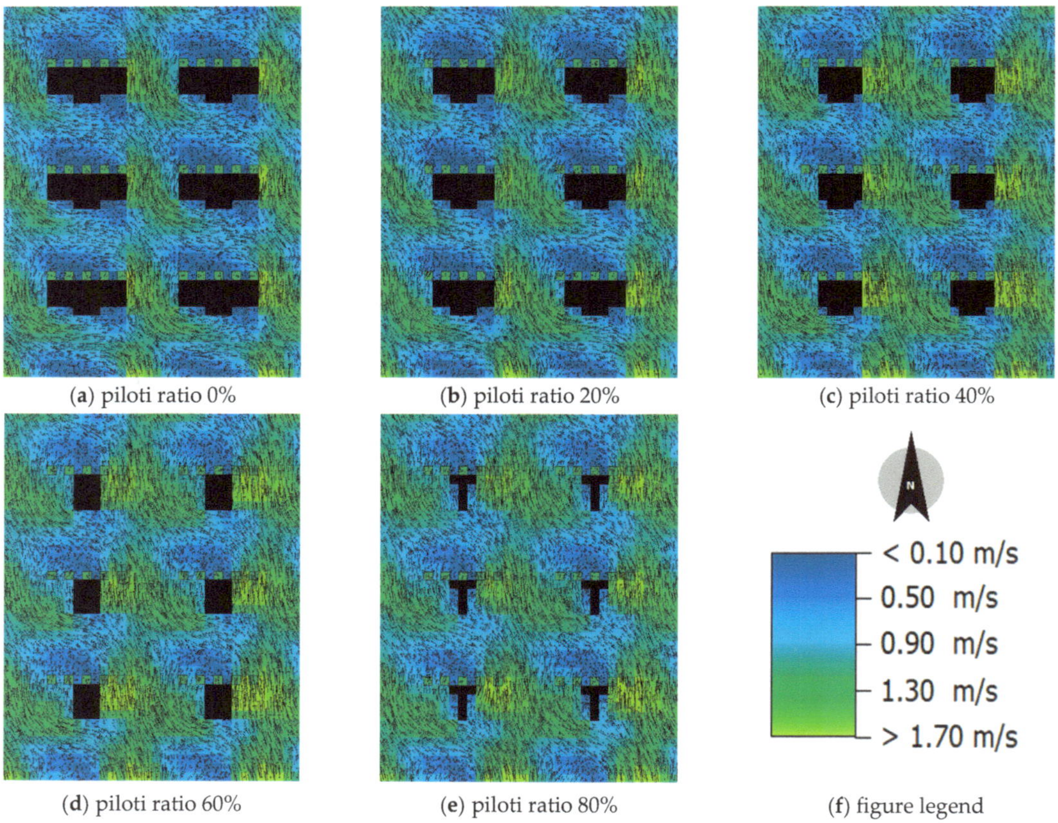

Figure A1. Vector wind speed cloud chart (8:00) of target area (Different piloti ratio model).

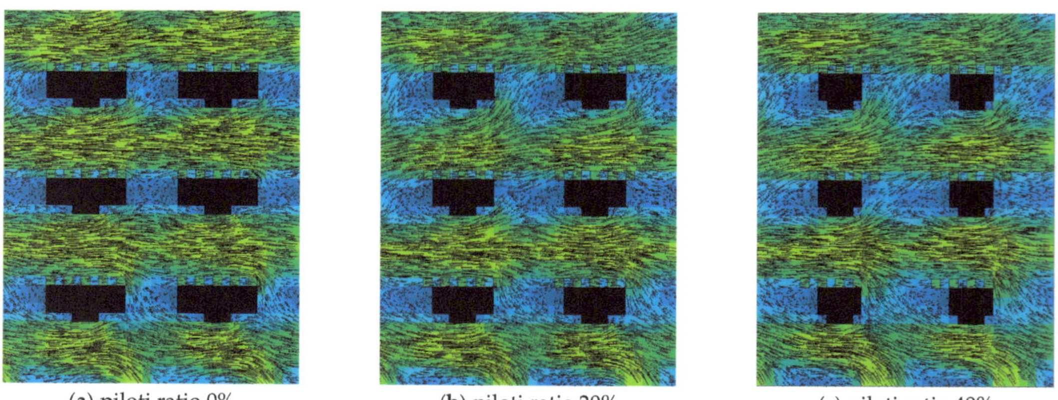

Figure A2. *Cont.*

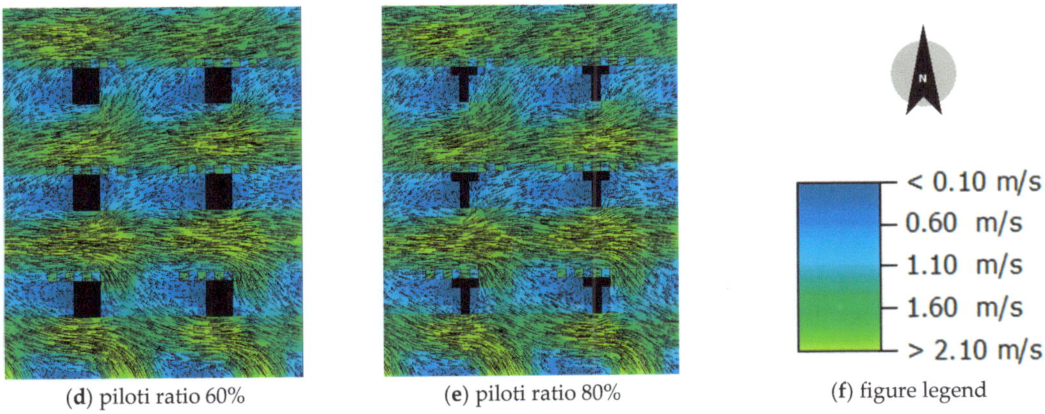

(**d**) piloti ratio 60% (**e**) piloti ratio 80% (**f**) figure legend

Figure A2. Vector wind speed cloud chart (14:00) of target area (Different piloti ratio model).

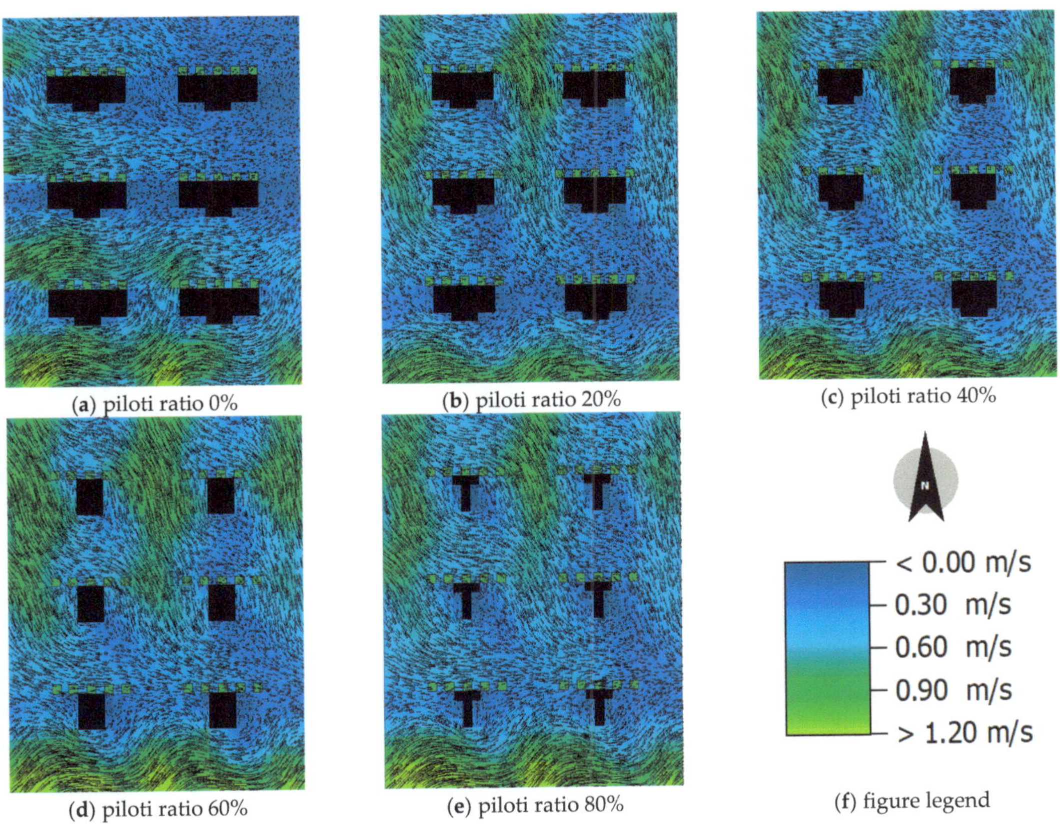

(**a**) piloti ratio 0% (**b**) piloti ratio 20% (**c**) piloti ratio 40%

(**d**) piloti ratio 60% (**e**) piloti ratio 80% (**f**) figure legend

Figure A3. Vector wind speed cloud chart (18:00) of target area (Different piloti ratio model).

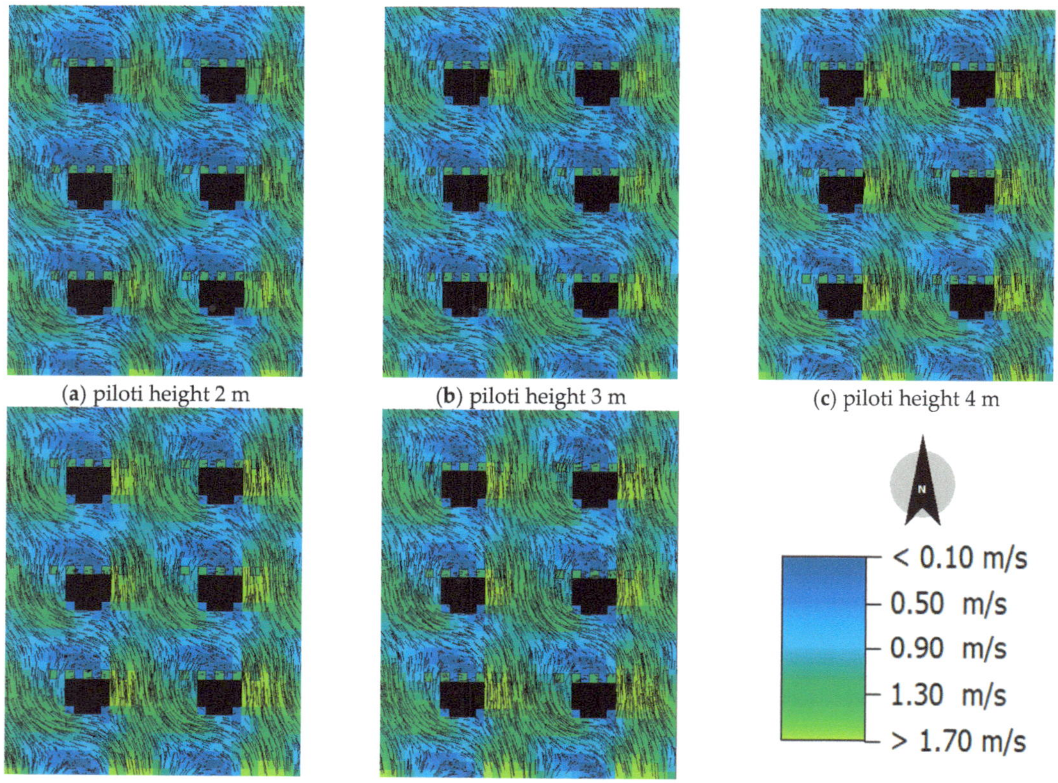

(a) piloti height 2 m (b) piloti height 3 m (c) piloti height 4 m
(d) piloti height 5 m (e) piloti height 6 m (f) figure legend

Figure A4. Vector wind speed cloud chart (8:00) of target area (Different piloti height model).

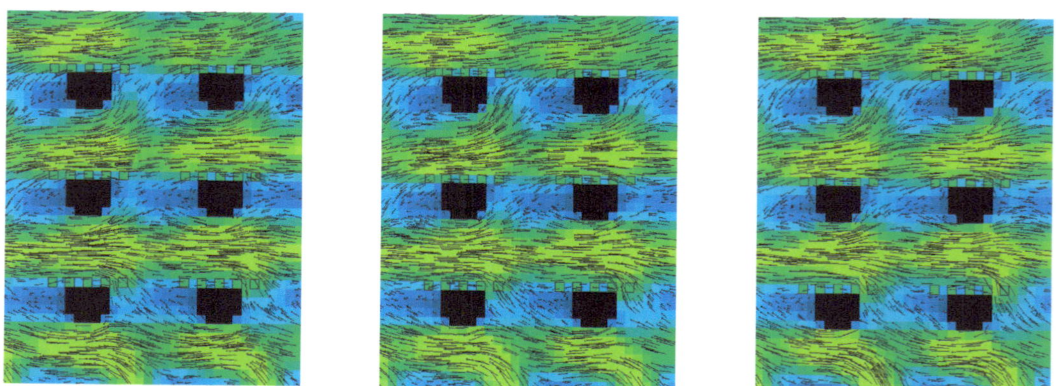

(a) piloti height 2 m (b) piloti height 3 m (c) piloti height 4 m

Figure A5. *Cont.*

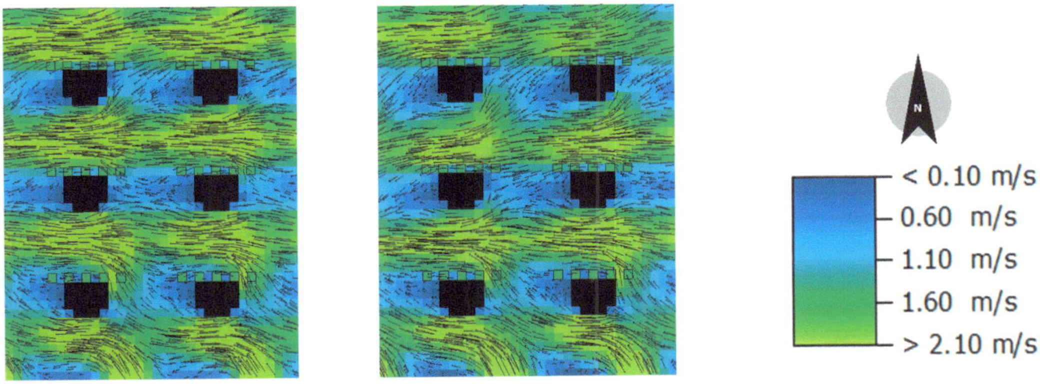

(**d**) piloti height 5 m (**e**) piloti height 6 m (**f**) figure legend

Figure A5. Vector wind speed cloud chart (14:00) of target area (Different piloti height model).

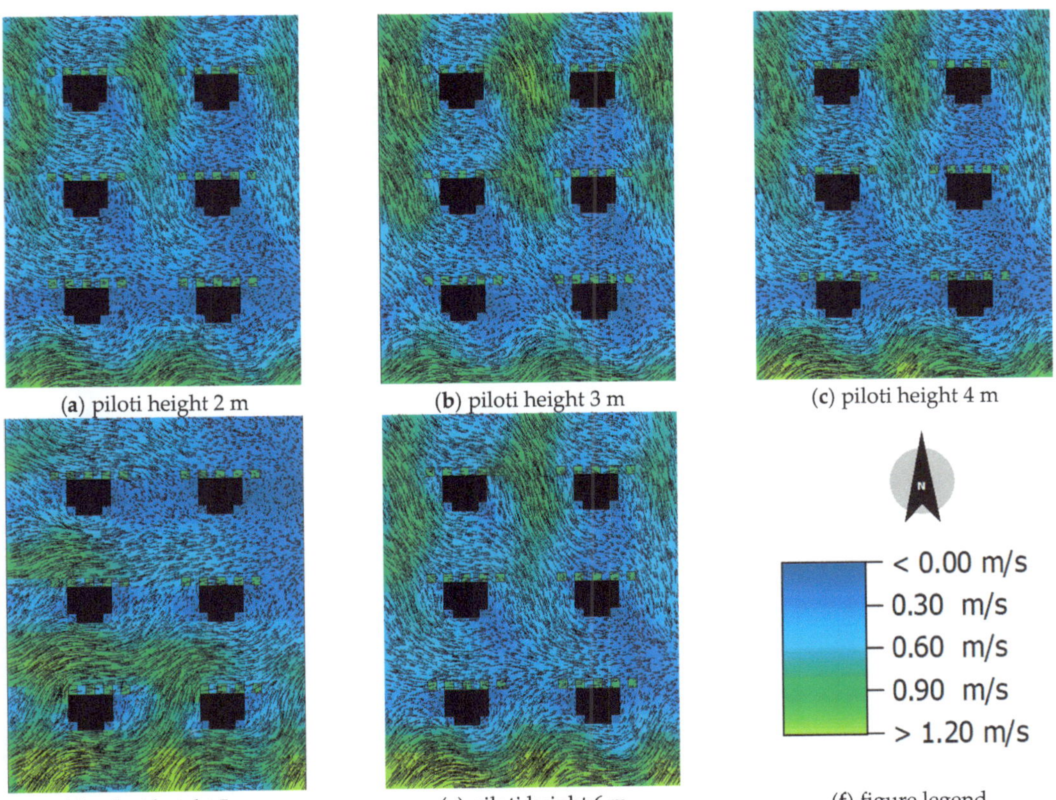

(**a**) piloti height 2 m (**b**) piloti height 3 m (**c**) piloti height 4 m

(**d**) piloti height 5 m (**e**) piloti height 6 m (**f**) figure legend

Figure A6. Vector wind speed cloud chart (18:00) of target area (Different piloti height model).

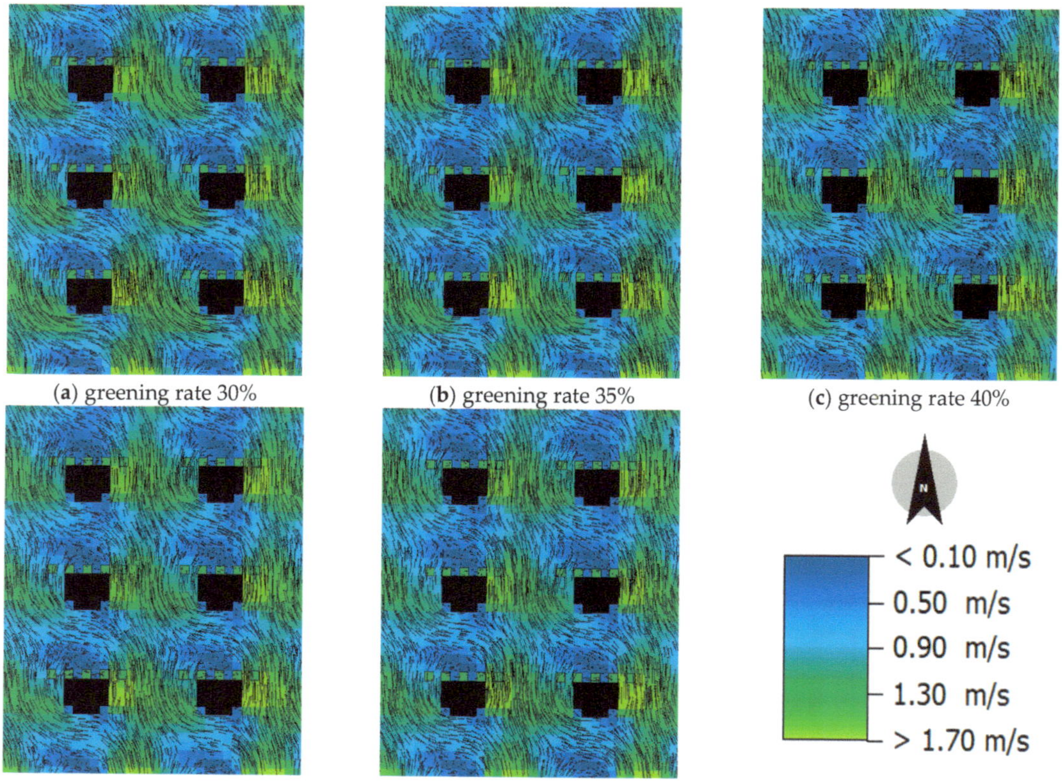

(**a**) greening rate 30% (**b**) greening rate 35% (**c**) greening rate 40%

(**d**) greening rate 45% (**e**) greening rate 50% (**f**) figure legend

Figure A7. Vector wind speed cloud chart (8:00) of target area (Different greening rate model).

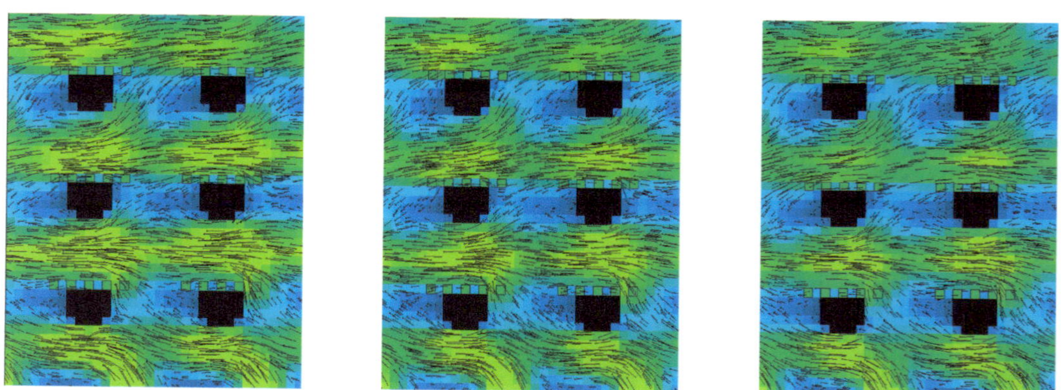

(**a**) greening rate 30% (**b**) greening rate 35% (**c**) greening rate 40%

Figure A8. *Cont.*

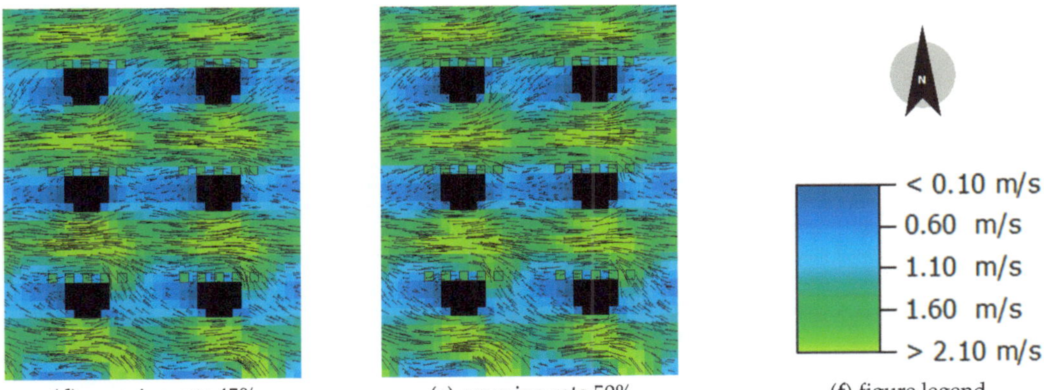

(**d**) greening rate 45% (**e**) greening rate 50% (**f**) figure legend

Figure A8. Vector wind speed cloud chart (14:00) of target area (Different greening rate model).

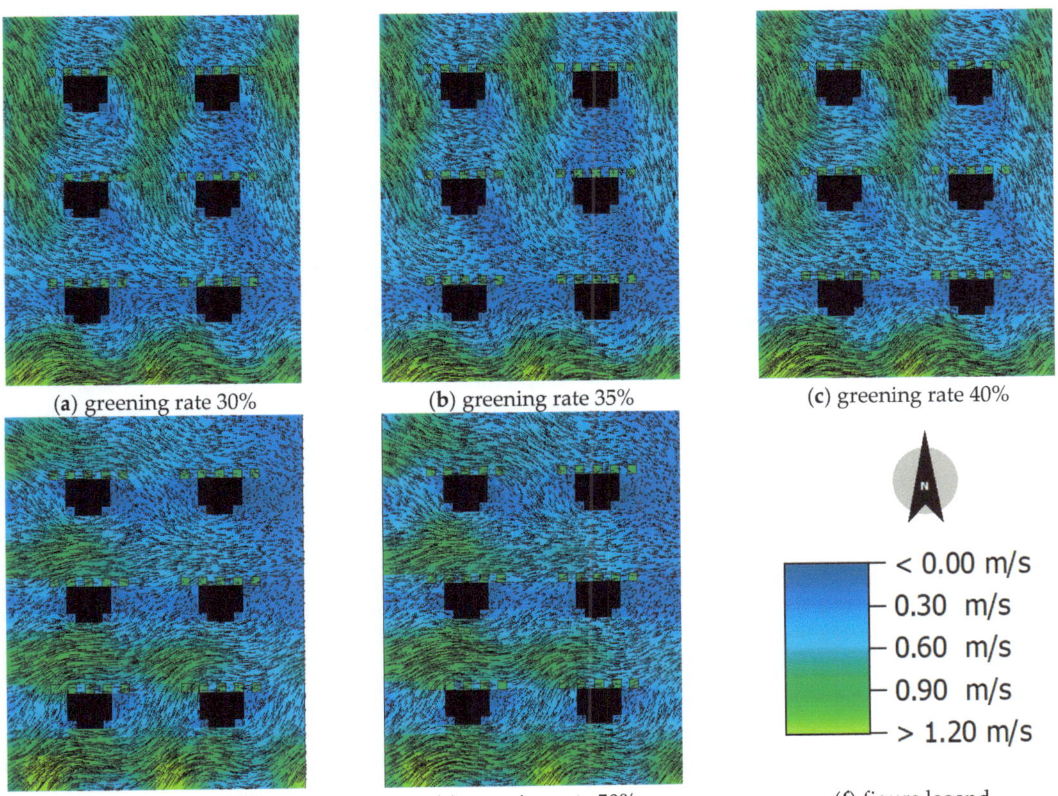

(**a**) greening rate 30% (**b**) greening rate 35% (**c**) greening rate 40%

(**d**) greening rate 45% (**e**) greening rate 50% (**f**) figure legend

Figure A9. Vector wind speed cloud chart (18:00) of target area (Different greening rate model).

References

1. Jin, H.; Liu, Z.; Jin, Y.; Kang, J.; Liu, J. The effects of residential area building layout on outdoor wind environment at the pedestrian level in severe cold regions of China. *Sustainability* **2017**, *9*, 2310. [CrossRef]
2. Yu, Z.; Yang, G.; Zuo, S.; Jørgensen, G.; Koga, M.; Vejre, H. Critical review on the cooling effect of urban blue-green space: A threshold-size perspective. *Urban For. Urban Green.* **2020**, *49*, 126630. [CrossRef]
3. Lin, L.; Gao, T.; Luo, M.; Ge, E.; Yang, Y.; Liu, Z.; Zhao, Y.; Ning, G. Contribution of urbanization to the changes in extreme climate events in urban agglomerations across China. *Sci. Total Environ.* **2020**, *744*, 140264. [CrossRef]
4. Zhou, L.; Dickinson, R.E.; Tian, Y.; Fang, J.; Li, Q.; Kaufmann, R.K.; Tucker, C.J.; Myneni, R.B. Evidence for a significant urbanization effect on climate in China. *Proc. Natl. Acad. Sci. USA* **2004**, *101*, 9540–9544. [CrossRef]
5. Yu, Z.; Chen, T.; Yang, G.; Sun, R.; Xie, W.; Vejre, H. Quantifying seasonal and diurnal contributions of urban landscapes to heat energy dynamics. *Appl. Energy* **2020**, *264*, 114724. [CrossRef]
6. Tuholske, C.; Caylor, K.; Funk, C.; Verdin, A.; Sweeney, S.; Grace, K.; Peterson, P.; Evans, T. Global urban population exposure to extreme heat. *Proc. Natl. Acad. Sci. USA* **2021**, *118*, e2024792118. [CrossRef] [PubMed]
7. Watkins, R.; Palmer, J.; Kolokotroni, M.; Littlefair, P. The balance of the annual heating and cooling demand within the London urban heat island. *Build. Serv. Eng. Res. Technol.* **2002**, *23*, 207–213. [CrossRef]
8. Santamouris, M.; Papanikolaou, N.; Livada, I.; Koronakis, I.; Georgakis, C.; Argiriou, A.; Assimakopoulos, D. On the impact of urban climate on the energy consumption of buildings. *Sol. Energy* **2001**, *70*, 201–216. [CrossRef]
9. Ko, Y. Trees and vegetation for residential energy conservation: A critical review for evidence-based urban greening in North America. *Urban Forest. Urban Green.* **2018**, *34*, 318–335. [CrossRef]
10. Zhu, S.; Yang, Y.; Yan, Y.; Causone, F.; Jin, X.; Zhou, X.; Shi, X. An evidence-based framework for designing urban green infrastructure morphology to reduce urban building energy use in a hot-humid climate. *Build. Environ.* **2022**, *219*, 109181. [CrossRef]
11. Huang, J.; Zhou, C.; Zhuo, Y.; Xu, L.; Jiang, Y. Outdoor thermal environments and activities in open space: An experiment study in humid subtropical climates. *Build. Environ.* **2016**, *103*, 238–249. [CrossRef]
12. Gu, C.; Guan, W.; Liu, H. Chinese urbanization 2050: SD modeling and process simulation. *Sci. China Earth Sci.* **2017**, *60*, 1067–1082. [CrossRef]
13. Du, Y.; Mak, C.M.; Li, Y. Application of a multi-variable optimization method to determine lift-up design for optimum wind comfort. *Build. Environ.* **2018**, *131*, 242–254. [CrossRef]
14. Li, J.; Wang, W.; Jin, H.; Li, Y.; Bu, N. Thermal responses of people exhibiting high metabolic rates when exercising in piloti spaces in hot and humid areas. *J. Build. Eng.* **2022**, *48*, 103930. [CrossRef]
15. Du, Y.; Mak, C.M.; Liu, J.; Xia, Q.; Niu, J.; Kwok, K. Effects of lift-up design on pedestrian level wind comfort in different building configurations under three wind directions. *Build. Environ.* **2017**, *117*, 84–99. [CrossRef]
16. Huang, T.; Niu, J.; Xie, Y.; Li, J.; Mak, C.M. Assessment of "lift-up" design's impact on thermal perceptions in the transition process from indoor to outdoor. *Sustain. Cities Soc.* **2020**, *56*, 102081. [CrossRef]
17. Sha, C.; Wang, X.; Lin, Y.; Fan, Y.; Chen, X.; Hang, J. The impact of urban open space and 'lift-up'building design on building intake fraction and daily pollutant exposure in idealized urban models. *Sci. Total Environ.* **2018**, *633*, 1314–1328. [CrossRef] [PubMed]
18. Tse, K.T.; Zhang, X.; Weerasuriya, A.U.; Li, S.W.; Kwok, K.C.; Mak, C.M.; Niu, J. Adopting 'lift-up'building design to improve the surrounding pedestrian-level wind environment. *Build. Environ.* **2017**, *117*, 154–165. [CrossRef] [PubMed]
19. Chen, L.; Mak, C.M. Integrated impacts of building height and upstream building on pedestrian comfort around ideal lift-up buildings in a weak wind environment. *Build. Environ.* **2021**, *200*, 107963. [CrossRef]
20. Lai, D.; Liu, W.; Gan, T.; Liu, K.; Chen, Q. A review of mitigating strategies to improve the thermal environment and thermal comfort in urban outdoor spaces. *Sci. Total Environ.* **2019**, *661*, 337–353. [CrossRef]
21. Xi, T.; Ding, J.; Jin, H.; Mochida, A. Study on the influence of piloti ratio on thermal comfort of residential blocks by local thermal comfort adaptation survey and CFD simulations. *Energy Procedia* **2017**, *134*, 712–722. [CrossRef]
22. Xi, T.; Li, Q.; Mochida, A.; Meng, Q. Study on the outdoor thermal environment and thermal comfort around campus clusters in subtropical urban areas. *Build. Environ.* **2012**, *52*, 162–170. [CrossRef]
23. Zhou, Z.; Deng, Q.; Yang, G.; Lin, Y. Quantitative study of using piloti for passive climate adaptability in a hot-summer and cold-winter city in China. *Int. J. Environ. Res. Public Health* **2018**, *15*, 2202. [CrossRef] [PubMed]
24. Chen, G.; Zhao, L.H.; Li, Q. Simulation Study on Effect of Piloti on Outdoor Thermal Environment of Campus Cluster in Guangzhou. *Appl. Mech. Mater.* **2013**, *368*, 1888–1893. [CrossRef]
25. Jiao, M.; Zhou, W.; Zheng, Z.; Yan, J.; Wang, J. Optimizing the shade potential of trees by accounting for landscape context. *Sustain. Cities Soc.* **2021**, *70*, 102905. [CrossRef]
26. Wong, N.H.; Tan, C.L.; Kolokotsa, D.D.; Takebayashi, H. Greenery as a mitigation and adaptation strategy to urban heat. *Nat. Rev. Earth Environ.* **2021**, *2*, 166–181. [CrossRef]
27. Yu, Z.; Zhang, J.; Yang, G. How to build a heat network to alleviate surface heat island effect? *Sustain. Cities Soc.* **2021**, *74*, 103135. [CrossRef]
28. Battista, G.; de Lieto Vollaro, R.; Zinzi, M. Assessment of urban overheating mitigation strategies in a square in Rome, Italy. *Sol. Energy* **2019**, *180*, 608–621. [CrossRef]

29. Georgescu, M.; Chow, W.T.; Wang, Z.H.; Brazel, A.; Trapido-Lurie, B.; Roth, M.; Benson-Lira, V. Prioritizing urban sustainability solutions: Coordinated approaches must incorporate scale-dependent built environment induced effects. *Environ. Res. Lett.* **2015**, *10*, 061001. [CrossRef]
30. Liu, H.; Lim, J.Y.; Thet, B.W.H.; Lai, P.Y.; Koh, W.S. Evaluating the impact of tree morphologies and planting densities on outdoor thermal comfort in tropical residential precincts in Singapore. *Build. Environ.* **2022**, *221*, 109268. [CrossRef]
31. Zhou, W.; Huang, G.; Pickett, S.T.; Wang, J.; Cadenasso, M.; McPhearson, T.; Grove, J.M.; Wang, J. Urban tree canopy has greater cooling effects in socially vulnerable communities in the US. *One Earth* **2021**, *4*, 1764–1775. [CrossRef]
32. Yang, Y.; Zhou, D.; Wang, Y.; Meng, X.; Gu, Z.; Xu, D.; Han, X. Planning method of centralized greening in high-rise residential blocks based on improvement of thermal comfort in summer. *Sustain. Cities Soc.* **2022**, *80*, 103802. [CrossRef]
33. Middel, A.; Chhetri, N.; Quay, R. Urban forestry and cool roofs: Assessment of heat mitigation strategies in Phoenix residential neighborhoods. *Urban For. Urban Green.* **2015**, *14*, 178–186. [CrossRef]
34. Lee, H.; Mayer, H.; Chen, L. Contribution of trees and grasslands to the mitigation of human heat stress in a residential district of Freiburg, Southwest Germany. *Landsc. Urban Plan.* **2016**, *148*, 37–50. [CrossRef]
35. Mayer, H.; Höppe, P. Thermal comfort of man in different urban environments. *Theor. Appl. Climatol.* **1987**, *38*, 43–49. [CrossRef]
36. Murakami, S. Criteria for assessing wind-induced discomfort considering temperature effect. *J. Archit. Plann. Environ. Eng. AIJ* **1985**, *358*, 9–17.
37. Li, Q.; Meng, Q.L.; Zhao, L.H.; Yang, X.S.; Shu, L.F. Experimental study on outdoor thermal environment of campus building clusters in tropical climate. In Proceedings of the 7th International Conference on Urban Climate (ICUC-7), Okohama, Japan, June 29–3 July 2009.
38. Zhang, L.; Zhan, Q.; Lan, Y. Effects of the tree distribution and species on outdoor environment conditions in a hot summer and cold winter zone: A case study in Wuhan residential quarters. *Build. Environ.* **2018**, *130*, 27–39. [CrossRef]
39. Willmott, C.J. Some comments on the evaluation of model performance. *Bull. Am. Meteorol. Soc.* **1982**, *63*, 1309–1313. [CrossRef]

buildings

MDPI

Article

Analysis of Influencing Factors on Solid Waste Generation of Public Buildings in Tropical Monsoon Climate Region

Tingwei Meng [1], Xiaofang Shan [1,2], Zhigang Ren [1,2] and Qinli Deng [1,2,*]

[1] School of Civil Engineering and Architecture, Wuhan University of Technology, Wuhan 430070, China; tingweimeng@whut.edu.cn (T.M.); xfshan@whut.edu.cn (X.S.); renzg@whut.edu.cn (Z.R.)
[2] Hainan Institute, Wuhan University of Technology, No. 5 Chuangxin Road, Sanya 572024, China
* Correspondence: deng4213@whut.edu.cn

Abstract: Environmental problems including the depletion of natural resources and energy have drawn a lot of attention from all sectors of society in the context of high-quality global development, and solid waste generated by the construction industry accounts for 36% of the total amount of municipal waste. The generation of large amounts of construction waste not only causes a waste of resources, but also causes great damage to the environment. Reducing the quantity of solid waste produced during a building's new construction period can be greatly aided by construction site solid waste statistics and forecasts. Based on the statistical data of 61 public construction projects in Hainan Province, China, this study uses the Random Forest algorithm to rank the importance of possible factors affecting the amount of solid waste generated, and linearly fits the data to achieve the prediction of solid waste at construction sites. The findings indicate that building area, building height, concrete usage, steel usage and assembly rate are the main factors affecting solid waste in construction sites. In office buildings and exhibition buildings, an increase in ground area, building height, concrete usage, and steel usage increases the generation of each type of solid waste (inorganic non-metallic solid waste, metallic solid waste), with the exception of an increase in concrete usage, which results in a decrease in the generation of metallic solid waste. Furthermore, a higher assembly rate can substantially lower the production of all waste types. These results offer a theoretical foundation for the implementation of assembly construction to support the high-quality development of the construction industry, as well as partial design inspiration for the architectural design stage.

Keywords: construction site; solid waste; random forest; prediction

Citation: Meng, T.; Shan, X.; Ren, Z.; Deng, Q. Analysis of Influencing Factors on Solid Waste Generation of Public Buildings in Tropical Monsoon Climate Region. *Buildings* **2024**, *14*, 513. https://doi.org/10.3390/buildings14020513

Academic Editors: Jan Fořt and Haoxin Li

Received: 22 December 2023
Revised: 9 February 2024
Accepted: 9 February 2024
Published: 13 February 2024

1. Introduction

The building area in China has been steadily increasing during the "13th Five-Year Plan" period, indicating the rapid development of the country's construction industry. By 2020, the country's housing construction area is expected to reach 14.947 billion square meters [1]. China is the largest nation in the field of new construction, and Zhu et al. noted that the country produces a significant amount of construction solid waste, which makes up 30–40% of all municipal waste [2]. China's rapid urbanization over the past few decades has resulted in the generation of a significant amount of solid waste from construction sites [3,4]. As the construction industry becomes more important as the country's pillar industry, the quantity of solid waste produced at construction sites will unavoidably rise along with the number of new buildings that are constructed.

Landfills are used to dispose of a lot of solid waste currently [5], but in most places, the amount of landfill material generated exceeds the landfill's capacity [4]. The random stacking of a large amount of solid waste not only causes a large amount of material waste, but also leads to air and soil pollution, seriously harming the ecosystem [6–8]. Recycling solid garbage can significantly lessen its environmental impact [9]. Nonetheless, compared to the 76% recycling rate in the United States and the average 90% recycling rate in the

28 nations that make up the European Union, China's recycling percentage for construction solid waste is extremely low [10–12]; it is only about 5% [13].

Many academics have previously studied the topic of recycling construction solid waste (hereinafter referred to as solid waste). China and the United States have made exceptional contributions in this area [14] and have investigated a variety of recycling techniques for it. In place of natural aggregate resources, recycled aggregates from construction solid waste are added to cement and concrete [15–19]. Strategies to improve their performance are also investigated. Some scholars use solid waste as recycled aggregates as sub-base materials in road construction [20–22].

Reducing the generation of solid waste from construction sites is the best strategy to manage it for new buildings, yet scholars have paid relatively little attention to this aspect of research [23–25]. A portion of the academics used software simulation and the project construction personnel questionnaire to analyze the current solid waste management conundrum, propose fundamental guidelines for the development and management of solid waste at construction sites, and offer initiatives and suggestions to minimize construction solid waste during the construction phase [24–29]. Other scholars have analyzed different solid waste treatment methods in the whole life cycle from both economic and environmental aspects, and provided suggestions for policy makers and practitioners on how to improve waste management through the analysis of research findings [30]. Duan et al. anticipated a significant increase in the generation of medical waste during the COVID-19 public health crisis, and predicted that the treatment system of this medical waste had serious challenges, so they conducted research to improve its treatment efficiency. Their findings not only reveal the changes brought about by the outbreak, but also provide effective measures to improve the efficiency of medical waste management in emergencies, which can provide guidance for policy makers and practitioners on how to improve management during public health crises [31]. In an effort to reduce the quantity of solid waste that may be produced during building construction from the design stage, several academics have also investigated the influence of various influencing factors on the amount of construction solid waste generated. Many elements, such as floor area, building type, building height, assembly rate, and construction environment influence the quantity of solid waste generated during a building's construction phase [32]. Prefabricated structures have been shown to significantly reduce the generation of solid waste [33,34].

China's Hainan Province has a tropical monsoon climate, which means that there are increased demands for building materials due to the region's high seasonal temperatures, high humidity, and high salinity levels. High-performance materials mean that their cost is higher than that of ordinary building materials, and the solid waste generated at the construction site will lead to more serious economic losses and environmental pollution. Therefore, under such climate conditions, It is very important to study the influencing factors of the generation rate of solid waste at the construction site of public buildings and their action rules for the future development of the construction industry, which can avoid environmental pollution and resource waste caused by construction solid waste from the source. In order to achieve this goal, this paper made accurate data collection of 61 public buildings in Hainan Province in the construction stage, to a certain extent, to make up for the lack of data in the current research. At the same time, the influence of various factors on the production of solid waste in the construction site of new buildings is studied, and the importance of each factor is analyzed by the Random Forest algorithm. According to the results of the importance analysis, the main influencing factors of solid waste production rate were determined, and the correlation analysis was carried out with the measured data to explore the connotation between the changes of influencing factors and the solid waste production rate.

2. Research Objectives and Methods

2.1. Research Objects

China has not provided a precise definition of construction solid waste. Construction waste disposal technical standards will be defined as: engineering residue, engineering slurry, engineering rubbish, demolition waste, decoration waste, etc., including new construction, expansion, alteration and demolition of all kinds of buildings, structures, pipe networks, etc., as well as residents to decorate the process of house decoration of the discarded soil, discarded materials and other waste, excluding the inspection, identification of hazardous waste construction waste [35]. Considering the different contents of the study, this paper defines construction solid waste as follows: the generic term for engineering sludge, engineering mud, engineering rubbish, and decorating garbage, including new construction, expansion, alteration and demolition of all kinds of buildings, structures, pipe networks, etc., as well as residents to decorate the process of house decoration of the discarded soil, discarded materials and other waste, including solid waste tested as hazardous waste. This paper focuses on the solid waste generated at the construction site of the main stage of the building, so the solid waste mentioned below refers to all of the solid waste generated at the construction site of the main stage of the building.

It is necessary to classify the solid waste in order to better analyze the solid waste generated during the construction site of the main stage of the building [7]. The source classification method, the physical composition classification method, the chemical composition classification method, and other methods are currently used to classify solid waste. The classification method of Chen Lei et al. [36] is cited in this paper. It divided construction solid waste into five categories: inorganic non-metallic solid waste (INSW), organic non-metallic solid waste (ONSW), metallic solid waste (MSW), composite solid waste (CSW) and hazardous solid waste (HSW). The specific classification is as follows Table 1.

Table 1. Classification of new construction solid waste.

Sort	Description and Contained Component
INSW	Stands for inorganic non-metallic solid waste, including natural stone, burnt earth products, cement, concrete and silicate products, etc.
ONSW	Stands for inorganic non-metallic solid waste, including waste plastic, waste paint, waste adhesive and other plant materials, synthetic polymer materials and asphalt materials, etc.
MSW	Stands for metallic solid waste, including scrap steel bars, scrap copper pipes and other ferrous and non-ferrous materials, etc.
CSW	Stands for composite solid waste, including lightweight metal sandwich board, gypsum board, etc.
HSW	Stands for hazardous solid waste, including rock wool, asbestos, glass glue, etc.

2.2. Research Boundary

Determining reasonable research boundaries plays a crucial role in determining the accuracy of conclusions [37]. This study focuses on the production of solid waste at construction sites. In order to more precisely quantify the solid waste, this document gives defined parameters for the study scope, such as time boundaries, object boundaries, field pair boundaries, element boundaries, and other requirements [38]. Table 2 shows the methodology used to determine the exact limitations.

Table 2. Research Boundary Delineation Table.

Boundary Name	Boundary Range
Time boundaries	From the beginning of construction to the completion and acceptance of the entire construction project.
Object boundaries	Take the buildings in the project as the research object, only the solid waste generated in the underground, main, decoration and electromechanical stages of the new building is calculated, and the solid waste generated in any other construction stage is not included in the calculation.
Site boundaries	All kinds of building materials start to be calculated when they enter the boundary of the construction site, and solid waste generated by transport and processing outside the site is not calculated.
Element boundaries	Solid waste generated from the daily lives of construction workers is not counted.
Other provisions	Materials that can be reused directly, such as waste soil, are not counted as solid waste, and materials that need to be processed before they can be used, such as waste steel pipes, need to be counted.

2.3. Data Sources

Data collection is a very important part of the whole research. In order to better complete the data collection during the study and ensure the accuracy of the data, data collection was carried out around a certain construction phase of 61 public buildings in Hainan Province, including 29 office buildings and 32 exhibition buildings. The classification outlined in 1.1 is used to statistically weigh the solid trash produced at the construction site. Figure 1: Project Distribution Map displays the projects' distribution.

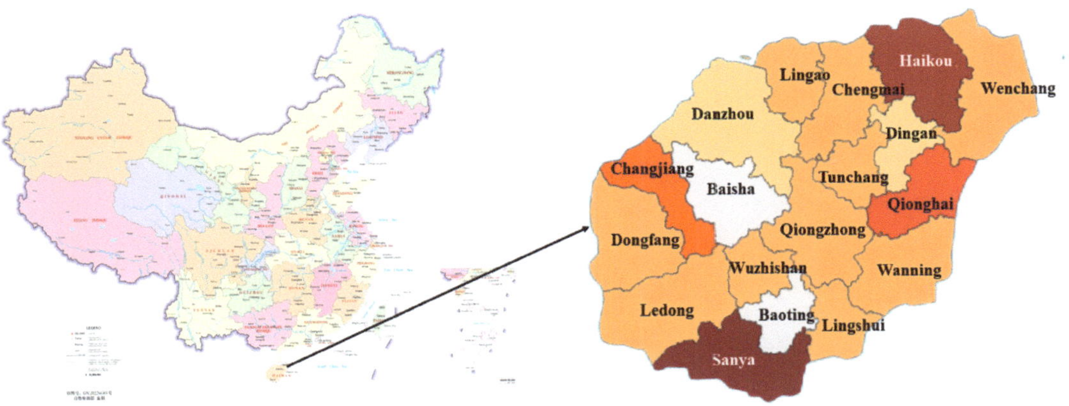

Place	Total	Exhibition Buildings	Office Buildings	Place	Total	Exhibition Buildings	Office Buildings	Place	Total	Exhibition Buildings	Office Buildings
Baisha	1	1	0	Ledong	2	2	0	Wenchang	3	1	2
Danzhou	1	0	1	Wanning	3	2	1	Changjiang	4	2	2
Dingan	1	0	1	Chengmai	3	2	1	Qionghai	6	4	2
Wuzhishan	2	2	0	Lingao	3	1	2	Haikou	9	4	5
Dongfang	2	0	2	Lingshui	3	2	1	Sanya	13	7	6
Tunchang	2	0	2	Qiongzhong	3	2	1				

Figure 1. Project distribution map.

2.4. Random Forest

Random Forest is a classifier containing many decision trees proposed by Leo Breiman, which is a non-parametric machine learning method. Random Forest is an effective method for handling regression and classification issues, and it may also be used to evaluate the relative significance of explanatory factors. Random Forest works effectively, and provides precise variable importance prediction even when there are a lot of features and few data.

The algorithm of Random Forest consists of the following steps: 1. Use sampling method with put-back to draw n samples from the sample set as a training set. 2. Randomly draw M (M is less than the total number of features) features from the sampled samples as inputs to the training decision tree, to find the optimal delineation of features, and to construct the decision tree. 3. Repeat the steps 1–2 for K times, to generate K decision trees that constitute a random forest. 4. Finally, the classification results of the K decision trees are integrated, and the samples are predicted and analyzed. By quantifying each feature's contribution to the classification performance of the K decision tree construction process, Random Forest evaluates the importance of features. The evaluation index used in this general calculation is the out-of-bag error rate. Feature importance measures (FIM) are the measures used to assess the importance of features.

Defining Indicator Functions $I(x, y)$.

$$I(x,y) = \begin{cases} 1, x = y \\ 0, x \neq y \end{cases} \tag{1}$$

The $FIM_{km}^{(OOB)}$ of the mth feature Fm in the kth tree is

$$FIM_{km}^{(OOB)} = \frac{\sum_{p=1}^{n_0^k} I(Y_p, Y_p^k)}{n_0^k} - \frac{\sum_{p=1}^{n_0^k} I(Y_p, Y_{p,\pi_m}^k)}{n_0^k} \tag{2}$$

In this equation: n_0^k is a sample of observations from the kth tree, Y_p is the true classification label corresponding to the Pth sample, and Y_{p,π_m}^k is the classification result of the Kth decision tree for the Pth sample after the ensuing replacement Fm, where the Kth decision tree after the replacement Fm needs to be retrained. When the feature F does not appear in the kth bare tree, $FIM_{km}^{(OOB)} = 0$.

The importance score of feature F in the whole random forest is defined as follows:

$$FIM_m^{(OOB)} = \frac{\sum_{p=1}^{K} FIM_{km}^{(OOB)}}{K\sigma} \tag{3}$$

where K denotes the number of decision trees in the random forest and σ denotes the standard deviation of $FIM_{km}^{(OOB)}$. The importance score $FIM_m^{(OOB)}$ of feature Fm characterizes the contribution of Fm to the correct classification rate.

3. Results

3.1. Influence Factor Screening

Pre-processing the data is necessary to make better use of the sample data. There are instances of missing data in the samples that were gathered for the study. In a statistical sample, a missing value can be filled in using techniques like mean replacement and predictive simulation filling. However, the total amount of measured data in this study is rather small, so adopting this type of data cleaning method will have a greater impact on the subsequent impact analysis; so we choose to delete the statistical sample. Furthermore, the gathered data samples suffer from significant variations in the dataset's characteristic interval range. As a result, the data must be normalized to ensure that the influencing elements with varying scales are measured on the same scale.

Following the preprocessing of the data, the importance of each factor that might have an impact on the rate of solid waste generation during the construction phase was determined. The findings indicated that certain factors, including pile diameter ratio, pit support and structural type, had very little bearing on the rate of solid waste generation. Based on this, Random Forest was used to analyze the major factors influencing the rate of solid waste generation during the main phase and explore the importance of each factor (importance is measured on a scale of 0 to 1, with 1 representing the highest level of importance for a feature). The specific findings are displayed in Figure 2.

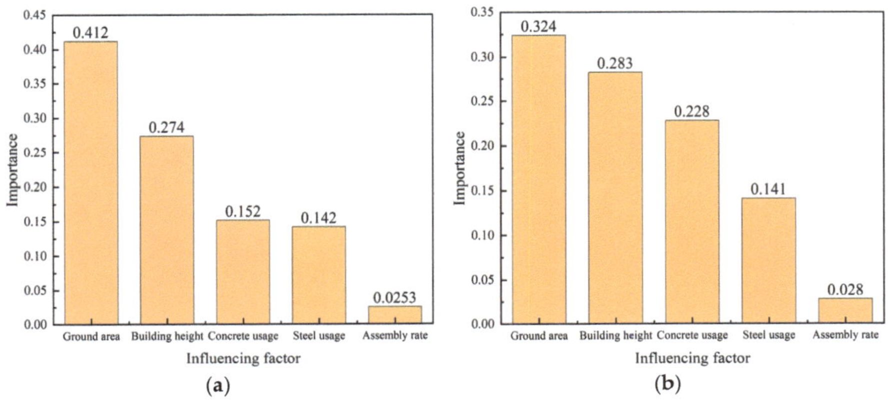

Figure 2. Importance analysis of factors affecting solid waste generation rate at construction sites. (**a**) Office buildings; (**b**) Exhibition buildings.

A number of influencing elements, including ground area, building height, concrete usage, steel usage, and assembly rate, were chosen for the importance analysis of influencing factors. The above figure shows that ground area (0.412) and building height (importance factor: 0.274) have the greatest effects on the construction site solid waste generation rate in the new construction phase of office buildings. There is not much difference in the influence of the amount of concrete usage and steel usage, with an influence factor of 0.152 and 0.142, respectively. With only 0.0253, the assembly rate has the least significant influence. When it comes to exhibition buildings, several influencing factors on the production rate of solid waste on the construction site are, in order of importance, ground area, building height, concrete usage, steel usage and assembly rate, and their influencing factors are 0.324, 0.283, 0.228, 0.141 and 0.028, respectively.

In addition to the assembly rate, the results are basically consistent with the conclusions of Hu et al., that is, that the construction area and building height are the main factors affecting the solid waste generation rate at the construction site; but Hu et al. paid more attention to the impact of the process on the average solid waste generation rate [32]. The construction stage assembly rate for various types of construction solid waste generation rate is not large in the results of the analysis of the importance of the impact factors. This is because the sample data's assembly rate distribution is more centralized, and there are only a few sample data points where the assembly rate is different. As a result, the assembly rate's impact on the analysis of the impact is smaller. Analyzing the gathered samples further demonstrates that in real projects, the solid waste generation rate of buildings with assembled components is significantly lower than that of buildings without assembled components. This has been amply demonstrated by other researchers.

3.2. Influence of Ground Area

The building area had the biggest influence on solid waste at the building construction site, based on the Random Forest analysis results. According to the type of building use, the sample data from the main stage is divided into office buildings and exhibition buildings. The more concentrated portion of the data is then chosen, a scatter plot is created, and it is linearly fitted to yield a fitting curve and the correlation coefficient R^2 value. The results are shown in Figure 3.

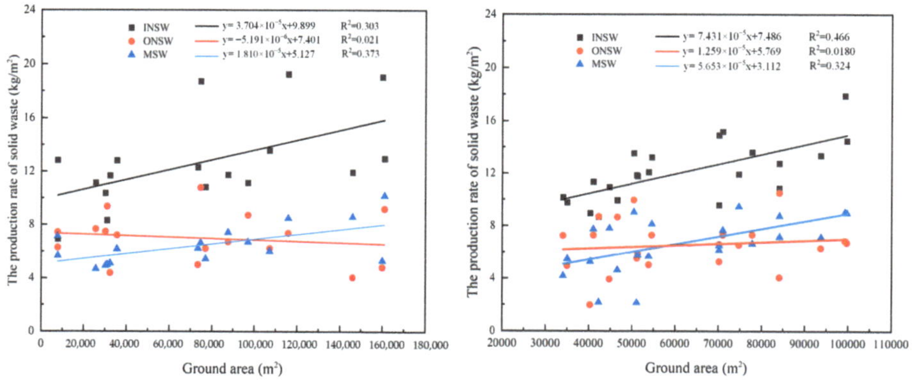

(a) (b)

Figure 3. Effect of ground area on solid waste generation rate. (**a**) Office buildings; (**b**) Exhibition buildings.

While ONSW was fitted poorly, INSW and MSW were fitted better. As ground area increases, there is a trend toward an increase in the generation rate of INSW and MSW at construction sites. In office buildings, the generation rate of INSW and MSW increases by 0.370 kg/m^2 and 0.181 kg/m^2, respectively, for every 10,000 m^2 increase in ground area. As for exhibition buildings, the generation rate of INSW and MSW increases by 0.743 kg/m^2 and 0.565 kg/m^2, respectively, for every 10,000 m^2 increase in ground area.

3.3. *Influence of Building Height*

The height of the building has a large impact on the solid waste of the construction site. The same method in Section 3.2 was followed to fit the curve. The results are shown in Figure 4.

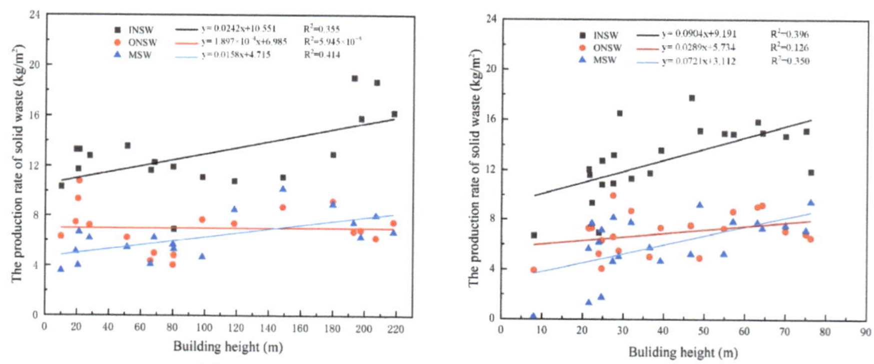

(a) (b)

Figure 4. Effect of building height on solid waste generation rate. (**a**) Office buildings; (**b**) Exhibition buildings.

The results showed that the goodness of fit was equally high for INSW and MSW, and poorer for ONSW. The generation rate of all types of solid waste in the construction stage of new buildings shows an increasing trend with the increase of building height. For office buildings, every 10 m increase in building height, the generation rate of INSW and MSW increases by 0.242 kg/m^2 and 0.158 kg/m^2, respectively. As for exhibition buildings, the generation rate of INSW and MSW increases by 0.904 kg/m^2 and 0.721 kg/m^2, respectively, for every 10 m increase in building height.

3.4. Influence of the Amount of Concrete Usage

The amount of concrete usage has a large impact on the solid waste generation rate of the building construction site. The results are shown in Figure 5.

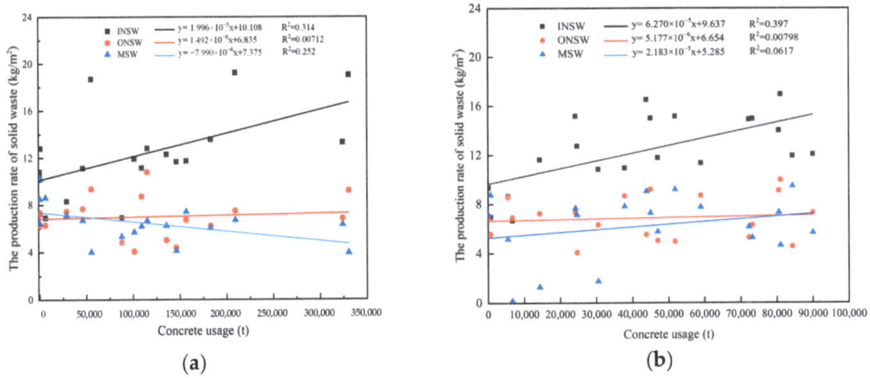

Figure 5. Effect of the amount of concrete usage on solid waste generation rate. (**a**) Office buildings; (**b**) Exhibition buildings.

The fitting results show that the generation rate of ONSW is not affected by the amount of concrete usage, but the amount of concrete usage and the generation rate of INSW and MSW in the construction stage of office buildings show a certain correlation. For office buildings, for every 10,000 t increase in the amount of concrete usage, INSW increases by 0.199 kg/m^2, and MSW generation rate decreases by 0.0799 kg/m^2. As for exhibition buildings, for every 10,000 t increase in the amount of concrete usage, INSW and MSW generation rate increases by 0.627 kg/m^2 and 0.218 kg/m^2, respectively.

3.5. Influence of the Amount of Steel Usage

The amount of steel usage is also one of the main influencing factors of the solid waste generation rate. The results are shown in Figure 6.

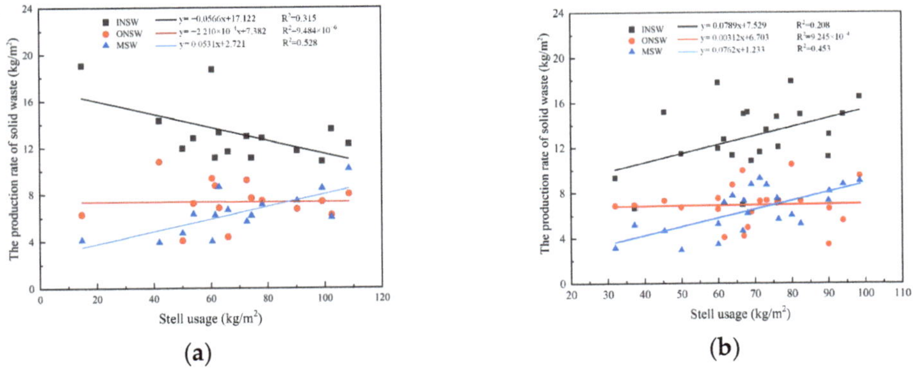

Figure 6. Effect of the amount of steel usage on solid waste generation rate. (**a**) Office buildings; (**b**) Exhibition buildings.

The fitting findings indicate that the amount of steel usage has a significant effect on the generation rate of MSW, and also influences the generation rate of INSW; however, the amount of steel usage has no effect on the generation rate of ONSW. In office buildings, for every 10 kg/m^2 increase in the amount of steel used, MSW increases by 0.531 kg/m^2, and the generation rate of INSW decreases by 0.566 kg/m^2. While exhibition buildings, the use of steel increased by 10 kg/m^2, INSW and MSW generation rate increases by 0.789 kg/m^2 and 0.762 kg/m^2, respectively.

3.6. The Impact of Assembly Rate

Because the data samples of the assembly rate are more concentrated, the typical values of measured data under different assembly rates are selected to make the fitting curve of solid waste generation under different assembly rates, and the values of the fitting curve and the correlation coefficient R^2 are found. Specific results are shown in Figure 7.

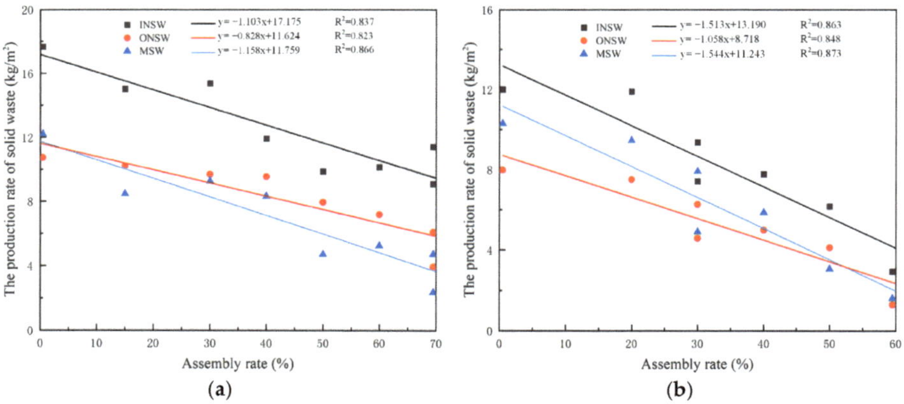

Figure 7. Effect of the assembly rate on solid waste generation rate. (**a**) Office buildings; (**b**) Exhibition buildings.

Analyzing the fitted curves shows that an increase in the assembly rate during the construction phase of new buildings leads to a decrease in the generation rate of each type of waste. For office buildings, for every 10% increase in the assembly rate, the generation rates of INSW, ONSW and MSW are reduced by 1.103 kg/m^2, 0.828 kg/m^2 and 1.158 kg/m^2, respectively. And for every 10% increase in the assembly rate of exhibition buildings, the generation rates of INSW, ONSW and MSW decreased by 1.513 kg/m^2, 1.058 kg/m^2 and 1.544 kg/m^2, respectively.

4. Conclusions and Perspective

By measuring the amount of solid waste generated during the construction stage of 61 projects in Hainan Province, this paper uses the Random Forest algorithm to explore the relationship between influencing factors and solid waste generation rate. The following conclusions are summed up based on the examination of the project's assessed data processing:

(1) The impact of building area: in office buildings, the generation rate of INSW and MSW increases by 0.370 kg/m^2 and 0.181 kg/m^2, respectively, for every 10,000 m^2 increase in building area. For exhibition buildings, every increase of 10,000 m^2 in building area, the generation rate of INSW and MSW increases by 0.743 kg/m^2 and 0.565 kg/m^2, respectively.

(2) The impact of building height: in office buildings, every 10 m increase in building height, INSW and MSW generation rate increases by 0.242 kg/m^2 and 0.158 kg/m^2, respectively. For exhibition buildings, for every 10 m increase in building height, the generation rate of INSW and MSW increases by 0.904 kg/m^2 and 0.721 kg/m^2, respectively.

(3) The impact of concrete usage: for office buildings, the rate of MSW generation decreases by 0.0799 kg/m^2 and INSW increases by 0.199 kg/m^2 for every 10,000 t increase in concrete usage. For exhibition buildings, the rate of MSW and INSW

generation increases by 0.218 kg/m^2 and 0.627 kg/m^2 for every 10,000 t increase in concrete usage.

(4) The impact of steel usage: for office buildings, the steel usage increases by 10 kg/m^2, MSW increases by 0.531 kg/m^2, and INSW generation rate decreases by 0.566 kg/m^2. For exhibition buildings, for every kg/m^2 increase in steel usage, the generation rate of INSW and MSW increases by 0.789 kg/m^2 and 0.762 kg/m^2, respectively.

(5) The impact of assembly rate: for office buildings, assembly rate increases by 10%, and INSW, ONSW and MSW generation rate decreases by 1.103 kg/m^2, 0.828 kg/m^2 and 1.158 kg/m^2, respectively. For exhibition buildings, for every 10% increase in the assembly rate, the generation rate of INSW ONSW and MSW decreases by 1.513 kg/m^2, 1.058 kg/m^2 and 1.544 kg/m^2, respectively.

After summarizing the article, the following suggestions are made:

(1) For all types of buildings, increasing the rate of assembly and controlling the building height and floor area during design can reduce the generation of solid waste.

(2) For visiting buildings, increasing the rate of assembly has a significant impact on reducing the rate of solid waste generation. The design of these buildings ought to prioritize the highest possible rate of assembly.

(3) Construction contractors ought to fortify their oversight and management protocols, enhance the caliber of their workforce, and ameliorate the construction site's environmental conditions to curtail the production of solid waste.

The production of a large amount of solid waste will not only cause economic waste, but also have a huge impact on the environment. In the process of data collection, we found that developers and construction teams are reluctant to pay attention to solid waste recycling. On the one hand, solid waste recycling will occupy a large part of human resources, but on the other hand, solid waste recycling cannot bring economic benefits to the construction side. In further studies, we expect to explore the relationship between solid waste production rates and influencing factors in other climate types. Following this study, it is expected to investigate the suggestions of major participants in the construction site on solid waste management by issuing on-site questionnaires, explore feasible solid waste management schemes for construction sites, and provide suggestions for decision makers.

Author Contributions: Conceptualization, Q.D. and T.M.; methodology, X.S.; resources, Q.D. and Z.R.; data curation, T.M.; writing—original draft preparation, T.M.; writing—review and editing, T.M.; visualization, Q.D. All authors have read and agreed to the published version of the manuscript.

Funding: This work was supported by Hainan Province Science and Technology Special Fund (ZDKJ2021024).

Data Availability Statement: The data presented in this study are available on request from the corresponding author.

Conflicts of Interest: The authors declare no conflicts of interest.

References

1. Ministry of Housing and Urban-Rural Development of the People's Republic of China. *14th Five-Year Plan' for the Development of the Construction Industry*; Ministry of Housing and Urban-Rural Development of the People's Republic of China: Beijing, China, 2022.
2. Zhu, L.; Hao, W. A review of construction waste treatment in China. *Adv. Mater. Res.* **2011**, *280*, 62–66. [CrossRef]
3. Duan, H.; Li, J. Construction and demolition waste management: China's lessons. *Waste Manag. Res.* **2016**, *34*, 397–398. [CrossRef]
4. Jin, R.; Li, B.; Zhou, T.; Wanatowski, D.; Piroozfar, P. An empirical study of perceptions towards construction and demolition waste recycling and reuse in China. *Resour. Conserv. Recycl.* **2017**, *126*, 86–98. [CrossRef]
5. Safiuddin, M.; Jumaat, M.Z.; Salam, M.A.; Islam, M.S.; Hashim, R. Utilization of solid wastes in construction materials. *Int. J. Phys. Sci.* **2010**, *5*, 1952–1963.
6. Wu, Z.; Yu, A.T.W.; Shen, L. Investigating the determinants of contractor's construction and demolition waste management behavior in Mainland China. *Waste Manag.* **2017**, *60*, 290–300. [CrossRef]
7. Lu, W.; Yuan, H. Exploring critical success factors for waste management in construction projects of China. *Resour. Conserv. Recycl.* **2010**, *55*, 201–208. [CrossRef]

8. Poon, C.; Yu, A.T.; Ng, L. On-site sorting of construction and demolition waste in Hong Kong. *Resour. Conserv. Recycl.* **2001**, *32*, 157–172. [CrossRef]
9. Zambrana-Vasquez, D.; Zabalza-Bribián, I.; Jáñez, A.; Aranda-Usón, A. Analysis of the environmental performance of life-cycle building waste management strategies in tertiary buildings. *J. Clean. Prod.* **2016**, *130*, 143–154. [CrossRef]
10. Zhang, C.; Hu, M.; Di Maio, F.; Sprecher, B.; Yang, X.; Tukker, A. An overview of the waste hierarchy framework for analyzing the circularity in construction and demolition waste management in Europe. *Sci. Total. Environ.* **2021**, *803*, 149892. [CrossRef] [PubMed]
11. Gálvez-Martos, J.-L.; Styles, D.; Schoenberger, H.; Zeschmar-Lahl, B. Construction and demolition waste best management practice in Europe. *Resour. Conserv. Recycl.* **2018**, *136*, 166–178. [CrossRef]
12. Nunes, K.; Mahler, C. Comparison of construction and demolition waste management between Brazil, European Union and USA. *Waste Manag. Res. J. Sustain. Circ. Econ.* **2020**, *38*, 415–422. [CrossRef]
13. Huang, B.; Wang, X.; Kua, H.; Geng, Y.; Bleischwitz, R.; Ren, J. Construction and demolition waste management in China through the 3R principle. *Resour. Conserv. Recycl.* **2018**, *129*, 36–44. [CrossRef]
14. Zhang, X.; Xiao, Y.; Xue, Y.; Liu, J.; Chen, Z.; Zhang, R. Global Bibliometric Developments on Solid Waste Recycling in Concrete Construction Engineering. *Materials* **2022**, *15*, 4142. [CrossRef]
15. Pacheco, J.; de Brito, J. Recycled aggregates produced from construction and demolition waste for structural concrete: Constituents, properties and production. *Materials* **2021**, *14*, 5748. [CrossRef]
16. Shi, C.; Li, Y.; Zhang, J.; Li, W.; Chong, L.; Xie, Z. Performance enhancement of recycled concrete aggregate—A review. *J. Clean. Prod.* **2016**, *112*, 466–472. [CrossRef]
17. Wang, B.; Yan, L.; Fu, Q.; Kasal, B. A Comprehensive Review on Recycled Aggregate and Recycled Aggregate Concrete. *Resour. Conserv. Recycl.* **2021**, *171*, 105565. [CrossRef]
18. Ginga, C.P.; Ongpeng, J.M.C.; Daly, M.K.M. Circular economy on construction and demolition waste: A literature review on material recovery and production. *Materials* **2020**, *13*, 2970. [CrossRef] [PubMed]
19. Borrachero, M.V.; Payá, J.; Brito, S.; Segura, Y.P.; Soriano, L.; Tashima, M.M.; Monzó, J.M. Reusing Construction and Demolition Waste to Prepare Alkali-Activated Cement. *Materials* **2022**, *15*, 3437. [CrossRef] [PubMed]
20. Poon, C.S.; Chan, D. Feasible use of recycled concrete aggregates and crushed clay brick as unbound road sub-base. *Constr. Build. Mater.* **2006**, *20*, 578–585. [CrossRef]
21. Xuan, D.X.; Molenaar, A.A.A.; Houben, L.J.M. Evaluation of cement treatment of reclaimed construction and demolition waste as road bases. *J. Clean. Prod.* **2015**, *100*, 77–83. [CrossRef]
22. Contrafatto, L.; Cosenza, R.; Barbagallo, R.; Ognibene, S. Use of recycled aggregates in road sub-base construction and concrete manufacturing. *Ann. Geophys.* **2018**, *61*, 223. [CrossRef]
23. Poon, C.S. Reducing construction waste. *Waste Manag.* **2007**, *27*, 1715–1716. [CrossRef]
24. Marzouk, M.; Azab, S. Environmental and economic impact assessment of construction and demolition waste disposal using system dynamics. *Resour. Conserv. Recycl.* **2014**, *82*, 41–49. [CrossRef]
25. Chen, X.; Lu, W. Identifying factors influencing demolition waste generation in Hong Kong. *J. Clean. Prod.* **2017**, *141*, 799–811. [CrossRef]
26. Esa, M.R.; Halog, A.; Rigamonti, L. Strategies for minimizing construction and demolition wastes in Malaysia. *Resour. Conserv. Recycl.* **2017**, *120*, 219–229. [CrossRef]
27. Nikmehr, B.; Hosseini, M.R.; Rameezdeen, R.; Chileshe, N.; Ghoddousi, P.; Arashpour, M. An integrated model for factors affecting construction and demolition waste management in Iran. *Eng. Constr. Arch. Manag.* **2017**, *24*, 1246–1268. [CrossRef]
28. Peng, C.L.; Scorpio, D.E.; Kibert, C.J. Strategies for successful construction and demolition waste recycling operations. *Constr. Manag. Econ.* **1997**, *15*, 49–58. [CrossRef]
29. Ma, W.; Liu, T.; Hao, J.L.; Wu, W.; Gu, X. Towards a circular economy for construction and demolition waste management in China: Critical success factors. *Sustain. Chem. Pharm.* **2023**, *35*, 101226. [CrossRef]
30. Lu, D.; Iqbal, A.; Zan, F.; Liu, X.; Chen, G. Life-cycle-based greenhouse gas, energy, and economic analysis of municipal solid waste management using system dynamics model. *Sustainability* **2021**, *13*, 1641. [CrossRef]
31. Lu, D.; Iqbal, A.; Zan, F.; Liu, X.; Dong, Z.; Jiang, C.; Chen, G. Integrated life cycle assessment with data envelopment analysis for enhancing medical waste management during a public health crisis. *J. Clean. Prod.* **2023**, *426*, 139074. [CrossRef]
32. Hu, R. Quantification and Prediction of Construction Waste Generation of New-Constructed Residential Buildings. Master's Thesis, Wuhan University of Technology, Wuhan, China, 2020.
33. Hu, R.; Chen, K.; Fang, W.; Zheng, L.; Xu, J. The technology-environment relationship revisited: Evidence from the impact of prefabrication on reducing construction waste. *J. Clean. Prod.* **2022**, *341*, 130883. [CrossRef]
34. Tam, V.W.Y.; Tam, C.M.; Zeng, S.X.; Ng, W.C.Y. Towards adoption of prefabrication in construction. *Build. Environ.* **2007**, *42*, 3642–3654. [CrossRef]
35. Ministry of Housing and Urban-Rural Development of the People's Republic of China. *Technical Standard for Construction Waste Disposal CJJT 134-2019*; China Architecture & Building Press: Beijing, China, 2019.
36. Chen, L. Quantification and Reduction of Solid Waste from Construction Sites and Resource Utilisation. Ph.D. Thesis, Wuhan University of Technology, Wuhan, China, 2020.

37. Iqbal, A.; Liu, X.; Chen, G.H. Municipal solid waste: Review of best practices in application of life cycle assessment and sustainable management techniques. *Sci. Total. Environ.* **2020**, *729*, 138622. [CrossRef] [PubMed]
38. Ministry of Industry and Information Technology of the People's Republic of China. *Technical Specification for Construction Solid Waste Treatment (JCT2744-2023)*; Ministry of Industry and Information Technology of the People's Republic of China: Beijing, China, 2023.

buildings

MDPI

Article

Source Location Identification in an Ideal Urban Street Canyon with Time-Varying Wind Conditions under a Coupled Indoor and Outdoor Environment

Yuwei Dai, Minzhang Hou, Haidong Wang * and Wanli Tu

School of Environment and Architecture, University of Shanghai for Science and Technology, 516 Jungong Rd., Shanghai 200093, China; ywdai@usst.edu.cn (Y.D.); 223382100@st.usst.edu.cn (M.H.); 233432047@st.usst.edu.cn (W.T.)
* Correspondence: whd@usst.edu.cn

Abstract: Source location identification methods are typically applied to steady-state conditions under pure indoor or outdoor environments, but under time-varying wind conditions and coupled indoor and outdoor environments, the applicability is not clear. In this study, we proposed an improved adjoint probability method to identify the pollutant source location with time-varying inflows in street canyons and used scaled outdoor experiment data to verify the accuracy. The change in inflow velocity will affect the airflow structure inside the street canyons. Outdoor wind with a lower temperature will exchange heat with the air with a higher temperature inside the street canyon, taking away part of the heat and reducing the heat of the air inside the street canyons. Moreover, the room opening will produce some air disturbance, which is conducive to the heat exchange between the air near the opening and the outdoor wind. Furthermore, the fluctuations of the upper wind will influence the diffusion of the tracer gas. We conducted three cases to verify the accuracy of the source identification method. The results showed that the conditioned adjoint location probability (CALP) of each case was 0.06, 0.32, and 0.28. It implies that with limited pollutant information, the improved adjoint probability method can successfully identify the source location in the dynamic wind environments under coupled indoor and outdoor conditions.

Keywords: inverse CFD modeling; scaled outdoor experiment; dynamic inflow conditions; source identification; coupled indoor and outdoor environment

Citation: Dai, Y.; Hou, M.; Wang, H.; Tu, W. Source Location Identification in an Ideal Urban Street Canyon with Time-Varying Wind Conditions under a Coupled Indoor and Outdoor Environment. *Buildings* **2023**, *13*, 3121. https://doi.org/10.3390/buildings13123121

Academic Editor: Ricardo M. S. F. Almeida

Received: 27 October 2023
Revised: 10 December 2023
Accepted: 13 December 2023
Published: 15 December 2023

1. Introduction

Urban street canyons, often used for daily activities, are typical urban microclimate environments [1]. The dynamic and thermal properties of street canyons affect the microclimate environment within the urban canopy, making them a crucial area of research in urban ventilation and pollutant dispersion [2]. The indoor and outdoor airflows in the urban microclimate environment are closely related and interact with each other under natural ventilation, which is a common ventilation method used in urban residential areas [3,4]. It may increase the exposure risk of urban residents to fine particulate pollutants, such as aerosols [5,6]. To prevent the spread of contaminants or viruses in urban street canyons, it is essential to comprehend the dynamic characteristics of airborne transmission and locate sources in coupled indoor and outdoor environments.

Many previous studies have investigated the flow regime and pollutant dispersion in urban street canyons, including in situ measurements, wind tunnel experiments, computational fluid dynamics (CFD) simulations, and scaled outdoor experiments. Data monitoring and in situ experiments in urban environments are important for understanding the characteristics of temperature and airflow within and above urban environments. These experiments can provide data from realistic atmospheric conditions (e.g., dynamic velocities and heterogeneous temperature distributions of building facades). Nakamura and Oke [7] studied the wind and temperature characteristics in an east–west street canyon under various stability conditions. They observed an approximately linear correlation between wind velocity above and inside a street canyon. Santamouris et al. [8] investigated airflow patterns and temperature distribution in a deep street canyon under hot weather conditions. According to their observations, the airflow in the canyon was characterized by either one main vortex or a double vortex when the ambient wind blew vertically toward the street. Eliasson et al. [9] conducted long-term measurements within an urban canyon and observed a high degree of vertical mixing within the canyon. Niachou et al. [10] assessed the wind speed, air temperature, and surface temperature inside an urban street canyon under hot summer weather conditions. They observed that airflow patterns in street canyons were primarily affected by ground thermal conditions. The results of these studies imply that in street canyons, the airflow fluctuated continuously but was not as large as that shown in the numerical simulations [11,12].

In addition to in situ measurements, numerous studies have used numerical simulations to investigate the transmission characteristics of pollutants and viruses in urban street canyons [13–17]. Yang et al. [18] used a steady Reynold-averaged Navier–Stokes (RANS) model to simulate the effects of outdoor pollutants on indoor air quality and the characteristics of outdoor pollutant dispersion in street canyons. They observed that airflow from upstream buildings alters the flow field with windows in street canyons and the distribution of pollutants from downstream buildings. Fan et al. [19] adopted the steady-state standard k–ε model to simulate the influence of aerosol transmission distance on residents' health in street canyons. The results of the study revealed that aerosol transmission over short distances poses a health risk to the residents of street canyons. Hang et al. [20] investigated the effect of wind speed on ventilation performance and pollutant concentration in street canyons. They observed that low wind speeds change ventilation performance and pollutant concentrations in street canyons. The results of these studies indicated that pollutant dispersion routes are influenced by changes in the airflow field in street canyons.

More recently, scaled outdoor experiments, an effective alternative, have gained popularity for investigating airflow fields [21], pollutant dispersion [22], urban ventilation [23], and thermal conditions [24–26] in urban environments. In our previous studies [27,28], we conducted scaled outdoor experiments in two-dimensional street canyons to investigate airflow fields and pollutant dispersion problems. In this experiment, we investigated the ventilation effectiveness and pollutant dispersion among rooms in street canyons.

Most previous studies have considered or simplified the urban boundary layer airflow as a time-independent flow combined with turbulence. However, actual urban airflow is frequently gusty, with time-varying wind speeds and directions [29,30]. Studies have revealed that varying inflow wind speeds and directions can considerably affect the characteristics of airflow and pollution dispersion [31,32], which cannot be ignored. Zhang et al. [31] compared the impact of a time series of incoming winds (time-scale of 1 min) and a constant inflow velocity on the airflow fields and pollutant dispersion. They observed that the airflow circulation inside the street canyon was interrupted by time-varying inflows, which dramatically increased the dispersion efficiency of pollutants. Wang et al. [33] adopted a periodic sinusoidal fluctuating wind as the inlet flow to investigate the effect of a dynamic airflow field on the ventilation and pollutant dispersion characteristics of street canyons. According to their findings, an airflow field with periodic sinusoidal fluctuations could hasten the decay of pollutant concentrations inside street canyons. More recently, Zhang et al. [2] summarized the effects of time-varying wind conditions on airflow and pollutant dispersion in street canyons. They indicated that dynamic inflow is an essential factor that influences the airflow structure and pollutant dispersion characteristics inside urban street canyons.

Prior research has suggested that time-varying inflows lead to more complicated airflow and pollutant dispersion fields in urban street canyons compared to constant inflows. However, these studies primarily focused on pure outdoor environments, overlooking the prevalence of indoor and outdoor environments in urban settings. Therefore, it is crucial to extend this investigation to include such hybrid environments.

For airborne transmitted viruses, identifying a certain source location or infection point in a street canyon environment is crucial. Liu and Zhai [34] summarized the methods to locate pollutant sources into three categories: forward, backward, and probability methods. Among these methods, when information about the pollutant source is insufficient, particularly without a likely source site, the probability method is frequently used, which is also suitable for identifying the source location in a built environment [35]. Based on the adjoint method proposed by Neupauer and Wilson [36], Zhai et al. [37] proposed a standard adjoint location probability (SALP) method for locating a dynamic source in an indoor environment. Subsequently, Wang et al. [38] improved the adjoint probability model and identified the indoor source locations under a dynamic airflow field. Chen et al. [39] combined the adjoint probability method with CFD to establish an inverse model of radionuclide diffusion under ventilation conditions to estimate the probability of locating radioactive source leakage. Instead of focusing primarily on pollutant source identification in pure indoor or outdoor environments, few studies have considered pollutant dispersion in coupled indoor and outdoor environments. More recently, our previous study [35] employed the adjoint probability method in inverse CFD modeling to locate a pollutant source with single-sided natural ventilation in an isolated building. We suggested that the inverse modeling method with the adjoint probability method can successfully locate pollutant sources under coupled indoor and outdoor conditions. However, the former study investigated only the time-independent inflow condition. Considering the more complex airflow and pollutant dispersion fields with the time-carrying inflow, the applicability of the inverse CFD simulation should also be tested with the dynamic inflow.

In this study, we intend to develop an improved method for identifying the location of pollutant sources in urban street canyons with time-varying inflow in a coupled indoor and outdoor environment. For this purpose, we conducted a scaled outdoor experiment to determine the actual urban boundary layer airflow and pollutant dispersion in street canyon rooms. We then reconstructed the dynamic airflow and pollutant dispersion fields using unsteady CFD simulations. Finally, we employed the adjoint probability method with inverse CFD simulations to identify pollutant source locations. This research can fill the gap for both the inverse modeling with dynamic wind conditions and source identifications in coupled indoor and outdoor environments. This research can also help to locate the pollutant source in a real urban environment quickly and accurately. The remainder of this paper is organized as follows: Section 2 briefly introduces the scaled outdoor experiment; Section 3 introduces CFD methods with inverse modeling under time-varying inflow; Section 4 validates the CFD simulation with the results of the scaled outdoor experiment; Section 5 presents the results and discussions of the CFD simulations; and Section 6 concludes the study.

2. Scaled Outdoor Experiment

2.1. Description of the Experiment Setup

Scaled outdoor experiments were conducted in idealized two-dimensional (2D) street canyons at Sun Yat-sen University [27,28]. The experiment field was located in the southern area of Guangzhou, China (23°01′ N, 113°24′ E). The experimental field was built on a concrete foundation measuring 57 m × 57.5 m. As shown in Figure 1a, the entire north–south street canyon field has dimensions of width × length = 44.4 m × 12 m. The street canyon deviates from the northern direction by approximately 30°. Customized as a hollow cuboid, each concrete model had a size of length × width × height = 0.5 m × 0.5 m × 1.2 m. As shown in Figure 1b, the walls of the concrete model were painted dark grey to resemble typical urban buildings. Four aspect ratios (H/W), 1, 2, 3, and 6 are included in the street canyons. Each aspect ratio contained six rows, with each row containing 24 concrete models.

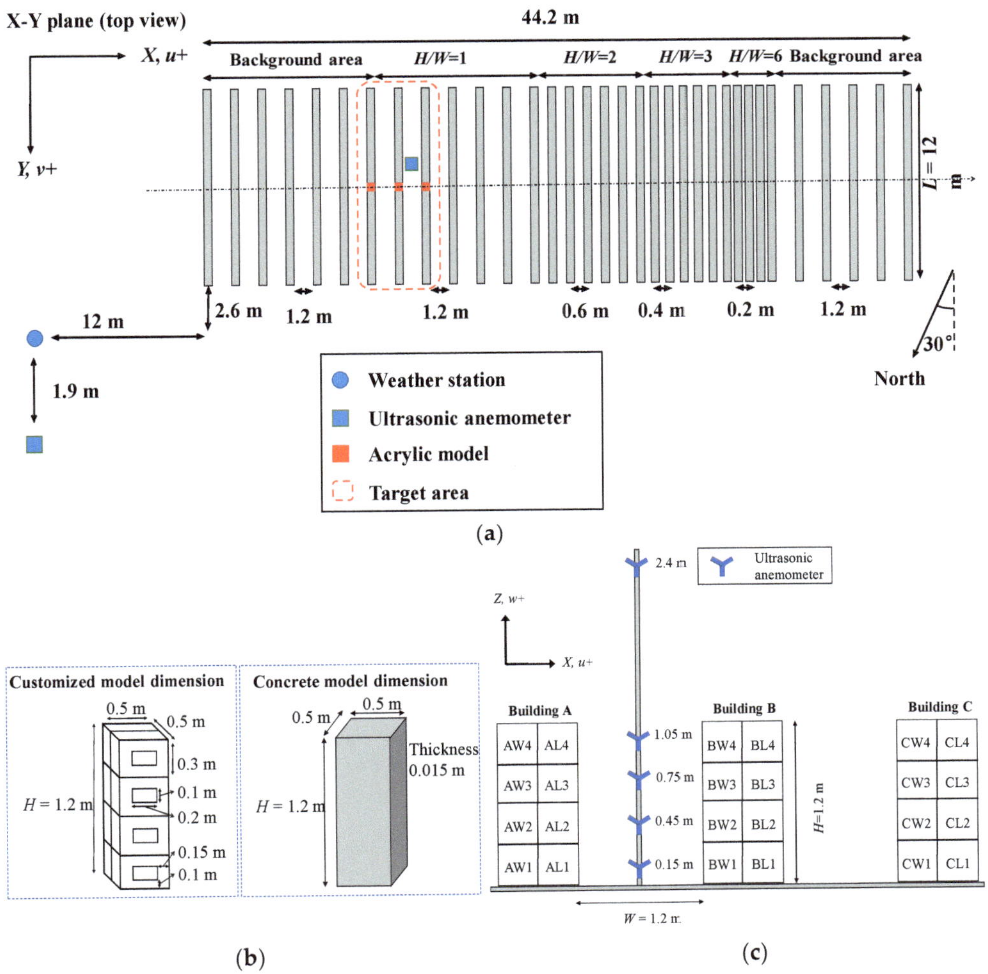

Figure 1. Descriptions of the experiment setup: (**a**) schematic view of the experiment field on the X–Y plane and instrument positions; (**b**) dimensions of the customized and concrete models; (**c**) schematic view of buildings and rooms in the street canyon.

To simulate the coupled indoor and outdoor environment in the urban street canyon, we customized an acrylic model (5 mm thick) with four floors, and each floor contained two rooms with opposite openings, for a total of eight rooms. The dimensions of the acrylic and concrete models were identical. As shown in Figure 1b, the height and width of each opening were 0.1 and 0.2 m, respectively. This study selected a 1:1 street canyon as the main research area, and three acrylic models were positioned in the middle of three consecutive street canyons with a 1:1 aspect ratio. In Figure 1c, *W* and *L* represent the windward and leeward sides of the building model, respectively.

During the experimental period, wind speed was measured using 3D ultrasonic anemometers (Gill WindMaster, Leamington, UK, 1.5%, in the range of 0–50 m/s). The indoor and outdoor air/wall temperature was measured using K-type fine-wire thermocouples (Omega, TT-K-36-SLE, Φ 0.127 mm, 1.1 °C or 0.4% in a range of −200–260 °C, Agilent Technologies Inc., Santa Clara, CA, USA). A CO_2 sensor (HR International Co., Taiwan, China, ±40 ppm in a range of 400–10,000 ppm) was used to detect CO_2 concentrations in the building models. Atmospheric data, such as background air temperature and wind speed, were obtained from an automatic weather station (RainWise Inc., Dover, DE, USA, ±0.25 in the range of −54–74 °C, ±2% in the range of 0–67 m/s, 3° in the range of 0°–360°). Two wind masts were installed in the experimental field, on which several ultrasonic anemometers were installed at different positions, as shown in Figure 2a,b. Figure 2c illustrates the positions of the thermocouples during the experiment. Figure 2c–e show the concentration measurement setup. Each room had a CO_2 sensor to measure the indoor concentration. A lengthy tube with a diameter of 8 mm was used to carry the tracer gas (CO_2) into the source room from a compressed gas cylinder. To reduce the injection velocity, we inserted a customized plastic ball (diameter of 30 mm) with six uniformly arranged holes (diameter of 5 mm) at the end of the tube close to the room center.

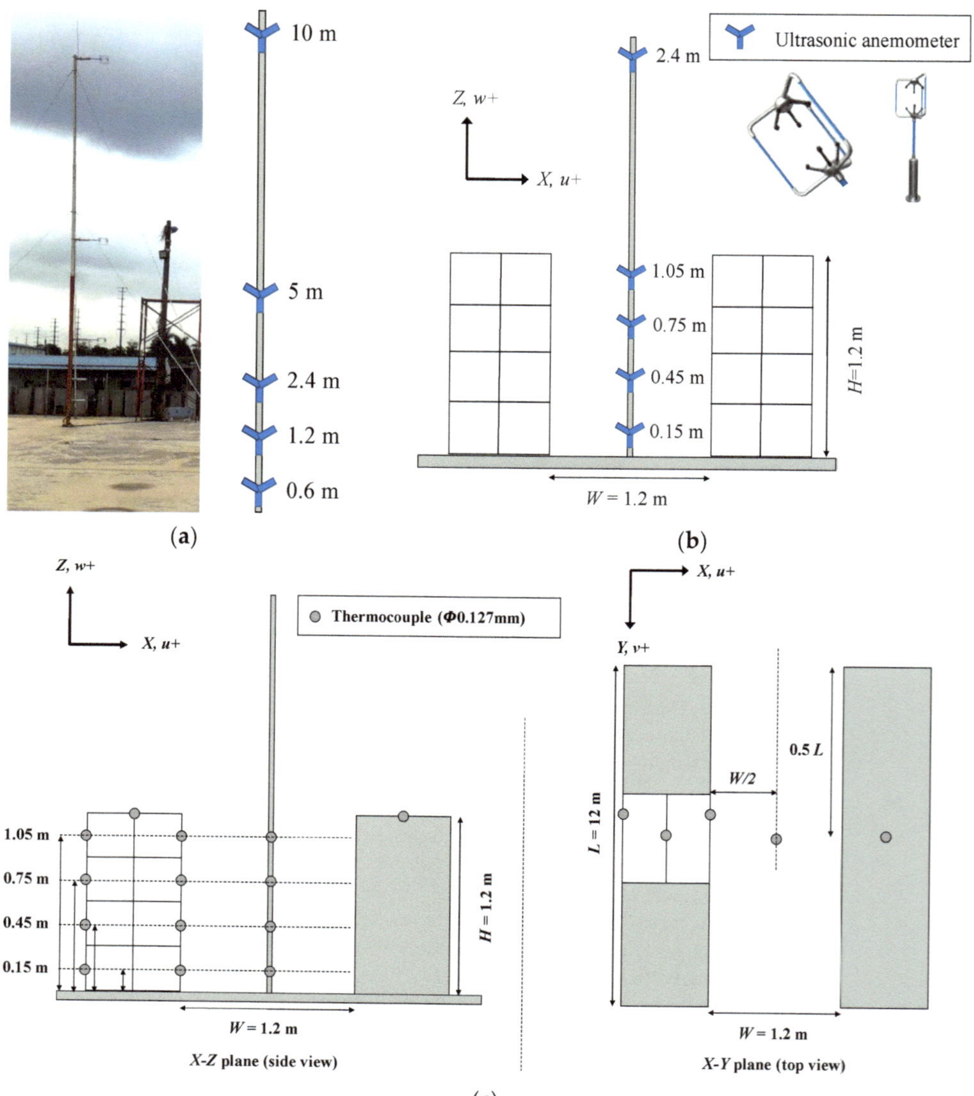

Figure 2. *Cont.*

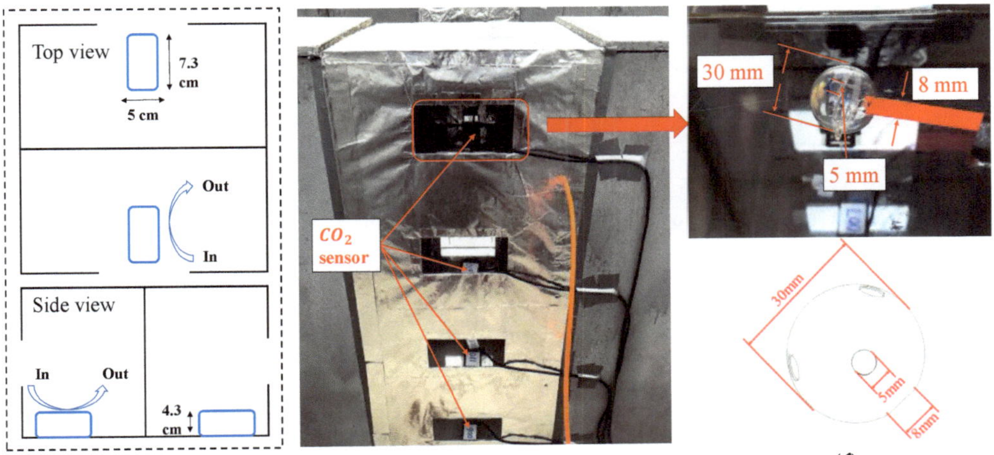

(d) (e) (f)

Figure 2. Setup for the wind masts, ultrasonic anemometer positions, tracer gas release, and concentration measurement: (**a**) 10 m mast at the side of the experiment field; (**b**) 2.4 m mast in the middle of the street canyon; (**c**) positions of the thermocouples; (**d**) positions of the CO_2 sensors in each room; In: inlet of the sensor, Out: outlet of the sensor; (**e**) measurement setup; (**f**) customized plastic ball.

In this experiment, each room served as a source room several times. The source gas concentration was 10^5 ppm. The output values were averaged over one second at a sampling frequency of 1 Hz. The flow rate of the tracer gas was 1.5 L/min, and it was continuously delivered for approximately 30 min. For the wind speed, wind direction, air and wall temperatures, and concentrations in each room, all instruments were used concurrently. The gas-releasing tube was removed after each test was completed. Before the subsequent test was started, the CO_2 concentration in the former source room had to reach background levels. Detailed information can be observed in our previous papers [27,28].

2.2. Experimental Results

During the measurement periods, the wind direction and speed were measured by an automatic weather station, as described in our previous paper [28]. The dominant wind direction was southwest (mostly between 43° and 50°), flowing in a nearly perpendicular direction toward the street canyons. In this study, we primarily considered the variations in the inflow velocities and set the wind direction to normal incidence.

We set the CO_2 concentration to more than 10 ppm (excluding the background concentration) as the threshold, considering the accuracy of the CO_2 sensor. In this study, we used the data from 10:40:00 to 17:40:00 on 9 June 2019. Table 1 shows the time and concentration of the tracer gas in the monitored room above the threshold, as well as the maximum and final concentrations detected during the entire experimental period.

Table 1. Concentrations of each room detected with different source rooms (excluding background concentration).

Period	Room	Time over 10 ppm	Concentration over 10 ppm	Maximum Value	Final Value
10:40:05–11:09:55 Source room: BW1	-	-	-	-	-
11:32:05–12:01:56 Source room: BW2	BW1	11:34:30	10.7 ppm	58.6 ppm	40 ppm

Table 1. *Cont.*

Period	Room	Time over 10 ppm	Concentration over 10 ppm	Maximum Value	Final Value
12:27:06–12:56:55 Source room: BW3	BW2	12:28:45	30.4 ppm	88 ppm	63.5 ppm
	BW1	12:31:35	10 ppm	15 ppm	13 ppm
13:19:05–13:48:56 Source room: BW4	BW3	13:21:00	30.4 ppm	49 ppm	37 ppm
	BW2	13:20:10	10.2 ppm	13 ppm	10.5 ppm
14:16:06–14:45:55 Source room: BL1	BL2	14:18:28	12.8 ppm	39.7 ppm	36.1 ppm
	BL3	14:22:20	10 ppm	16 ppm	12 ppm
	CW3	14:30:58	10 ppm	13 ppm	8 ppm
	CW1	14:31:34	10 ppm	12.5 ppm	7 ppm
	BW2	14:42:16	10 ppm	10 ppm	6 ppm
15:11:04–15:40:55 Source room: BL2	BL3	15:12:34	16 ppm	154.6 ppm	130 ppm
	BL4	15:15:05	10 ppm	26.8 ppm	21.5 ppm
16:13:04–16:42:56 Source room: BL3	BL4	16:15:44	10 ppm	90 ppm	66.3 ppm
17:09:04–17:39:00 Source room: BL4	-	-	-	-	-

For the thermal information, Table 2 lists the outdoor air temperature, the wall temperature at different heights of the acrylic model, the concrete model building wall temperature, and the roof temperature of both acrylic and concrete models. The thermal conditions were set as boundary conditions in the CFD simulations.

Table 2. Average temperature of different areas in the street canyons in different periods.

Temperature (°C) Period	Outdoor Air (1.8 m)	Acrylic Wall (0.15 m, 0.45 m, 0.75 m, 1.05 m)	Concrete Wall	Model Roof	Concrete Roof
10:40:05–11:09:55	35.0	35.4 36.0 35.5 35.4	38.5	41.8	51.5
11:32:05–12:01:56	35.1	35.9 36.4 35.9 35.7	39.5	40.6	45.1
12:27:06–12:56:55	36.4	38.1 37.8 37.1 37.0	40.9	42.9	49.4
13:19:05–13:48:56	36.0	38.2 38.3 37.5 37.3	42.2	41.2	51.2
14:16:06–14:45:55	36.1	39.0 38.9 37.9 37.6	42.8	41.0	49.1

Table 2. *Cont.*

Temperature (°C) / Period	Outdoor Air (1.8 m)	Acrylic Wall (0.15 m, 0.45 m, 0.75 m, 1.05 m)	Concrete Wall	Model Roof	Concrete Roof
15:11:04–15:40:55	35.7	39.3 39.5 38.7 38.2	44.1	40.8	48.5
16:13:04–16:42:56	34.6	36.9 37.6 36.8 36.2	42.5	38.4	44.5
17:09:04–17:39:00	33.5	34.4 34.7 34.3 33.8	37.4	35.5	38.3

3. CFD Methodology

3.1. CFD Models

Three-dimensional Navier–Stokes equations are solved to simulate the airflow field around a building, which is very complex and varies with time. The continuity and momentum equations are expressed as follows:

$$\frac{\partial u_i}{\partial t} = 0 \tag{1}$$

$$\frac{\partial u_i}{\partial t} + \frac{\partial}{\partial x_j}(u_i u_j) = -\frac{1}{\rho}\frac{\partial p}{\partial x_i} + \frac{\partial}{\partial x_j}(2v s_{ij}) \tag{2}$$

where u_i is the velocity vector in x_i direction, p is the pressure, ρ is the density, v is the molecular viscosity, and s_{ij} is the stain-rate tensor as defined in Equation (3).

$$s_{ij} = \frac{1}{2}\left(\frac{\partial u_i}{\partial x_i} + \frac{\partial x_i}{\partial x_j}\right) \tag{3}$$

The contaminant dispersion equation is used to describe the characteristics of contaminant dispersion in a street canyon environment. The pollutant diffusion equation assumes that no reaction or mass exchange occurs between the ambient air and pollutants during transport and that the flow is stable.

$$\frac{\partial C}{\partial t} + \frac{\partial}{\partial x_j}(c u_j) = \frac{\partial}{\partial x_j}\left(D\frac{\partial c}{\partial x_j}\right) \tag{4}$$

where t denotes the time, c is the transient concentration, and D is the molecular diffusion coefficient. The renormalization group (RNG) $k-\varepsilon$ model is derived from a rigorous statistical technique, which was proposed by Yakhot et al. [40].

$$R_\varepsilon = \frac{C_u \rho \eta^3 (1-\eta/\eta_0)}{1+\xi\eta^3} \cdot \frac{\varepsilon^2}{k} \tag{5}$$

where C_μ, η_0, and ξ are model constants, and $\eta \equiv Sk/\varepsilon$, where S is the scale of strain rate.

3.2. Inverse Identification of Source Location Method in Dynamic Airflow Fields

Our previous research applied the inverse CFD method to steady-state airflow fields [35]. In this study, we primarily investigated the application of inverse CFD modeling to the dynamic airflow fields. Based on the adjoint probability method, the major difference between dynamic and steady-state airflow fields is that the airflow field changes with time. In dynamic airflow fields, the backward time and multiple inverse groups of airflow fields must be considered. In the forward simulation, multiple groups of forward airflow fields $\vec{V}_j$ (j represents different time points) can be obtained, and the remaining forward process is consistent with the operating process under the steady-state airflow field.

In the inverse simulation process, the velocity field at the last time step of the forward simulation process is initially reversed. The concentration is calculated at the location of the monitoring point to obtain the first concentration field in the reverse time step. Subsequently, all the concentration information in the computational domain is extracted and imported into the reverse velocity field under the penultimate time step in the forward process to continue the calculation. This is performed until the first time step is calculated in the forward process, and the obtained concentration field is the SALP in the dynamic flow field. The remaining processes in the reversal process under a dynamic airflow field are consistent with those under a steady-state airflow field, which can be observed in our previous paper [35]. A flowchart of the specific process used to identify the pollutant sources under dynamic airflow fields is shown in Figure 3a.

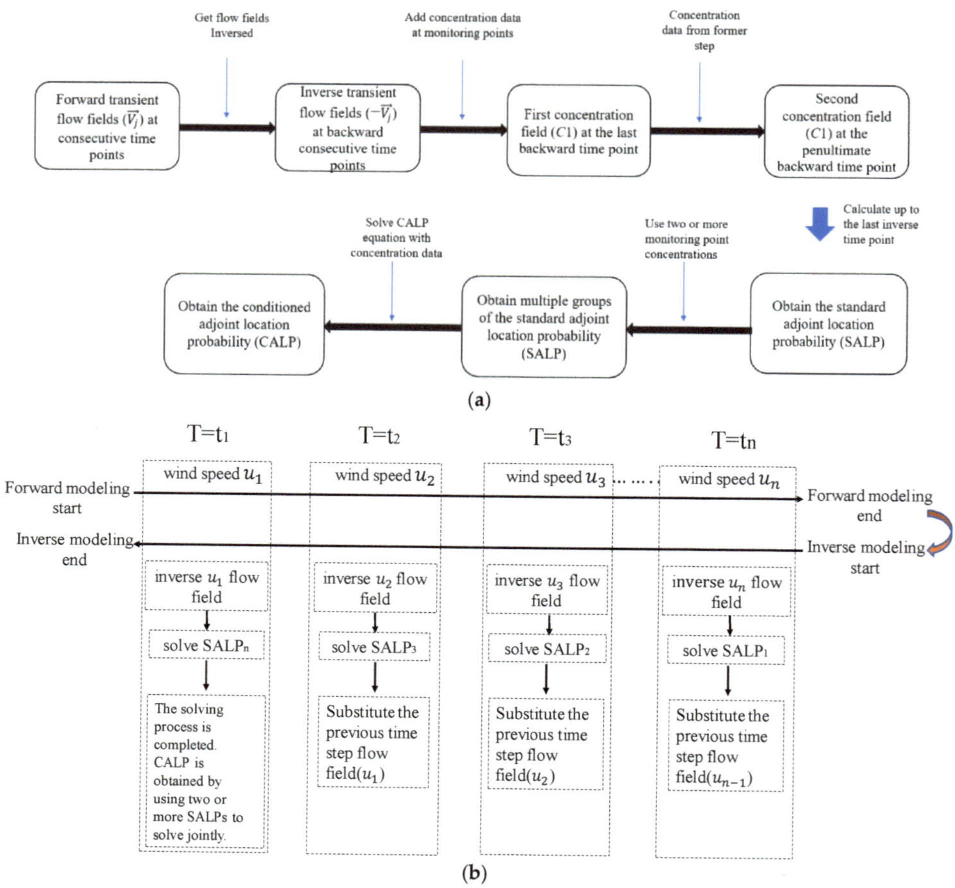

Figure 3. Illustrations of the source identification method with dynamic wind conditions: (**a**) flow chart of pollutant source identification under dynamic airflow fields; (**b**) steps of source identification under dynamic airflow field with a period of t_n.

Figure 3b shows the detailed source identification steps under the dynamic airflow field, which are as follows:

1. The unsteady airflow fields at t_n min in the forward simulation are used. In the forward simulation, the airflow field is calculated every 1 min. During t_n min, the airflow fields are simulated 30 times to obtain 30 airflow fields in this period. The start time of the forward simulation is set to $T = t_1$. Subsequently, using the 1 min average outdoor wind speed as the inlet boundary condition, the airflow field u_1 can be obtained. The final airflow field u_n is obtained after 30 forward simulations at $T = t_n$. Subsequently, the forward simulation is completed.
2. Probability-based inverse modeling starts from the backward time step and the final airflow field u_n. Release one unit of tracer gas CO_2 in the monitored room that exceeds the concentration limit and obtain SALP1 of the final time step t_n.
3. The calculation is stopped immediately after the convergence of SALP1 is achieved. The SALP1 values of all the grid nodes in the calculation domain are exported. Subsequently, by inverting the airflow field u_{n-1}, airflow field u_{n-1} is imported to SALP1 and continuously calculated to obtain SALP2.
4. Each SALP is obtained by inverting 30 airflow fields in the forward simulation. The adjoint probability of the final pollution SALPn under a dynamic airflow field is then obtained.
5. Finally, the CALP was calculated by combining the SALP with the concentration data from multiple monitoring points.

4. Accuracy of the CFD Simulation

Numerical simulations of the urban two-dimensional street canyon airflow field were performed based on actual outdoor atmospheric conditions obtained from experiments. The results of the numerical simulations are compared with the experimental data to verify the accuracy of the simulations.

4.1. Simulation Setting

To simulate the airflow and pollutant dispersion fields in street canyons, we built a simplified T-shaped computational domain with the upstream, downstream, and height of the computational domain as 3H, where H is the building height (1.2 m), as shown in Figure 4a,b. According to Ai et al. [41], the use of a T-type computational domain can effectively predict airflow fields in street canyons. The T-shaped computational domain significantly reduces the number of grids, which also reduces the time required for the simulation, particularly for unsteady simulations.

The dimensions of the building models used in the simulation were established based on a scaled outdoor experiment, as shown in Figure 4a. The simulation areas of this study focused on the target building of the experiment, which had a height-to-width ratio of 1:1 in a street canyon. The buildings marked in red in Figure 4a are Buildings A, B, and C from left to right.

The boundary conditions used in this paper are shown in Table 3. A uniform velocity was used as the inlet velocity boundary condition. Ai et al. [41] verified that the simulation results of the uniform velocity boundary conditions in a T-shaped street canyon model were similar to those of the atmospheric boundary layer under the logarithmic law. This is because the roughness height at the top of the building is much smaller than that above the urban ground, where the buildings themselves are rough elements. Therefore, uniform inflow boundary conditions can be used under these circumstances.

The experimental results showed that the wind speed and direction changed constantly during the experimental period. We selected experimental data from 10:40 a.m. to 5:40 p.m. on 9 June 2019 as the inlet boundary condition. During the experimental period, the wind direction ranged from 43° to 50°, which could be approximated as blowing vertically toward the street canyons. We selected wind speed data monitored by the ultrasonic anemometer at a height of 1.2 m on the 10 m wind mast during this period as the inflow velocity, which was also the height of the building top. It is appropriate to select the wind speed value at the top of the building for the boundary conditions of the inlet velocity in the T-shaped calculation domain [41].

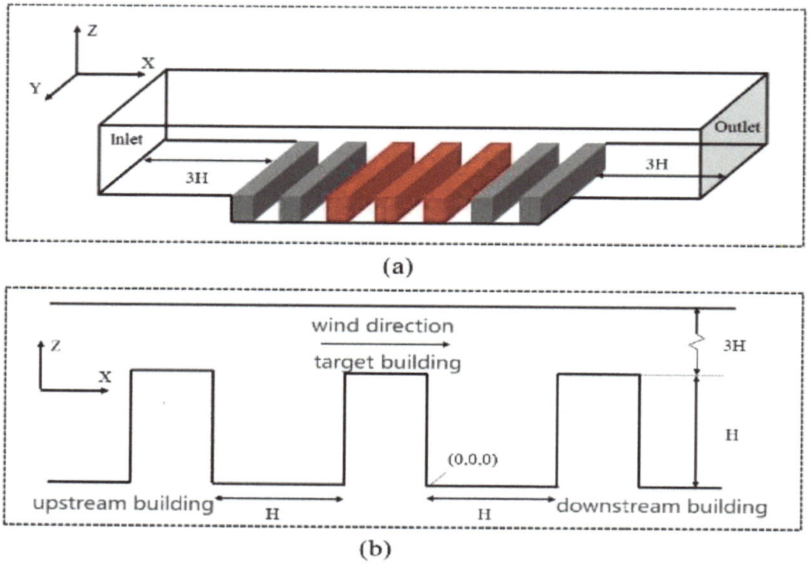

Figure 4. Model settings of the CFD simulation: (**a**) overall schematic of the model; (**b**) enlarged schematic of target buildings in the street canyon.

Table 3. Boundary conditions.

Inlet	Velocity Inlet (Wind Speed Values at 1.2 m from the 10 m Wind Mast on 9 June 2019, 14:16:05–14:31:05, as Shown in Figure 5)
Outlet	Pressure outlet
Near-Wall Treatment	Standard Wall Functions
Temperature	Average temperature of different areas in the street canyons in different periods (as shown in Table 2)
Building walls	Non-slip for wall shear stress
Turbulent kinetic energy	$k = \left(U_{ref} \times T_i\right)^2$
Turbulent viscous dissipation rate	$\varepsilon = \frac{C_\mu^{3/4} k^{3/2}}{l_t}$
Constants	$C_\mu = 0.09$ $T_i = 3\%$ $l_t = 0.7$ m $L_w = 0.9$ m $A_f = 0.6$ $U_{ref} = 6.7$ m/s

During the experimental period, data monitored by the ultrasonic anemometer at 1.2 m on the 10 m wind mast were recorded at 20 Hz. For the inflow boundary conditions, we considered an average wind speed every 1 s as a uniform inlet velocity. The processed velocity boundary conditions were written into a user-defined profile in Fluent as the inlet velocity. Figure 5 shows the velocities measured by the ultrasonic anemometer at 1.2 m on the 10 m wind mast during this period.

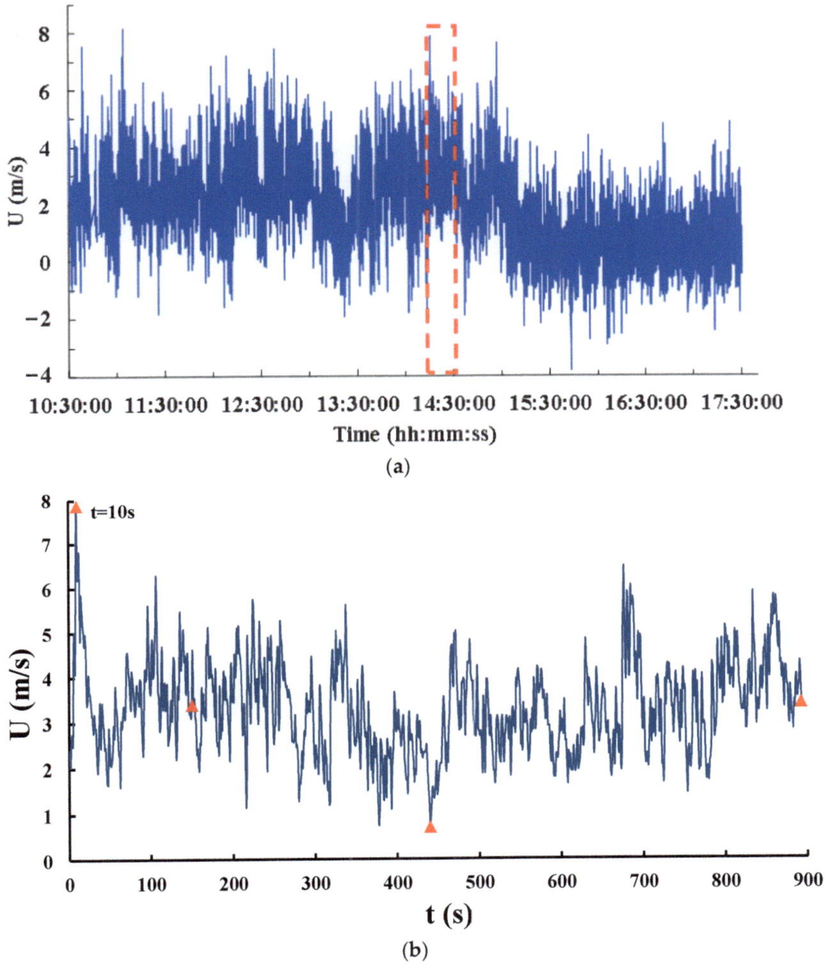

Figure 5. Wind speed values at 1.2 m from the 10 m wind mast on 9 June 2019; (**a**) wind speeds from 10:40 a.m. to 5:40 p.m.; (**b**) enlargement of 900 s wind speed from 14:16:05 to 14:31:05. (Dashed box represents a period of time selected as the inlet velocity boundary condition; Four red triangle represent the wind speed at each of the four time points (t = 10 s, t = 150 s, t = 440 s, t = 900 s).)

The $RNG\ k-\varepsilon$ model was selected as turbulence model in the numerical simulations, and the formula for inlet turbulence kinetic energy (k) and turbulence dissipation rate (ε) is expressed as follows [39]:

$$k = \left(U_{ref} \times T_i\right)^2 \tag{6}$$

$$\varepsilon = \frac{C_\mu^{3/4} k^{3/2}}{l_t} \tag{7}$$

where C_μ is a constant, C_μ = 0.09; T_i is the turbulence intensity, $T_i = 0.16Re^{-1/8} = 3\%$; l_t is turbulence characteristic size, $l_t = 4A_f/L_w = 0.7$ m; A_f is a cross-sectional area of fluid, A_f = 0.6 m; and L_w is wet perimeter length, L_w = 0.9 m.

The surface temperatures of the building models in the numerical simulations were based on the data measured during the experiment. During the experimental period, from

10:40 a.m. to 5:40 p.m., the intensity of solar radiation received by the experimental site varied. Therefore, the average temperature of each building surface during each period, as listed in Table 2, was considered the simulated temperature boundary condition.

4.2. Grid Sensitivity Test

We established scaled urban two-dimensional street canyons of 1:20, which were consistent with the experimental model. To reconstruct the full-scale airflow field, the Reynolds-number-independent similarity criterion was to be satisfied. The Reynolds number based on the building height was 75,600, which was greater than the reference value (Re = 11,000) of Reynolds independence [42].

In this study, the minimum grid sizes of the building model were 0.01, 0.005, and 0.0025 m, as shown in Figure 6a, with total grid numbers of 3.5 million, 5.6 million, and 7.2 million, respectively. A steady airflow field was used for comparison in the grid sensitivity test. A power law was fitted to the experimental data to determine the inlet velocity wind profile. By fitting the wind speed data from 10:00 a.m. to 3:00 p.m. on 9 June 2019, the formula can be expressed as follows [28]:

$$\frac{U}{U_H} = \left(\frac{Z}{Z_H}\right)^{\alpha} \tag{8}$$

where U is the wind speed at height H (m/s); U_H is the wind speed at the reference height of the street canyons (2.46 m/s); and Z_H is the height of the street canyon (1.2 m).

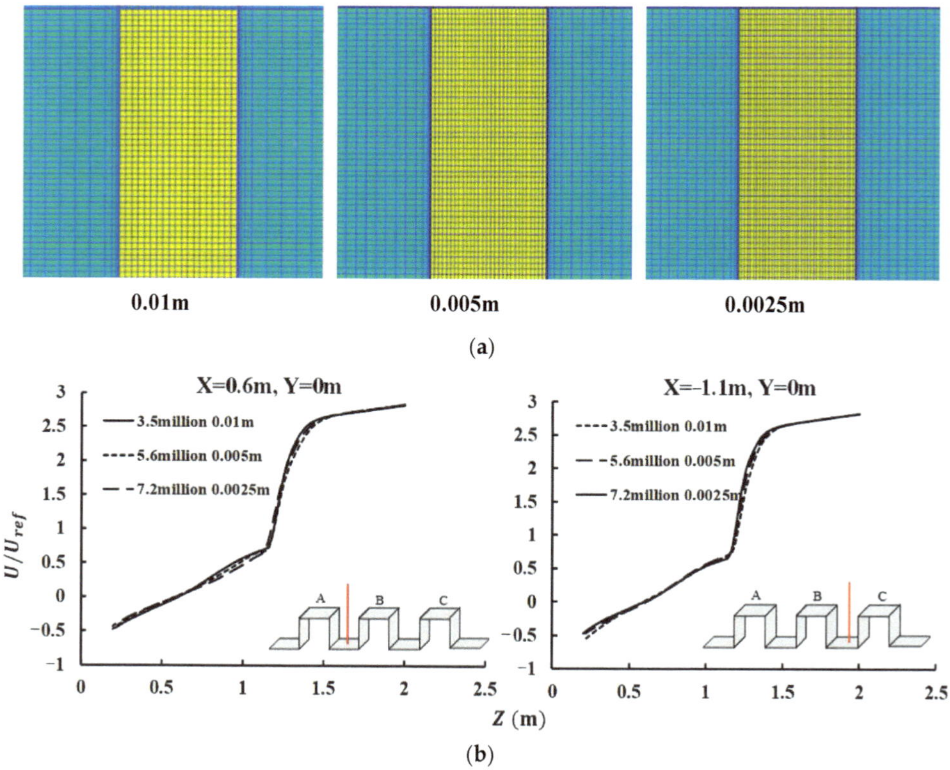

Figure 6. Sensitivity tests for the CFD simulations: (**a**) near wall grid sizing of the three systems; (**b**) sensitivity tests for the three mesh systems at lines X = −1.1 m, Y = 0 m, Z = 0–2 m; X = 0.6 m, Y = 0 m, Z = 0–2 m.

To verify the independence of the grid, we selected two vertical lines (X = −1.1 m, Y = 0 m, Z = 0–2 m and X = 0.6 m, Y = 0 m, Z = 0–2 m) at the center of the target street canyons for comparison with the normalized mean u-velocity (U/Uref). Figure 6b shows a comparison of the dimensionless velocity results of the three grid systems at the center of the street canyons. The speed prediction results of the three grid systems were very similar. Owing to the negligible deviation of the three grid systems, a coarse grid system was selected for the subsequent simulations in this study [35], which reduced the simulation resources for dynamic modeling.

4.3. Model Validation

We used wind speed values from 14:20:00 to 14:43:20 on 9 June 2019 to validate the accuracy of the CFD simulations. Figure 7 shows a comparison of the dimensionless velocity in the X-direction in the street canyon center between the CFD simulation and experimental measurements. The experimental data were obtained from the ultrasonic anemometer measurement values at a height of 1.05 m on a 2.4 m wind mast. Both simulated and experimental speeds underwent intermittent changes over time. A fifteen percent error bar was incorporated into the comparisons between the experiment data and the CFD results. This figure illustrates that the majority of the CFD simulation results fall within the 15% range of the experimental data. Furthermore, the trends of the simulations align with the fluctuating characteristics of outdoor environmental wind speeds.

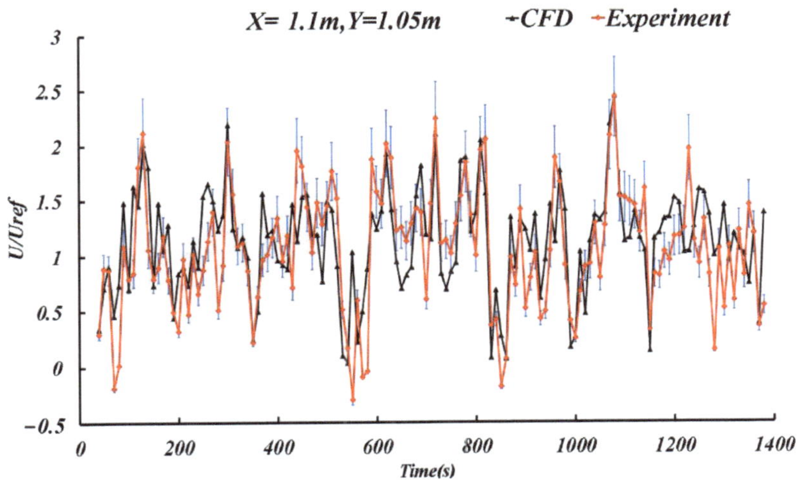

Figure 7. Comparisons of the dimensionless U velocity of the CFD simulation and experimental measurement (from 14:20:00 to 14:43:20 on 9 June 2019) in the street canyon.

However, the simulated velocities were not identical to the experimentally measured values. These deviations can be attributed to several factors. First, we used the 1 s average value to process the inlet velocity, which may have resulted in differences in wind velocities. Subsequently, the inlet wind profile was based on the experimental data from the 10 m wind mast, whereas the experimental data for comparisons were based on the 2.4 m wind mast. The distance between these two wind masts may have resulted in a 1–2 s delay in the wind speed, which may also have led to deviations in wind velocity results.

Generally, the overall trends of wind velocity in street canyons were similar, and the airflow characteristics, including sudden increases or decreases in velocity values, were reconstructed. Therefore, the simulated velocity field was considered to be in agreement with the velocity field in the actual environment.

5. Results and Discussions

5.1. Airflow Fields

We selected four time points to underline the airflow field inside and above the street canyons, as shown in Figure 5b. The velocities at $t = 10$ s (14:16:15) and $t = 440$ s (14:23:25) were the highest and lowest values of the boundary velocity, respectively; the velocities at $t = 10$ s (14:16:15) and $t = 900$ s (14:31:05) were the starting and ending values of the boundary velocity, respectively; and the velocities at $t = 150$ s (14:18:35) were the average values of the boundary velocity.

The influence of the time-varying wind conditions on the airflow fields inside and above the street canyon is shown in Figure 8 for four different time points ($t = 10, 150, 440,$ and 900 s). The results indicated that the upper-boundary time-varying wind conditions affected the vortex and shear layers in the street canyon. The airflow movement in the street canyons was stable with a single vortex. This vortex is a typical flow structure in a street canyon in a given direction with a constant wind velocity [43,44]. As indicated in previous studies [45,46], with an unchangeable wind direction under steady and real-time boundary conditions, the basic airflow structures inside and above the street canyon are similar. However, under real-time boundary wind conditions, the airflow structure inside and above street canyons exhibits a highly intermittent process [31]. In this study, this phenomenon occurred in the airflow structures of street canyons under time-varying wind conditions. At the four time points, a horizontal shear layer formed at the top of the street canyon, and a clockwise vortex formed inside the street canyon. The vortex characteristics in the street canyons were different, which were the results of the changing upper wind velocities from $t = 10$ to 900 s. For $t = 10$ s, compared with $t = 150$ s, the upper wind speed increased significantly, and the overshoot of the upper boundary airflow led to the compression of the vortex inside the street canyon. For $t = 440$ s, compared with $t = 10$ s, the upper wind speed decreased, and the decay of the upper boundary airflow led to the expansion of the vortex inside the street canyon.

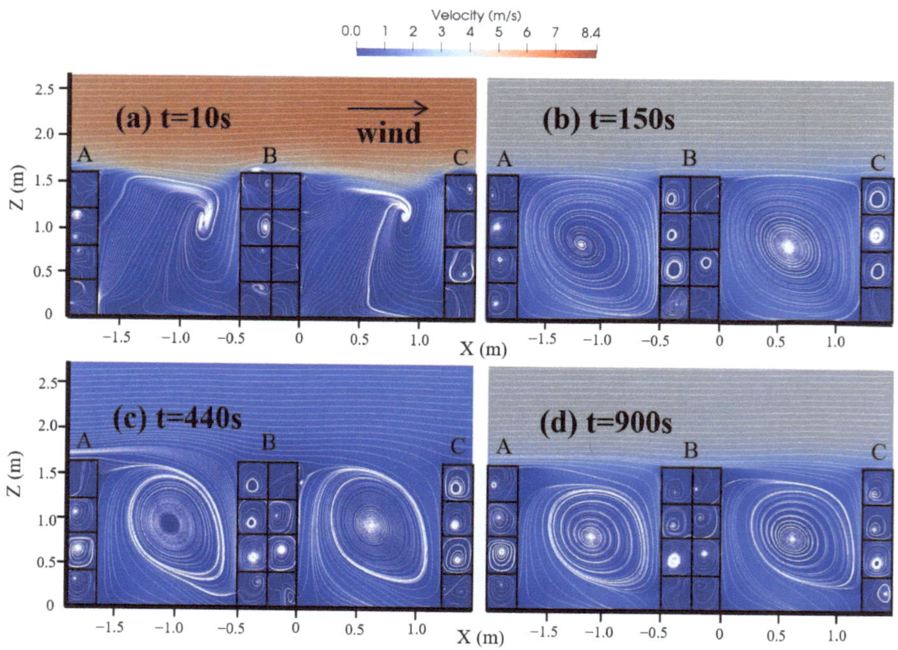

Figure 8. Velocity contours and vector distributions on the vertical plane ($Y = 0.25$ m) at four time points: (**a**) t = 10 s, (**b**) t = 150 s, (**c**) t = 440 s, (**d**) t = 900 s.

Moreover, the stability of the vortex structure within the street canyon is noteworthy. The 2D streamlines, as depicted in the streamwise flow profile (XZ-plane), offer a lucid depiction of how the airflow undergoes changes influenced by the street canyon's structure. The airflow follows a trajectory along the upper boundary of the street canyon, penetrates the interior, makes contact with the building walls and ground, thereby altering its course. The building walls and ground act as impediments to the airflow, resulting in a higher wind speed on the windward side in comparison to the leeward side. This discrepancy in wind speeds contributes to an elevated concentration of pollutants on the leeward side as opposed to the windward side during the dispersion of pollutants within the street canyon.

As shown in Figure 9, the wind speed was different at the vertical line (Z = 0 − 1.2 m) in the street canyon. Owing to the vortex movement and centrifugal effect, the wind speeds at the bottom and top of the street canyon were much higher than those at the middle height of the street canyon. From $t = 10$ to 900 s, the four wind speeds exhibited a fluctuating trend with changes in the time-varying wind conditions.

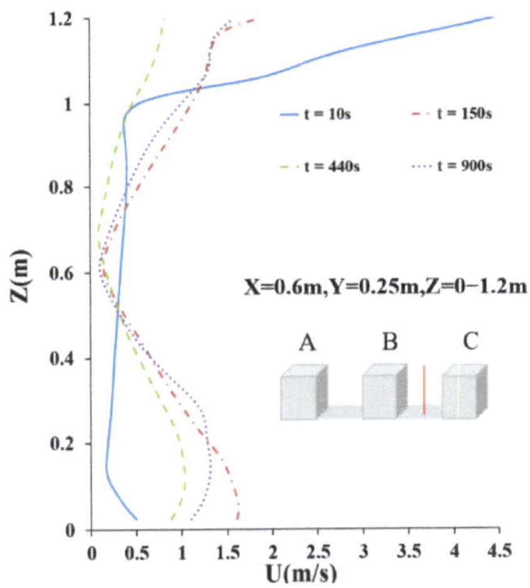

Figure 9. Velocities of the vertical line (Z = 0 − 1.2 m) in the street canyons at four time points.

The wind speed at the top of the street canyon was affected by the upper boundary wind conditions. The wind speeds at the bottom and middle parts were affected by both the upper boundary wind conditions and airflow structure in the street canyon. This conclusion was also reached by Zhang et al. [31] and Louka et al. [29], who experimentally determined the flapping of the shear layer. The shear layer at the top of the street canyon affects the vortex of the airflow inside the street canyon, reflecting the flapping of the shear layer.

In this study, the influence of time-varying wind conditions on the airflow field in a street canyon was investigated, and the dynamic wind conditions affected the diffusion of pollutants in the street canyon.

5.2. Temperature Fields

The effect of time-varying wind conditions on the temperature distribution in street canyon models was studied at three different time points ($t = 10$, 150, and 440 s). The velocities at $t = 10$ and 440 s were the highest and lowest values of the boundary velocity, respectively, and the velocity at $t = 150$ s was the average value of the boundary velocity. As shown in Figure 10, the leeward room temperature on top of the street canyon (room BL4) was higher than that of the windward room (room BW4), and the temperature on the leeward sidewall of the street canyon was slightly higher than that on the windward side. This may be attributed to the solar direction: the leeward side of the street canyon absorbed more solar energy than the windward side, as described in a scaled outdoor experiment [28]. The internal temperature distribution was analogous to that of the airflow vortex in a street canyon. When the low-temperature incident wind flowed over the street canyon, the incident wind and air in the street canyon underwent a heat exchange process, and the wind removed part of the heat in the street canyon, reducing the air temperature in the street canyon. Similar results were obtained by Li et al. [47], where the air temperature of the street canyon was directly related to the airflow temperature, and a lower wind speed was conducive to heat dispersion in the street canyon. Some studies have also shown that the ambient wind speed can affect the temperature of street canyons, and a decrease in ambient wind speed can result in an increase in temperature [48].

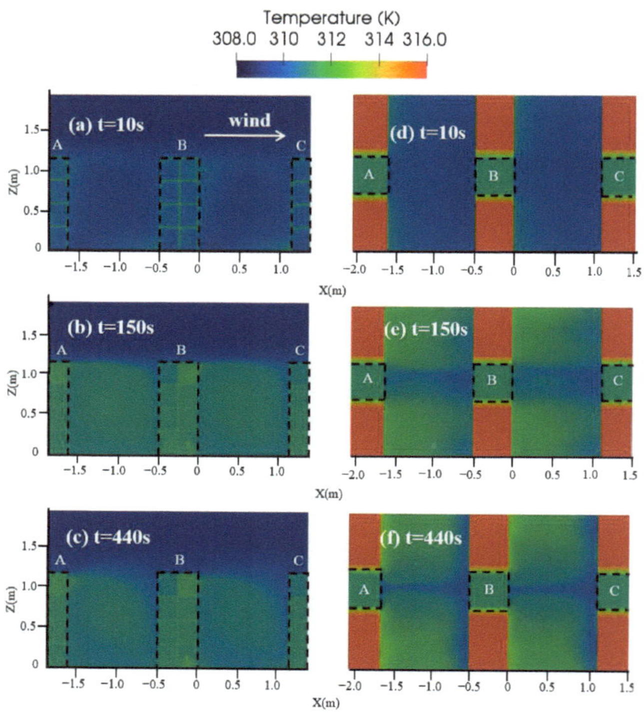

Figure 10. Temperature distributions on the vertical plane $Y = 0.25$ m and the horizontal plane $Z = 1.2$ m at three time points: (**a**,**d**) t = 10 s, (**b**,**e**) t = 150 s, (**c**,**f**) t = 440 s.

We observed that a street canyon room with openings can affect the temperature distribution and that the air temperature in a street canyon room with an opening is lower than that in a room without an opening. Owing to the room opening of street canyons, the low-temperature inlet wind enters each room, generates vortices, and exchanges heat with the air near the building opening, causing airflow disturbance and reducing the temperature. The temperatures at other locations in street canyons without openings are higher than those at the center of street canyons without air heat transfer.

5.3. Pollutant Dispersion inside the Street Canyons under Time-Varying Wind Conditions

5.3.1. Interunit Dispersion on the Leeward Side of Building B

This study investigated the influence of dynamic wind conditions on the dispersion of pollutants in a street canyon and identified the instant pollutant source locations using concentration data from a scaled outdoor experiment. We selected one room as the pollutant source to simulate pollutant dispersion in street canyons under time-varying wind conditions. Room BL1 was set as the source room and released carbon dioxide (CO_2) as the tracer gas, whereas the tracer gas concentrations in the other seven rooms (BL2, BL3, BL4, CW1, CW2, CW3, and CW4) were monitored.

In this study, continuous periods (t = 1, 5, 8, 12, and 15 min) were selected to observe concentration changes under time-varying wind conditions. As shown in Figure 11a, at t = 1 min, the CO_2 in room BL1 was released for 1 min, and the pollutants were primarily concentrated in the lower part of the room. At t = 5 min, the outdoor wind speed decreased, and the tracer gas entered the upper two rooms. Under the effect of dynamic wind, the tracer gas diffused from the opening of the source room to rooms BL2 and BL3; however, the concentration magnitude levels of these two rooms were much lower than those of the source room. At t = 8 min, the outdoor wind speed was the average wind speed during the entire process, which was slightly higher than that at t = 5 min. Here, the tracer gas concentration accumulated in the source room, and the concentration value was higher than that at t = 1 min. The tracer gas also spread across rooms; after entering BL2 and BL3 rooms successively, the tracer gas spread along the leeward side wall to the top of the street canyon. At t = 12 min, when the wind speed reached its maximum value, the air mass in the street canyons was compressed by the upper airflow, and the diffusion process of the tracer gas to the top of the street canyons was weakened. At t = 15 min, the re-entered tracer gas dispersed in rooms BL2 and BL3 was more evenly dispersed, owing to the dynamic wind effect.

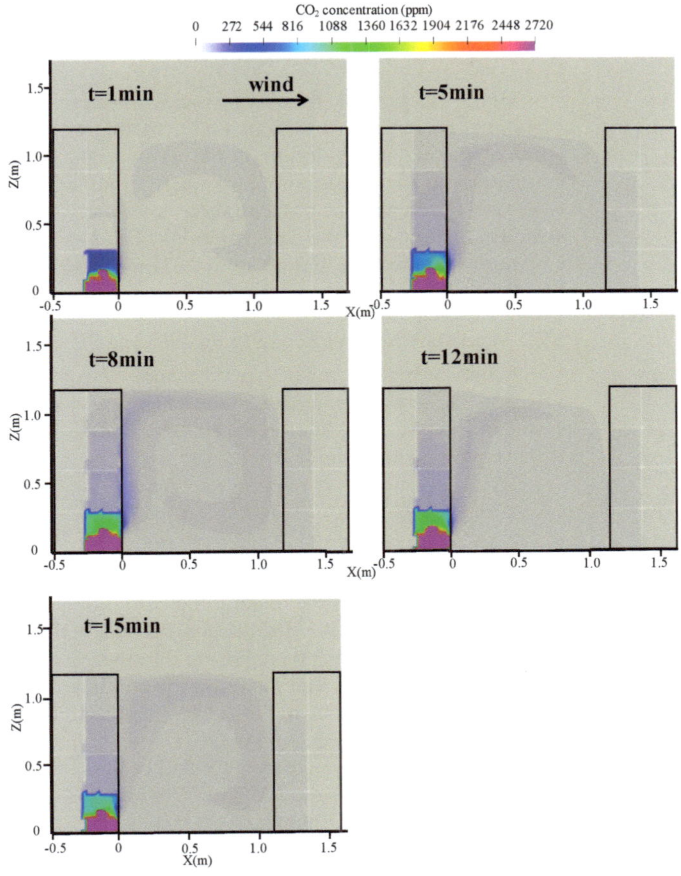

(a)

Figure 11. *Cont.*

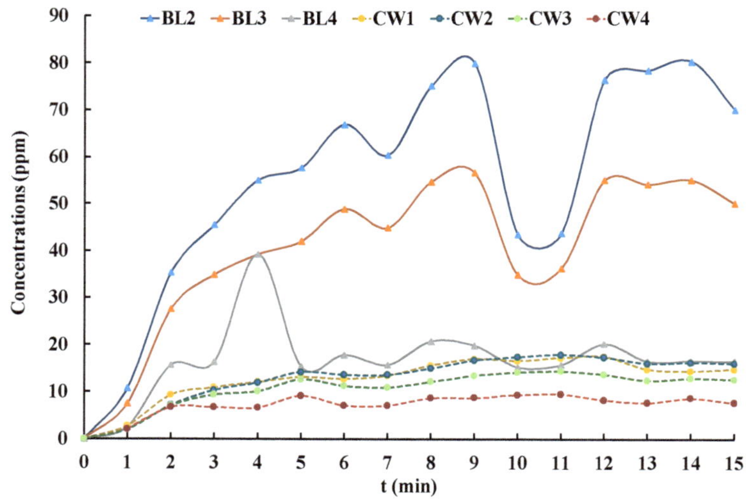

(b)

Figure 11. Concentration dispersions in the street canyon: (**a**) CO_2 dispersion in source room BL1 at five time points; (**b**) CO_2 concentrations of seven monitored rooms during gas releasing process for 15 min.

Figure 11b presents the 60 s averaged value of the CO_2 concentrations in the seven monitored rooms, which exhibited a fluctuating trend under the influence of dynamic wind. As shown in Figure 11b, of the seven rooms, BL2 and BL3 had the highest CO_2 concentrations. The concentration values of rooms BL2 and BL3 changed dramatically with a similar trend. The concentrations in rooms CW1, CW2, CW3, and CW4 changed steadily. At t = 0 min, the concentrations of BL2 and BL3 increased sharply, peaked at t = 9 min, decreased significantly until t = 11 min, and then increased to higher values after t = 11 min.

5.3.2. Ventilation Rates of Rooms in the Building B

In this study, the ventilation performances of eight rooms in Building B were analyzed using the tracer gas decay method. Assuming there was neither a background concentration nor an indoor tracer gas source, the air exchange per hour (ACH) (h^{-1}) can be calculated using a two-point method [49]:

$$ACH = \frac{\ln(\frac{C_{t_i}}{C_{t_{i+1}}})}{t_{i+1} - t_i} \times \frac{3600}{t} \tag{9}$$

Figure 12a presents the relationship between the tracer gas concentration and time with an exponential change. The attenuation rates of the tracer gas concentration in rooms BW2 and BW3 were higher than those in the other six rooms within the first 100 s after the tracer gas was released. The rooms on the windward side were directly affected by the wind, which resulted in the tracer gas decay rate of the room on the windward side being higher than that on the leeward side.

Figure 12b shows box charts of the ACH values of the eight rooms in the street canyon. The box edges represent the 25th and 75th percentiles. A line in the box represents the 50th percentile of a data group, and the symbol (□) indicates the mean value of each data group. When the room was on the windward side, the mean ACH of BL2 was slightly higher than those of the other three rooms. Under dynamic wind conditions, the ACH values of BW1 and BW4 fluctuated significantly, whereas the ACH values of BW2 and BW3 remained stable. When the room was located on the leeward side, the ACH values of the four rooms were consistent with those on the windward side. The ACH values of the highest and lowest rooms fluctuated significantly, and the ACH values of the middle room were relatively stable, which differed from the ventilation efficiency calculated using the integral method [50]. The tracer gas decay method can more precisely predict the ventilation distribution in rooms in street canyons under a transient airflow field.

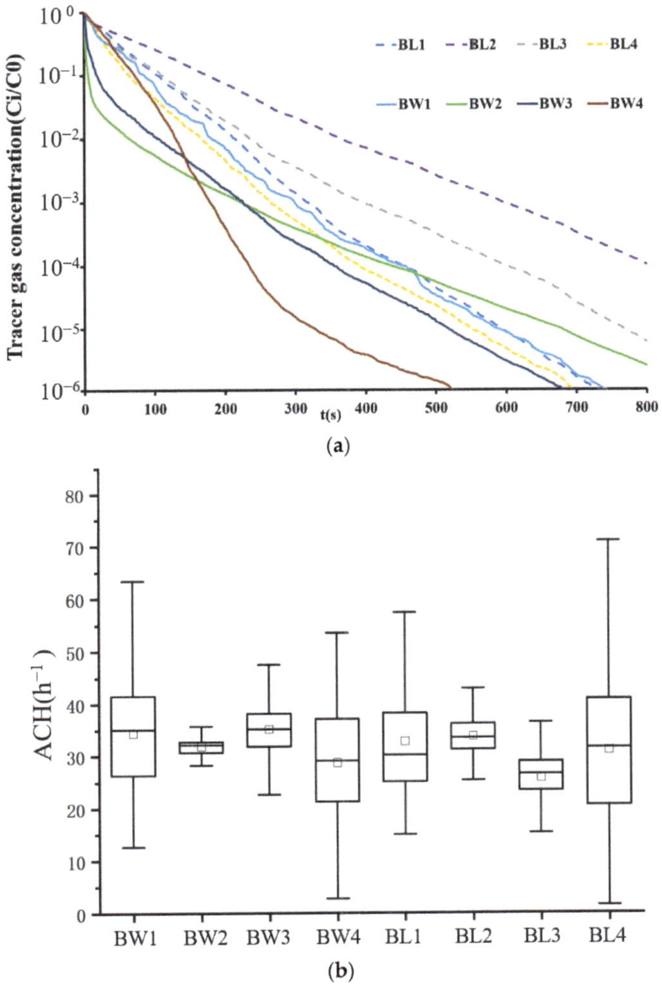

Figure 12. The results of concentration decays and ventilation rate of each room: (**a**) relationship between dimensionless tracer gas concentration and time with an exponential change; (**b**) box charts of ACH values of each room in the street canyon.

5.4. Results of Pollutant Source Identification

In this study, we selected three cases to apply probability-based inverse modeling to locate source rooms under real atmospheric conditions. The pollutant concentration was set to a limit of 10 ppm; when the pollutant concentration in the monitored room exceeded 10 ppm, an abnormal concentration alarm occurred immediately. Probability-based inverse modeling was then applied to identify the location of the source room in the street canyon building when alarms occurred in more than two rooms.

5.4.1. Case 1: Abnormal Concentration Values in Monitored Rooms BL2, BL3, and CW3

During the period from 14:18 to 14:31 p.m., the monitored rooms BL2, BL3, and CW3 in the street canyon building were alarmed, and the real-time concentration data monitored by the pollutant sensors are shown in Table 4. The highest concentration of pollutants in monitored room BL2 was 39.7 ppm and could still be maintained at the highest concentration at the end of the experimental period. Based on the obtained experimental data, a forward CFD simulation was first developed to obtain the ambient airflow field in street canyons during the period with real atmospheric boundary conditions. Room BL2, the first room to exceed the pollutant limit, was selected to simulate the SALP probability field using the source identification method under the dynamic airflow field proposed in Section 3.2. Similar operations were performed for rooms BL3 and CW3, which subsequently exceeded the pollutant limits, to obtain two sets of SALP probability fields. The final CALP probability field was obtained by solving the three sets of SALP fields jointly with the experimentally measured concentration values exceeding 10 ppm, as shown in Figure 13. The red part of the figure represents the maximum possible location of the pollutant source, and the black circle represents the location of the real pollutant source room in the experiment.

Table 4. Concentrations of room BL2, BL3, and CW3 over the limit (excluding background concentration, ppm).

	Room	Time over 10 ppm	Concentration Value over 10 ppm	Maximum Concentration Value	Final Concentration Value
Room 1	BL2	14:18:28	12.8 ppm	39.7 ppm	36.1 ppm
Room 2	BL3	14:22:20	10 ppm	16 ppm	12 ppm
Room 3	CW3	14:30:58	10 ppm	13 ppm	8 ppm

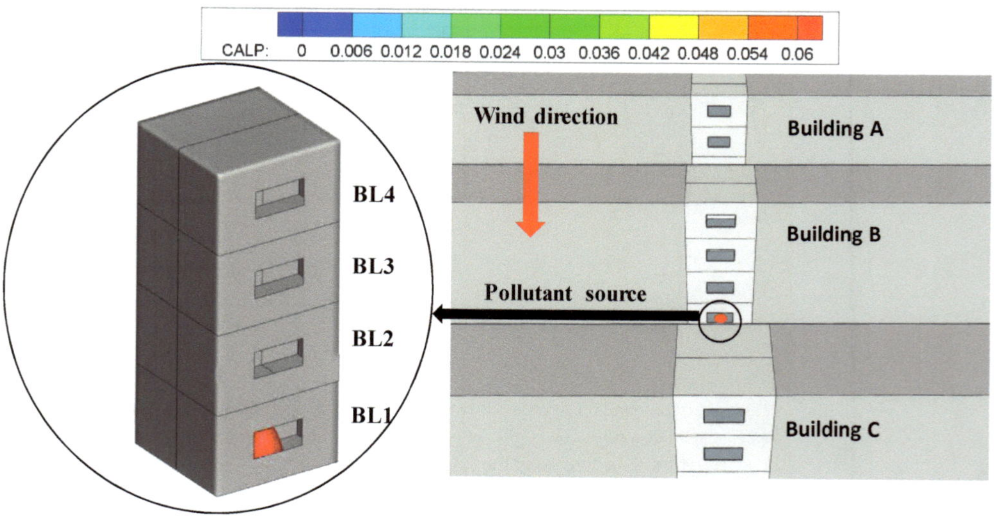

Figure 13. Predicted location probability of pollution source room BL1.

The figure shows that the maximum possible location of the predicted pollutant source was in the center of room BL1, and the value for the CALP is 0.06. The value of CALP for room BL1 is greater than the other rooms. Compared with the experimental settings, and in this period, the source room was also room BL1. Therefore, when the monitored concentration values in rooms BL2, BL3, and CW3 showed abnormalities, probability-based inverse modeling was initiated immediately to find the source location, which was successfully located in source room BL1. This case also implied that, in a street canyon environment, the pollutant in the room on the lower part of the leeward side has the greatest effect on the other rooms, and it is important to accurately identify the location of the source room.

Note that, from the experimental results in Table 1, five rooms were over the threshold, but in this case, we only used the first three rooms that exceeded the thresholds, which could already successfully locate the source room.

5.4.2. Case 2: Abnormal Concentration Values in Monitored Rooms BL3 and BL4

During the period from 15:12 to 15:19 p.m., monitoring rooms BL3, BL4, and AW3 of the street canyon building were alarmed, and the real-time concentration data monitored by the pollutant sensors are shown in Table 5. The pollutant concentrations in rooms BL3 and BL4 exceeded the limits approximately 1 and 4 min after the beginning of the experiment, respectively. The measured pollutant concentration data from these three rooms were used as inputs for inverse modeling. The forward velocity field was simulated based on the wind speed information obtained from the experiment during this period. Subsequently, the SALP probability fields of monitored rooms BL3 and BL4 could be obtained. The predicted pollutant source locations are shown in Figure 14.

Table 5. Concentrations of room BL3 and BL4 over the limit (excluding background concentration, ppm).

	Room	Time over 10 ppm	Concentration Value over 10 ppm	Maximum Concentration Value	Final Concentration Value
Room 1	BL3	15:12:34	16 ppm	154.6 ppm	130 ppm
Room 2	BL4	15:15:05	10 ppm	26.8 ppm	21.5 ppm

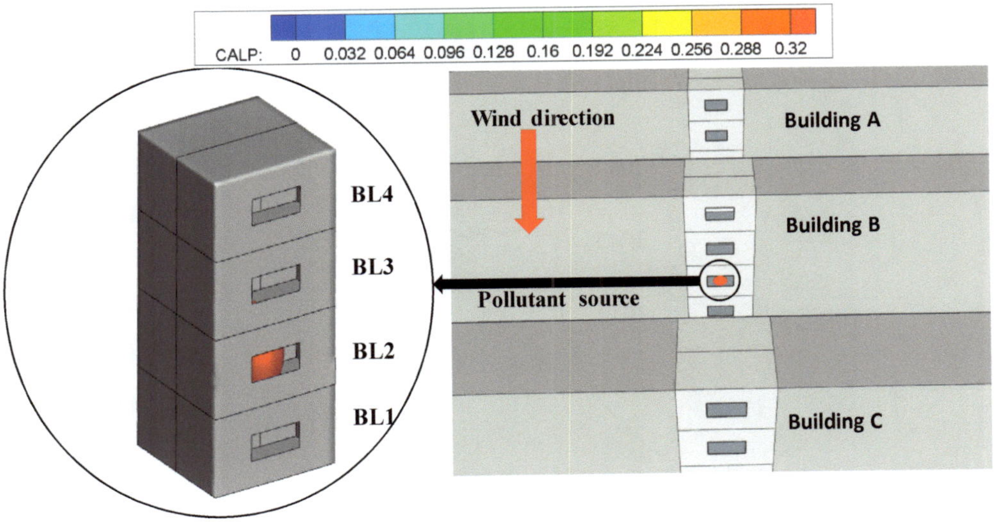

Figure 14. Predicted location probability of pollution source room BL2.

The figure shows that the predicted pollutant location was at the center of room BL2, and the value for the CALP is 0.32. The value of CALP for room BL2 is greater than the other rooms. For comparison with the experimental settings, the source room was room BL2. Furthermore, the highest concentration value occurred in room BL3, which could be up to 154.6 ppm, one order of magnitude beyond the concentration limit. This value was the maximum concentration detected in the monitored room throughout the experiment. This was sufficient to show that source room BL2 had a significant impact on the upper room.

5.4.3. Case 3: Abnormal Concentration Values in Monitored Rooms BW2 and BW1

During the period from 12:28 to 12:32 p.m., monitoring rooms BW2 and BW1 of the street canyon building were alarmed, and the real-time concentration data monitored by the pollutant sensors are shown in Table 6. Only the two monitored rooms in Case 3 obtained concentration data that exceeded the limit values during the experimental period. The pollutant concentration in room BW2 detected the over-limit value 1 min after the start of the experiment and directly reached 31.4 ppm, up to 88 ppm. Concentrations above the limit in room BW1 were detected after 4 min. Using the concentration data from these two monitored rooms and applying inverse modeling, we predicted that the maximum probability of the source was located on the lower side of room BW3 and the value for the CALP is 0.28, as shown in Figure 15. The value of CALP for room BW3 is greater than the other rooms. The experiment revealed that the source room was set in room BW3 during the period, which indicated the successful application of source identification.

Table 6. Concentrations of room BW2 and BW1 over the limit (excluding background concentration, ppm).

	Room	Time over 10 ppm	Concentration Value over 10 ppm	Maximum Concentration Value	Final Concentration Value
Room 1	BW2	12:28:45	31.4 ppm	88 ppm	63.5 ppm
Room 2	BW1	12:31:35	10 ppm	15 ppm	13 ppm

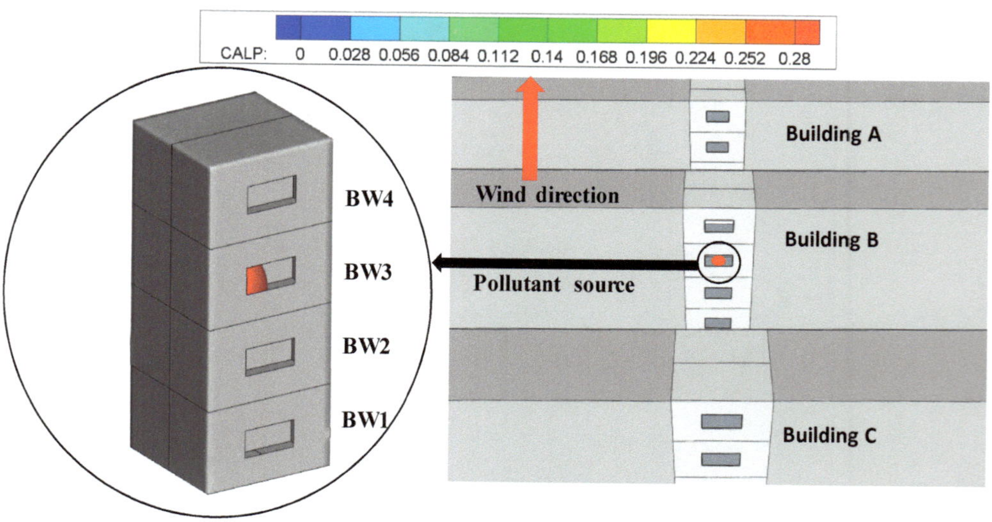

Figure 15. Probability of predicted location of pollution source room BW3.

6. Conclusions

In this study, an improved adjoint probability method based on CFD was developed to identify pollution source locations in street canyons under dynamic inflow conditions. We used scaled outdoor experimental data as the input conditions to simulate airflow, temperature, and concentration fields with time-varying wind conditions. Three experiments were conducted to verify the accuracy of the proposed source identification method. The following conclusions were drawn from the analysis of this study:

- The change in the boundary wind speed affects the airflow structure inside street canyons. The wind speed at each position in the street canyon fluctuates with changes in the boundary wind speed. Owing to vortex movement and the centrifugal effect, the wind speed at the top and bottom of the street canyons is greater than that in the middle.
- The outdoor wind with a lower temperature exchanges heat with the air at a higher temperature inside the street canyons, thereby removing part of the heat and reducing the heat of the air inside the street canyons. Moreover, the opening of the room produces some air disturbance, which is conducive to heat exchange between the air near the opening and the outdoor wind.
- The fluctuation of the upper wind speed influences the diffusion of the tracer gas, and the ventilation performance of the rooms in the middle of the street canyon is more stable than those in the top and bottom areas.
- We propose an improved adjoint probability method under dynamic wind conditions. Three cases were tested to verify the applicability of this method. The results showed that with limited pollutant information (two or more), all three cases could successfully identify the source location, indicating that this method can be used to locate pollutant sources in street canyons under time-varying inflows in coupled indoor and outdoor conditions.

This study acknowledges certain limitations. The model used in this study is based on a single-sided building with natural ventilation, whereas real-world buildings may have multiple openings. These differences in building types could potentially influence the results of pollutant source identification, an area that will be further explored in future research. Additionally, this study only considered time-varying wind velocities, while wind direction is also a dynamic factor. The impact of dynamic wind directions on source identifications will be considered in future studies.

Author Contributions: Conceptualization, Y.D.; methodology, Y.D.; software, M.H.; validation, M.H.; formal analysis, M.H.; investigation, M.H.; resources, Y.D.; data curation, M.H. and W.T.; writing—original draft preparation, M.H.; writing—review and editing, Y.D. and W.T.; visualization, M.H. and Y.D.; supervision, H.W.; project administration, H.W.; funding acquisition, Y.D. All authors have read and agreed to the published version of the manuscript.

Funding: This study was financially supported by the National Natural Science Foundation of China (No. 52200219), the Shanghai Sailing Program (No. 21YF1430700), and the Chen Guang project (No. 22CGA54), supported by the Shanghai Municipal Education Commission and Shanghai Education Development Foundation.

Data Availability Statement: The data presented in this study are available in the article.

Conflicts of Interest: The authors declare no competing interests relevant to the contents of this paper.

Nomenclature

Greek symbols

ε	Turbulent viscous dissipation rate (m^2/s^3)
η_0	Model constant
ξ	Model constant
κ	Von Karman constant, 0.41
ρ	Density (Kg/m^3)
ν	molecular viscosity (m^2/s)
φ	Adjoint probability

Nomenclature

A_f	Cross-sectional area (m^2)	t	Denotes time
C_u	Constant, 0.09	T_i	turbulence intensity, 4%
c	Transient concentration	u_i	Velocity component
D	Molecular diffusion coefficient (m^2/s)	α	Empirical coefficient (0.245)
p	Pressure (Pa)	U	Wind velocity at height H (m/s)
l_t	Mixing length, 0.4 m	U_H	Reference wind speed (2.46 m/s)
k	Turbulent kinetic energy (m^2/s^2)	Z_H	Height of the street canyon (1.2 m)
S	Scale of strain rate	$SALP$	Standard adjoint location probability
S_{ij}	Stain-rate tensor	$CALP$	Conditioned adjoint location probability

References

1. Zhou, Z.; Wang, P.; Deng, J.; Ouyang, C.; Xu, Y.; Jiang, W.; Ma, K. Numerical simulation of street canyon morphology and microclimate in hot summer and cold winter zone. *Buildings* **2023**, *13*, 2433. [CrossRef]
2. Zhang, Y.; Gu, Z.; Yu, C. Impact factors on airflow and pollutant dispersion in urban street canyons and comprehensive simulations: A review. *Curr. Pollut. Rep.* **2020**, *6*, 425–439. [CrossRef]
3. Albuquerque, D.P.; O'Sullivan, P.D.; da Graça, G.C. Effect of window geometry on wind driven single sided ventilation through opening. *Energy Build.* **2021**, *245*, 111060. [CrossRef]
4. Jiang, Z.; Kobayashi, T.; Yamanaka, T.; Sandberg, M.; Choi, N.; Kobayashi, N.; Sano, K.; Toyosawa, K. Wind tunnel experiment of wind-induced single-sided ventilation under generic shelter urban area. *Build. Environ.* **2023**, *242*, 110615. [CrossRef]
5. Mohammadi, M.; Calautit, J.; Owen, J. Inter-room pollutant transmission routes in naturally ventilated street canyon buildings. *J. Build. Eng.* **2023**, *79*, 107510. [CrossRef]
6. Li, Y.; Leung, G.; Tang, J.; Yang, X.; Chao, C.; Lin, J.; Lu, J.; Nielsen, P.; Niu, J.; Qian, H.; et al. Role of ventilation in airborne transmission of infectious agents in the built environment–A multidisciplinary systematic review. *Indoor Air* **2007**, *17*, 2–18. [CrossRef] [PubMed]
7. Nakamura, Y.; Oke, T. Wind, temperature and stability conditions in an east-west oriented urban canyon. *Atmos. Environ.* **1988**, *22*, 2691–2700. [CrossRef]
8. Dascalaki, E.; Santamouris, M.; Argiriou, A.; Helmis, C.; Asimakopoulos, D.N.; Papadopoulos, K.; Soilemes, A. On the combination of air velocity and flow measurements in single sided natural ventilation configurations. *Energy Build.* **1996**, *24*, 155–165. [CrossRef]
9. Eliasson, I.; Offerle, B.; Grimmond, C.S.B.; Lindqvist, S. Wind fields and turbulence statistics in an urban street canyon. *Atmos. Environ.* **2006**, *40*, 1–16. [CrossRef]
10. Niachou, K.; Livada, I.; Santamouris, M. Experimental study of temperature and airflow distribution inside an urban street canyon during hot summer weather conditions. Part II: Airflow analysis. *Build. Environ.* **2008**, *43*, 1393–1403. [CrossRef]
11. Xie, X.; Liu, C.; Leung, D. Impact of building facades and ground heating on wind flow and pollutant transport in street canyons. *Atmos. Environ.* **2007**, *41*, 9030–9049. [CrossRef]
12. Sini, J.; Anquetin, S.; Mestayer, P. Pollutant dispersion and thermal effects in urban street canyons. *Atmos. Environ.* **1996**, *30*, 2659–2677. [CrossRef]
13. Zheng, X.; Montazeri, H.; Blocken, B. Large-eddy simulation of pollutant dispersion in generic urban street canyons: Guidelines for domain size. *J. Wind. Eng. Ind. Aerodyn.* **2021**, *211*, 104527. [CrossRef]
14. Lv, W.; Wu, Y.; Zang, J. A review on the Dispersion and distribution characteristics of pollutants in street canyons and improvement measures. *Energies* **2021**, *14*, 6155. [CrossRef]
15. Hassan, S.; Akter, U.H.; Nag, P.; Molla, M.; Khan, A.; Hasan, F. Large-Eddy Simulation of airflow and pollutant dispersion in a model street canyon intersection of Dhaka city. *Atmosphere* **2022**, *13*, 1028. [CrossRef]
16. Jon, K.S.; Huang, Y.; Sin, C.; Cui, P.; Luo, Y. Influence of wind direction on the ventilation and pollutant dispersion in different 3D street canyon configurations: Numerical simulation and wind-tunnel experiment. *Environ. Sci. Pollut. Res.* **2022**, *30*, 31624–31652. [CrossRef] [PubMed]
17. Chen, G.; Hang, J.; Chen, L.; Lin, Y. Comparison of uniform and non-uniform surface heating effects on in-canyon airflow and ventilation by CFD simulations and scaled outdoor experiments. *Build. Environ.* **2023**, *244*, 110744. [CrossRef]
18. Yang, F.; Kang, Y.; Gao, Y.; Zhong, K. Numerical simulations of the effect of outdoor pollutants on indoor air quality of buildings next to a street canyon. *Build. Environ.* **2015**, *87*, 10–22. [CrossRef]
19. Fan, X.; Zhang, X.; Weerasuriya, A.; Hang, J.; Zeng, L.; Luo, Q.; Li, C.; Chen, Z. Numerical investigation of the effects of environmental conditions, droplet size, and social distancing on droplet transmission in a street canyon. *Build. Environ.* **2022**, *221*, 109261. [CrossRef]

20. Hang, J.; Liang, J.; Wang, X.; Zhang, X.; Wu, L.; Shao, M. Investigation of O3–NOx–VOCs chemistry and pollutant dispersion in street canyons with various aspect ratios by CFD simulations. *Build. Environ.* **2022**, *226*, 109667. [CrossRef]
21. Dallman, A.; Magnusson, S.; Britter, R.; Norford, L.; Entekhabi, D.; Fernando, H. Conditions for thermal circulation in urban street canyons. *Build. Environ.* **2014**, *80*, 184–191. [CrossRef]
22. Yee, E.; Biltoft, C. Concentration fluctuation measurements in a plume dispersing through a regular array of obstacles. *Boundary-Layer Meteorol.* **2004**, *111*, 363–415. [CrossRef]
23. Hang, J.; Chen, G. Experimental study of urban microclimate on scaled street canyons with various aspect ratios. *Urban Clim.* **2022**, *46*, 101299. [CrossRef]
24. Chen, G.; Yang, X.; Yang, H.; Hang, J.; Lin, Y.; Wang, X.; Wang, Q.; Liu, Y. The influence of aspect ratios and solar heating on flow and ventilation in 2D street canyons by scaled outdoor experiments. *Build. Environ.* **2020**, *185*, 107159. [CrossRef]
25. Chen, G.; Wang, D.; Wang, Q.; Li, Y.; Wang, X.; Hang, J.; Gao, P.; Ou, C.; Wang, K. Scaled outdoor experimental studies of urban thermal environment in street canyon models with various aspect ratios and thermal storage. *Sci. Total Environ.* **2020**, *726*, 138147. [CrossRef] [PubMed]
26. Hang, J.; Wang, D.; Zeng, L.; Ren, L.; Shi, Y.; Zhang, X. Scaled outdoor experimental investigation of thermal environment and surface energy balance in deep and shallow street canyons under various sky conditions. *Build. Environ.* **2022**, *225*, 109618. [CrossRef]
27. Dai, Y.; Mak, C.; Hang, J.; Zhang, F.; Ling, H. Scaled outdoor experimental analysis of ventilation and interunit dispersion with wind and buoyancy effects in street canyons. *Energy Build.* **2022**, *255*, 111688. [CrossRef]
28. Dai, Y.; Mak, C.; Zhang, Y.; Cui, D.; Hang, J. Investigation of interunit dispersion in 2D street canyons: A scaled outdoor experiment. *Build. Environ.* **2020**, *171*, 106673. [CrossRef]
29. Louka, P.; Belcher, S.; Harrison, R. Coupling between air flow in streets and the well-developed boundary layer aloft. *Atmos. Environ.* **2000**, *34*, 2613–2621. [CrossRef]
30. Cheng, X.; Zeng, Q.; Hu, F. Characteristics of gusty wind disturbances and turbulent fluctuations in windy atmospheric boundary layer behind cold fronts. *J. Geophys. Res. Atmos.* **2011**, *116*, D06101. [CrossRef]
31. Zhang, Y.; Gu, Z.; Cheng, Y.; Lee, S. Effect of real-time boundary wind conditions on the air flow and pollutant dispersion in an urban street canyon—Large eddy simulations. *Atmos. Environ.* **2011**, *45*, 3352–3359. [CrossRef]
32. Duan, G.; Ngan, K. Effects of Time-Dependent Inflow Perturbations on Turbulent Flow in a Street Canyon. *Boundary-Layer Meteorol.* **2018**, *167*, 257–284. [CrossRef]
33. Wang, H.; Wang, Q.; Yang, X.; Chen, T.; Lam, C.; Zhang, M.; Hang, J. Steady and unsteady turbulent flows and pollutant dispersion in 2D street canyons with novel boundary conditions and various Re numbers. *Urban Clim.* **2021**, *39*, 100973. [CrossRef]
34. Liu, X.; Zhai, Z. Inverse modeling methods for indoor airborne pollutant tracking: Literature review and fundamentals. *Indoor Air* **2007**, *17*, 419–438. [CrossRef]
35. Dai, Y.; Zhang, F.; Wang, H. Identification of source location in a single-sided building with natural ventilation: Case of interunit pollutant dispersion. *J. Build. Eng.* **2023**, *68*, 106049. [CrossRef]
36. Neupauer, R.; Wilson, J. Adjoint method for obtaining backward-in-time location and travel time probabilities of a conservative groundwater contaminant. *Water Resour. Res.* **1999**, *35*, 3389–3398. [CrossRef]
37. Zhai, Z.; Liu, X.; Wang, H.; Li, Y.; Liu, J. Experimental verification of tracking algorithm for dynamically-releasing single indoor contaminant. *Build. Simul.* **2012**, *5*, 5–14. [CrossRef] [PubMed]
38. Wang, H.; Lu, S.; Cheng, J.; Zhai, Z. Inverse modeling of indoor instantaneous airborne contaminant source location with adjoint probability-based method under dynamic airflow field. *Build. Environ.* **2017**, *117*, 178–190. [CrossRef]
39. Chen, L.; Chen, W.; Lin, J.; Chen, C.; Luo, Y.; Tao, L. Inverse model investigation of radionuclide dispersion in a ventilated room based on the adjoint probability method. *Ann. Nucl. Energy* **2022**, *169*, 108929. [CrossRef]
40. Yakhot, V.; Orszag, S. Renormalization-Group analysis of turbulence physical review letters. *J. Sci. Comput.* **1986**, *57*, 1722.
41. Ai, Z.; Mak, C. CFD simulation of flow in a long street canyon under a perpendicular wind direction: Evaluation of three computational settings. *Build. Environ.* **2017**, *114*, 293–306. [CrossRef]
42. Snyder, W. *Guideline for Fluid Modeling of Atmospheric Diffusion: Environmental Sciences Research Laboratory*; Office of Research and Development, US Environmental Protection Agency: Washington, DC, USA, 1981; Volume 81.
43. Chan, A.; So, E.; Samad, S. Strategic guidelines for street canyon geometry to achieve sustainable street air quality. *Atmos. Environ.* **2001**, *35*, 5681–5691. [CrossRef]
44. Gu, Z.; Zhang, Y.; Lei, K. Large eddy simulation of flow in a street canyon with tree planting under various atmospheric instability conditions. *Sci. China Technol. Sci.* **2010**, *53*, 1928–1937. [CrossRef]
45. Baik, J.; Park, R.; Chun, H.; Kim, J. A laboratory model of urban street-canyon flows. *J. Appl. Meteorol.* **2000**, *39*, 1592–1600. [CrossRef]
46. Walton, A.; Chen, A.; Yeung, W. Large-eddy simulation of pollution dispersion in an urban street canyon—Part I: Comparison with field data. *Atmos. Environ.* **2002**, *36*, 3602–3613. [CrossRef]
47. Li, Z.; Zhang, H.; Wen, C.; Yang, A.; Juan, Y. Effects of height-asymmetric street canyon configurations on outdoor air temperature and air quality. *Build. Environ.* **2020**, *183*, 107195. [CrossRef]
48. Memon, R.; Leung, D.; Liu, C. Effects of building aspect ratio and wind speed on air temperatures in urban-like street canyons. *Build. Environ.* **2010**, *45*, 176–188. [CrossRef]

49. Laussmann, D.; Helm, D. Air Change Measurements Using Tracer Gases: Methods and Results. Significance of Air Change for Indoor Air Quality. In *Chemistry, Emission Control, Radioactive Pollution and Indoor Air Quality*; IntechOpen: London, UK, 2011.
50. Ai, Z.; Mak, C. Wind-induced single-sided natural ventilation in buildings near a long street canyon: CFD evaluation of street configuration and envelope design. *J. Wind. Eng. Ind. Aerodyn.* **2018**, *172*, 96–106. [CrossRef]

buildings

MDPI

Article

The Numerical Simulation Study of Pumping Airflow Driven by Wind Pressure for Single- and Multi-Room Buildings

Xinpeng Qiu [1], Junli Zhou [1,*], Xue Xiao [1], Wenjun Zhu [1], Jiaji Zhang [1], Shuqiang Gui [2] and Yangdong Peng [2]

[1] School of Civil Engineering and Architecture, Wuhan University of Technology, Wuhan 430070, China; 274506@whut.edu.cn (X.Q.); 1363939726@whut.edu.cn (X.X.); 2538391456@whut.edu.cn (W.Z.); zhangjiaji@whut.edu.cn (J.Z.)

[2] WSGRI Engineering & Surveying Incorporation Limited, Wuhan 430080, China; 003117@wsgri.com (S.G.); 002687@wsgri.com (Y.P.)

* Correspondence: jlzhou@whut.edu.cn

Abstract: Pumping airflow occurs in single-sided natural ventilation buildings when the openings are located on the leeward side; the direction of airflow through the building then changes periodicity. In order to propose a calculation method of the ventilation rates for single-room buildings and analyze the pumping ventilation for multi-room buildings, the CFD method with an SST k-ω turbulence model was used to conduct numerical simulation in this study, which is verified by other experimental results. Firstly, the qualitative and quantitative characteristics of pumping ventilation were investigated for a single-room building with two openings. The results show that the steady-state method underestimates the ventilation flow rates, and the unsteady-state method captures the microstructures of the flow better. Secondly, the vortex-shedding and indoor airflow oscillation frequencies were analyzed based on transient simulation for a single-room building. It was found that both of them increase with air speed. Then, the factors affecting ventilation flow rates were analyzed. A calculation method for dimensionless ventilation rates is proposed. Finally, the pumping ventilation for a multi-room building was numerically simulated. The ventilation rate for a middle room is greater than that for a corner room, and the ventilation rate for any room in a multi-room building is greater than the single-room building given the same room size and the same incoming wind speed. The findings of this paper are helpful for the design and evaluation of natural ventilation.

Keywords: numerical simulation; pumping airflow; frequency; multi-room

Citation: Qiu, X.; Zhou, J.; Xiao, X.; Zhu, W.; Zhang, J.; Gui, S.; Peng, Y. The Numerical Simulation Study of Pumping Airflow Driven by Wind Pressure for Single- and Multi-Room Buildings. *Buildings* **2023**, *13*, 3066. https://doi.org/10.3390/buildings13123066

Academic Editor: Theodore Stathopoulos

Received: 31 October 2023
Revised: 1 December 2023
Accepted: 6 December 2023
Published: 8 December 2023

1. Introduction

Natural ventilation is an important passive technology for buildings, because its appropriate application can conserve energy and supply fresh outdoor air. Natural ventilation can be divided into single-sided ventilation and cross-ventilation. Many studies have shown that cross-ventilation is superior to single-side ventilation in the field of indoor ventilation [1,2], but in most cases, single-side ventilation is still very important in building design, as few buildings can be cross-ventilated due to internal zoning, fire codes, safety requirements and privacy issues.

Single-side ventilation systems can be categorized by their number of openings: a single opening (SS1) and multiple openings (SSn). In the former, air must enter and exit through the same openings, while in the latter, there is a clearer division of air entering and exiting the room [3]. It has been found that multiple openings provide a better ventilation effect than a single opening [4]. The unilateral natural ventilation model proposed by Warren [5] is suitable for the estimation of ventilation rates in most cases. In single-side ventilation, unsteady natural ventilation mechanisms are classified into continuous airflow, pulsating flow, eddy penetration and diffusion phenomena [6].

The ventilation is induced by the vortex shedding of the airflow around the building, which is referred to as the pumping airflow [7]. It can be considered as a special type of eddy penetration. In 2016, Daish et al. [7] first analyzed the pumping airflow of the SS2 model, which has two openings located on the leeward side of the structure. The ventilation airflow in this instance oscillated with the Strohal frequency for time T, exhibiting a unique periodic fluctuation in flow.

In 2018, the pumping airflow was also observed by Zhong [8] in a 2D CFD simulation, and the impact of the building aspect ratio, incoming wind speed, and opening spacing on the frequency and ventilation flow rate of the pumping flow was examined. The two-dimensional simulation method, however, was unable to accurately capture three-dimensional flow properties. In 2019, Zhong carried out quasi-3D numerical simulation, and they found that the increasing trend in the dimensionless ventilation flow rate disappears or even decreases as the wind speed increases to 5 m/s. The dimensionless ventilation flow rate increased as the aperture separation increased, but it had no effect on the vortex-shedding frequency [9]. In 2020, Daniel [10] pointed out that the studies proposed by Zhong et al. [8,9,11,12] did not consider the three-dimensional structure of the pumped ventilation flow in buildings. In 2021, Zhong [13] pointed out that the ventilation flow rate of the bottom and top floors of buildings is considerably different from that of the middle floors, because the ground and top floors of a building have a greater influence on vortex shedding. In 2021, Graça [14] found that there are particular wind angles that also cause pumping airflow, in addition to when the wind is perpendicular to the facade (single-sided rooms) or aligned with the building corner (corner rooms), when there were two openings in the same facade of a building.

In summary, pumping ventilation occurs when the two openings are located on the leeward side for single-room model buildings. The frequency change in the flow's direction is related to the shedding frequency of the vortex street. Moreover, it was found that the pumping ventilation volume was related to the aperture separation: for example, Daish [7] stated that the ventilation flow rate increased with $s'^{1/2}$ (s' is the aperture separation), while Kobayashi [15] and Daniel [10] concluded that the ventilation flow rate was linearly related to s'.

There are different conclusions about the relationship. In addition, previous research did not consider a multi-room building, which is the more realistic case. Does pumping ventilation still exist, and what is the difference for rooms in different locations? Aiming to answer these questions and further understand pumping ventilation, this paper analyzed the qualitative and quantitative characteristics of the pumping airflow of a single-room building, and it also investigated the pumping airflow of a multi-room building. Figure 1 shows the flow chart of this paper.

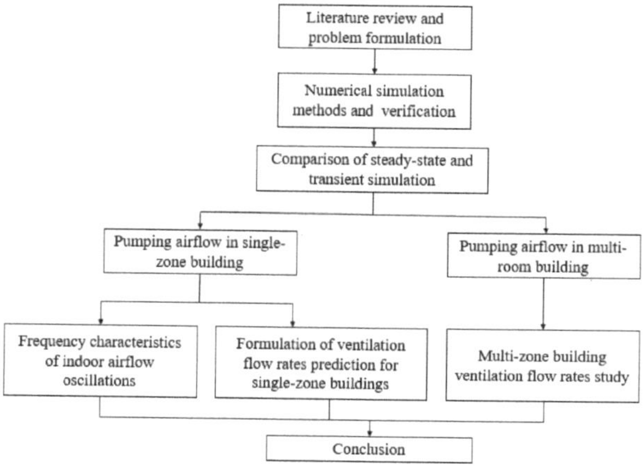

Figure 1. The flow chart of the study.

Revealing the principles of the natural ventilation is a prerequisite for further designing the thermal environment and evaluating the diffusion of pollutants. This study can provide references for understanding the mechanisms and effects of the natural ventilation.

2. Numerical Simulation Method

2.1. Model Simplification and Simulation Settings

2.1.1. Computational Geometry and Domain

A three-story building was taken as the model 1 (e.g., Figure 2). The total height of the building H is 9 m, and there is one room on each floor. The dimension of each room is $6 \text{ m} \times 6 \text{ m} \times 3 \text{ m}$ ($B \times B \times 1/3\,H$). The natural ventilation room is located on the second floor with two leeward openings of the same size: $1 \text{ m} \times 1 \text{ m}$. The centers of the two openings are 3 m apart and 1.5 m from the respective closer sides of the building. The three rooms in the building are not connected to each other and are separated by floors or ceilings.

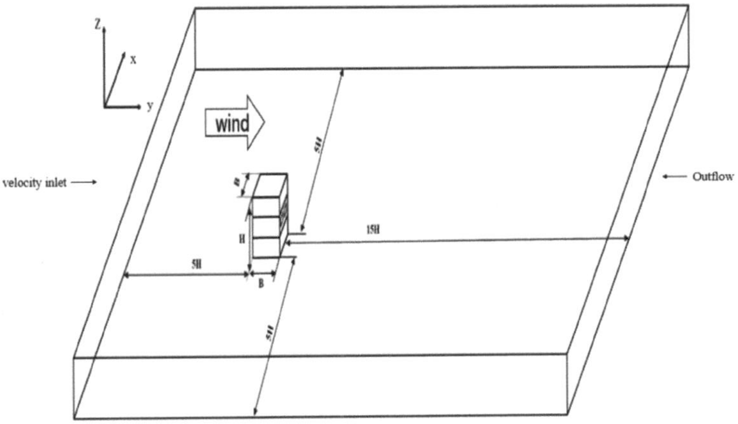

Figure 2. Building model 1 and computational domain.

The computational domain has a significant impact on the simulation results, and the blocking ratio is thought to be important [16], which meant that it should be less than 3% of the computational domain size. Additionally, for the building body, the computational domain's sides, the top, and the external exit boundary all need to be at least 5 H and 10 H away from the building, respectively. The computational domain is $186 \text{ m} \times 96 \text{ m} \times 9 \text{ m}$ in size because only the quasi-3D building case was taken into account in this study, which meant that the height of the computational domain was equal to the height of the building, the side distance is 5 H, and the distance from the leeward side to the exit is 15 H. The blockage ratios of it are less than 3%. Due to the fact that the pumping flow involved two apertures and that the spacing between the two openings has an impact on the airflow, three cases with varying spacing are adopted in this paper (e.g., Figure 3). The quoted dimensionless aperture distance is

$$s' = \frac{s}{B} \tag{1}$$

where s is the distance (in meters) between the centers of the two openings, and B is the building's width (in meters), using 0.25, 0.5, and 0.75 as examples.

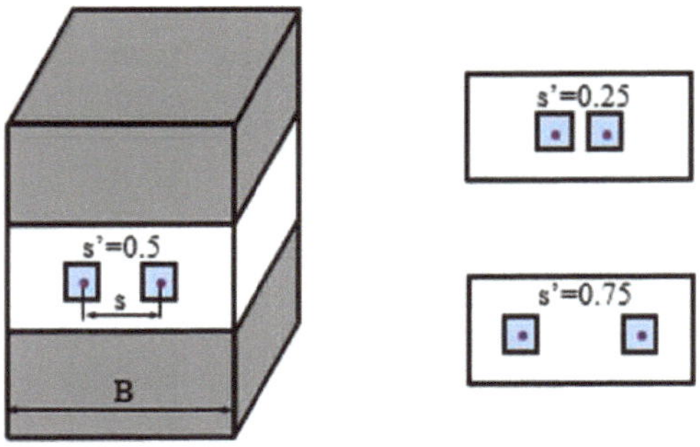

Figure 3. Schematic diagram of different opening distances.

2.1.2. Numerical Simulation Settings

Solution Settings

In this paper, the finite volume method was used to solve the governing equation. The governing equation was as follows:

$$\frac{\partial \rho}{\partial t} + \frac{\partial(\rho u)}{\partial x} + \frac{\partial(\rho v)}{\partial y} + \frac{\partial(\rho w)}{\partial z} = 0 \quad (2)$$

$$\rho\frac{\partial \vec{u}}{\partial t} + \rho\vec{u}\cdot\vec{\nabla}\cdot\vec{u} = -\vec{\nabla}p + \vec{\nabla}\cdot\tau + (\rho - \rho_\infty)g \quad (3)$$

$$\frac{\partial T}{\partial t} + u\frac{\partial T}{\partial x} + v\frac{\partial T}{\partial y} + w\frac{\partial T}{\partial z} = \frac{\lambda}{\rho c_p}\left(\frac{\partial^2 T}{\partial x^2} + \frac{\partial^2 T}{\partial y^2} + \frac{\partial^2 T}{\partial z^2}\right) + \frac{S_T}{\rho} \quad (4)$$

The three-dimensional RANS equations were solved together with the k-ω shear stress transport (SST k-ω) model:

$$\frac{\partial \rho k}{\partial t} + \frac{\partial(\rho u_i k)}{\partial x_i} = \tau_{ij}\frac{\partial u_i}{\partial x_j} - \beta^*\rho\omega k + \frac{\partial}{\partial x_j}\left[(\mu + \sigma_k\mu_t)\frac{\partial k}{\partial x_j}\right] \quad (5)$$

$$\frac{\partial \rho\omega}{\partial t} + \frac{\partial(\rho u_i\omega)}{\partial x_i} = \frac{\gamma}{\nu_t}\tau_{ij}\frac{\partial u_i}{\partial x_j} - \beta\rho\omega^2 + \frac{\partial}{\partial x_j}\left[(\mu + \sigma_\omega\mu_t)\frac{\partial \omega}{\partial x_j}\right] + 2(1 - F_1)\rho\sigma_{\omega 2}\frac{1}{\omega}\frac{\partial k}{\partial x_j}\frac{\partial \omega}{\partial x_j} \quad (6)$$

This paper assumed that the external wind speed was stable and the supply air parameters were constant. In addition, the air in the fluid region was considered to be an incompressible ideal gas. The SIMPLE method was chosen to the pressure–velocity coupling, and the PRESTO! pressure interpolation format was chosen. The controlling equations were discretized as the second-order windward terms because second-order discretization has a higher precision than the first-order discretization. With the exception of the energy equation residuals, which were 1×10^{-6}, the convergence criterion for the system of equations was 1×10^{-9} for all parameter residuals. The steady-state simulation was carried out firstly, and then the results were adopted as the initial conditions in the transient calculation.

For transient calculations, a time step which is too large or too small will affect the results. The *CFL* number can be used to determine the appropriate time step.

$$\Delta t = CFL \times \frac{l}{U} \quad (7)$$

where l is the distance from wall to the first grid, and the value of that is 0.01 m in this study. The *CFL* value for the explicit solver is close to 1, and for the implicit solver, it will be increased to a higher number [17]. The research [18] showed that the smaller time-step difference had a limited impact on the outcomes when the time-step length was substantially less than the vortex-shedding duration; hence, *CFL* may be considered as 10 [9]. When the incoming wind speed is 1 m/s, Δt is calculated as 0.1 s. Table 1 displays the time step changing with wind speed. A total of 10~20 iterations were run in each time step to ensure accurate calculation results.

Table 1. Time steps at different wind speeds.

U (m/s)	Re	Δt (s)
1	410,752	0.100
2	821,504	0.050
3	1,232,256	0.033
4	1,643,008	0.025
5	2,053,761	0.020

Boundary Conditions

The velocity inlet boundary condition was adopted to the inlet, with U_y values of 1 m/s, 2 m/s, 3 m/s, 4 m/s, 5 m/s, and the velocity in the other two directions was 0; the turbulence intensity at the inlet was set to 10%, and the turbulence integration length was set to 0.58 m [19]. The outlet was set as outflow; the sides, the top and the bottom were set as symmetry; the opening was set as the interior, and all the other wall surfaces were set as no-slip.

Grid Division and Independence Verification

The domain was constructed using the hexahedral gird (Figure 4). The grid was refined in the opening and the near-wall. For the case of $s' = 0.5$, three grid schemes with the grid numbers of 2,079,003, 3,167,446, and 4,851,384 were selected for independence verification. The corresponding y+ values for the three-grid method were 45, 30 and 15, and the minimum distances of the grids from the wall were 0.015 m, 0.01 m, and 0.005 m, respectively. The monitoring line $Y = 3$ m, and $Y = 9$ m in the plane with a Z value of 4.5 m is chosen for the sample set (Figure 5). The results with different numbers of grids are shown in Figure 6. As can be seen, the changing rules of the wind speed outside the building showed the same trend. Based on the results, the grid number of 3,167,446 (Figure 4) is finally selected for the case, as a further increase in the mesh density does not produce obvious changes in the numerical solutions.

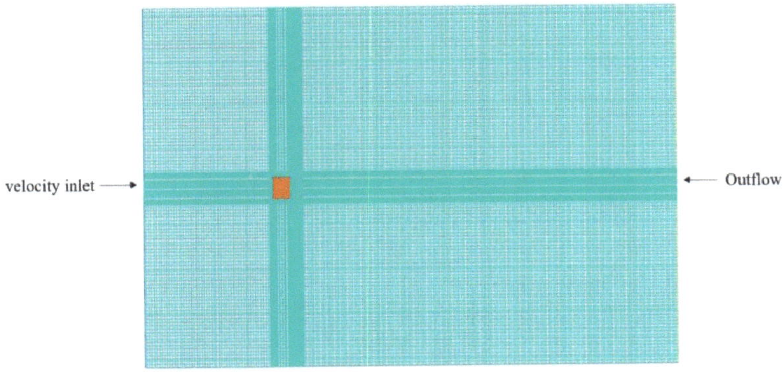

Figure 4. Schematic diagram of grid division.

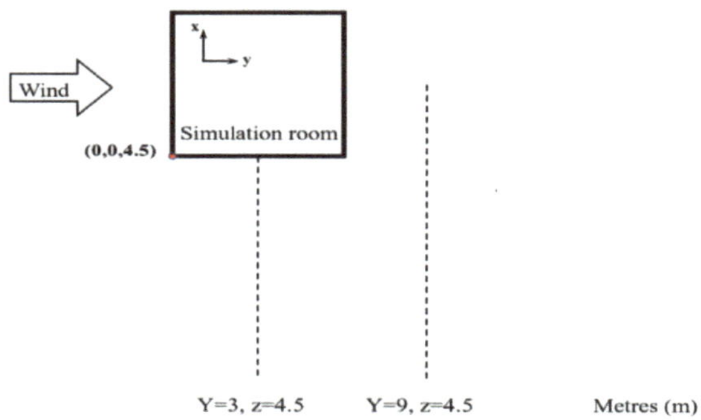

Figure 5. Monitoring lines for grid-independence verification.

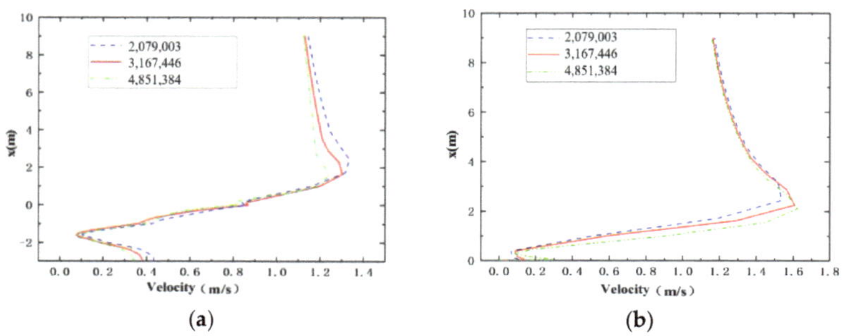

Figure 6. Airflow velocity distribution in the plane with $Z = 4.5$ m. (**a**) $Y = 3$ m. (**b**) $Y = 9$ m.

2.1.3. Validation of CFD Methodology

In 2003, Jiang et al. [20] conducted a wind tunnel experiment for natural ventilation. The study focused on a small (0.25 cubic) building model with a window opening size 0.084 m × 0.125 m (length × height), as shown in Figure 7. There is a workspace with the area of 2.0 m × 2.0 m, and the height 1.0 m, which is in the wind tunnel. The velocity measurements were conducted on a grid plane at the center of the building. In order to verify the CFD method used in this paper, Jiang's case was chosen for the numerical simulation. The downstream length was set as 8 H; the upstream length was 4 H; and the lateral length was 4 H on both sides of the building in the computational zone. The total number of the grids was 969,570. The boundary conditions and the wind speed were consistent with the literature, and the turbulence model was the SST k-ω. The numerical results are shown to have good agreement with the experimental data, and the relative error is less than 20% in the most regions (Figure 8). Therefore, the CFD method adopted in the paper was validated for the numerical simulation of the natural ventilation.

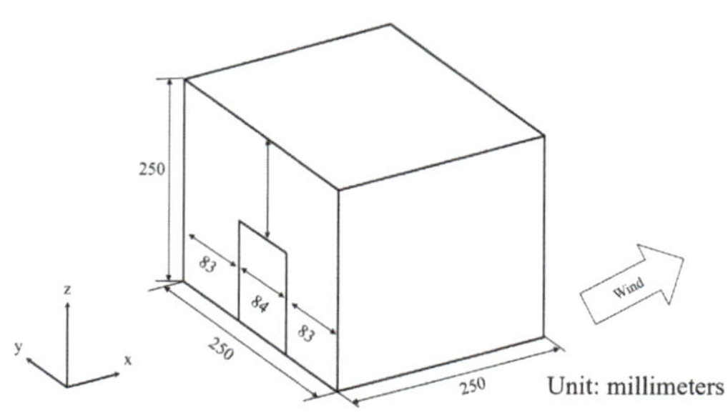

Figure 7. Schematic diagram of wind tunnel experiment room.

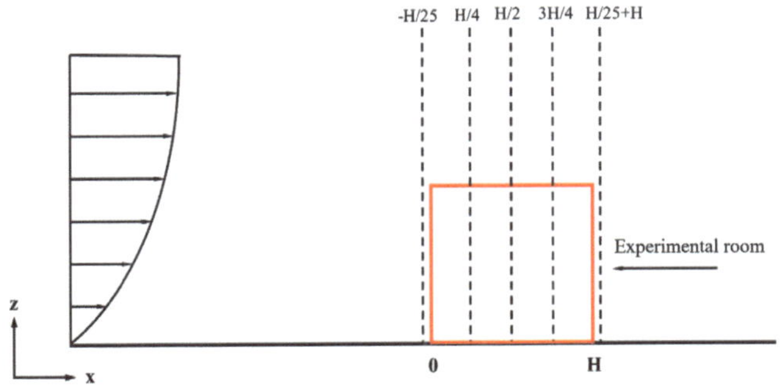

(a)

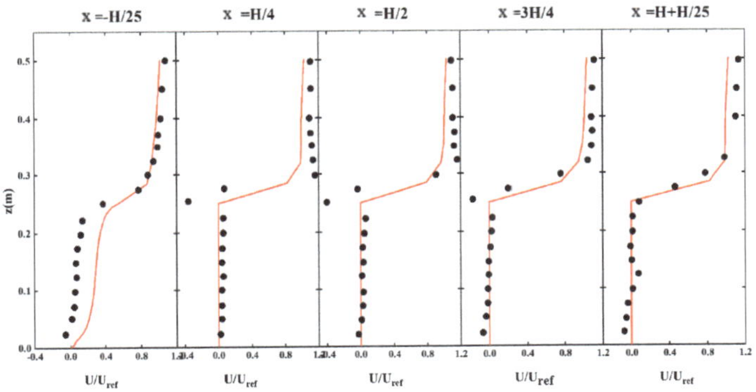

(b)

Figure 8. Validation of CFD methodology. (**a**) Schematic diagram of the location of wind tunnel experimental data measurements. (**b**) Numerical simulation of the velocity profile along the measurement line (red line) compared to the experimental values (black solid dots).

2.2. Comparison of Steady-State and Transient Simulation

2.2.1. The Flow Field

The transient simulation can describe the small structures in turbulent better than steady-state calculation. For example, as shown in Figure 9 (incoming wind speed was 1 m/s), there was an obvious vortex street in the building wake in the transient simulation compared to the steady-state results.

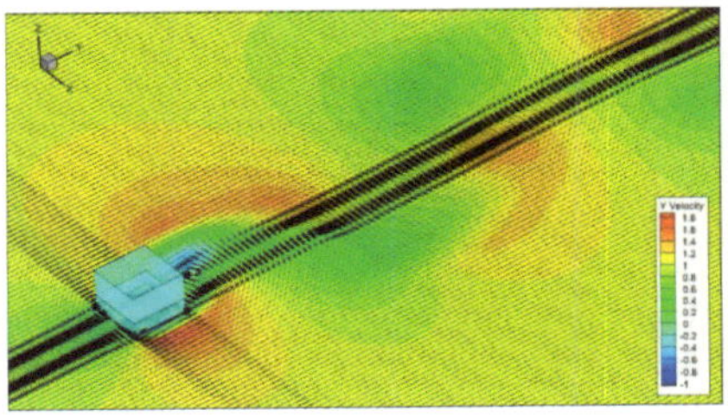

(a)

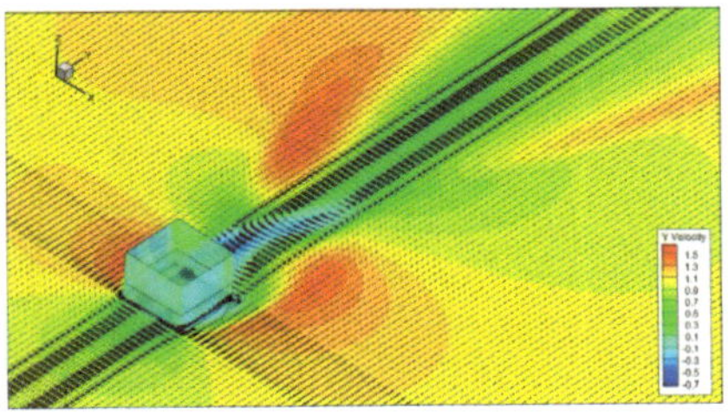

(b)

Figure 9. Flow field of the building exterior. (**a**) Transient calculations. (**b**) Steady-state calculations.

Figure 10 depicts the y-direction velocity cloud charts at the openings on the leeward side of the room, where the steady-state result did not have a symmetric image and the velocity distribution was rather uniform. The unsteady-state calculation revealed that the velocity distribution was symmetric about $z = 4.5$ m, where the minimum velocity was located at one of the openings and the maximum velocity was located near the wall. In general, there is a significant difference in the calculation results between the transient calculation and the steady-state calculation because the latter result is the average value.

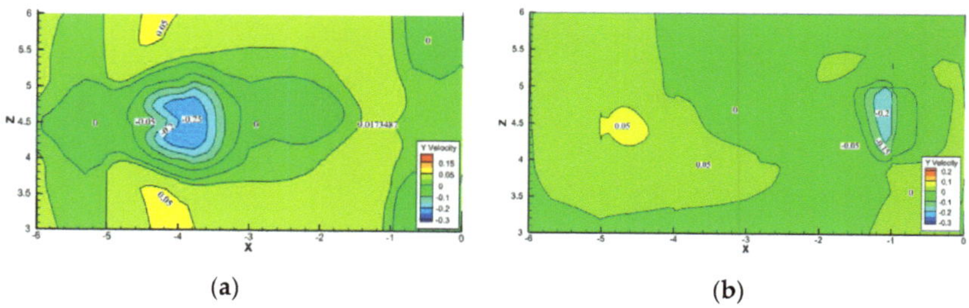

(a) (b)

Figure 10. Velocity distribution on the vertical surface of the opening. (**a**) Transient calculations. (**b**) Steady-state calculations.

2.2.2. The Ventilation Flow Rates

The ventilation flow rates are also compared here (Table 2). The steady-state case results could be obtained directly; however, the ventilation flow rate is the average of the total ventilation volume for each time step with the entire period in the non-stationary calculation.

Table 2. Results for different cases.

Wind Speed (m/s)	Q_(Steady State) (m^3/s)	Q_(Unsteady State) (m^3/s)
1	0.0778	0.1356
2	0.3018	0.3389
3	0.4207	0.5120

It can be seen that the steady-state calculation underestimates the ventilation flow rates of the building. The transient simulation method is more accurate because it takes into account the fluctuation of the wind around the buildings. Therefore, the method of the unsteady-state simulation was selected in the following calculation.

2.2.3. Periodic Characteristics of Pumping Airflow

The detailed flow field and the periodic characteristics of the typical pumping airflow are analyzed in this section. The case is the same as that in Section 2.2.1. Figure 11 displays the streamline distribution of the flow field with the cross-section situated at $Z = 4.5$ m.

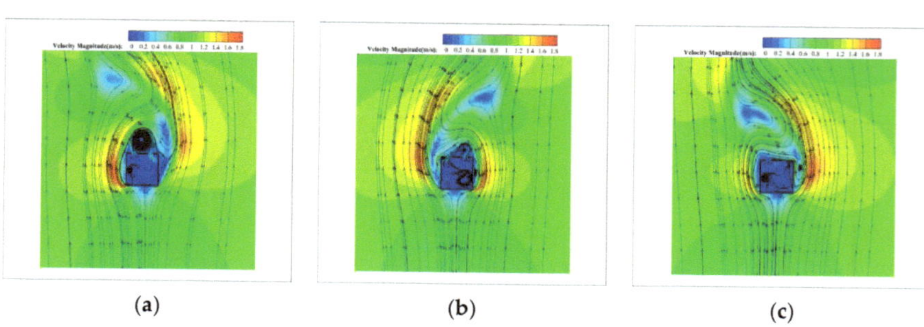

(a) (b) (c)

Figure 11. Streamline distribution of the flow field. (**a**) Moment t. (**b**) Moment t + 1/2t. (**c**) Moment 2t.

The flow around buildings in a natural wind field is similar to the flow around a blunt body. As shown in Figure 11, the incoming air flow separated on the building's windward surface and flowed to its two sides. As a result of the pressure difference, a clockwise vortex formed on the left side of the surface. Next, a separation was created at the backward corner of the left surface, and the vortex gradually grew to the entire leeward surface.

The airflow at the back of the building was forced to enter the building through the opening on the right side, create a circulating airflow that followed the vortex's direction, and exit through the opening on the left. The airflow on the opposite side, however, was exactly the opposite. It formed an anti-clockwise vortex on the right side surface. This caused the airflow to enter the room through the left side opening after separating and attaching itself to the leeward surface at the rear corner point of the lower right side. It then formed a circulating airflow that flowed out of the right side opening in the same direction as the vortex in an anti-clockwise direction.

Vortices were produced and shed alternatively on the either side of the square building, finally forming a regular vortex street in the wake flow. As a result, the fluctuating ventilation named as the pumping airflow was generated. Figure 12 describes the ventilation flow rates of the building changing with time. During the first 200 s, the changing amplitude of the ventilation flow rates was increasing gradually because the vortex shedding of the airflow was still in the developmental stage. After that, the change in amplitude decreased to form the regular fluctuations similar to the sinusoidal form. From Figure 13, it can be discovered that not all of the ventilation flow rates change with the sine wave, especially when the opening spacing was adjusted. However, all the cases clearly displayed the features of a regular fluctuation.

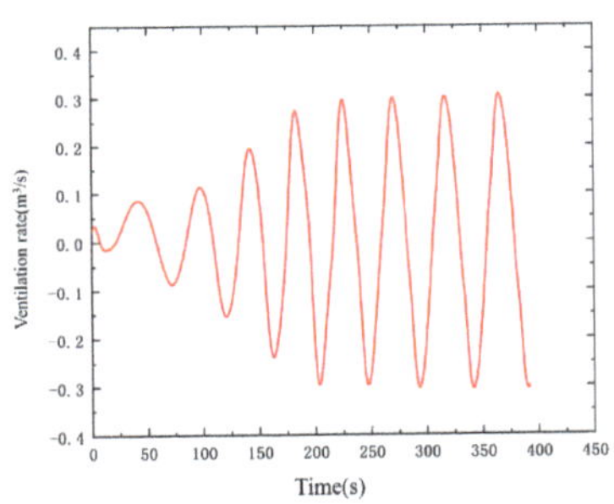

Figure 12. Time-range diagram of ventilation at $U_y = 1$ m/s, $s\prime = 0.5$.

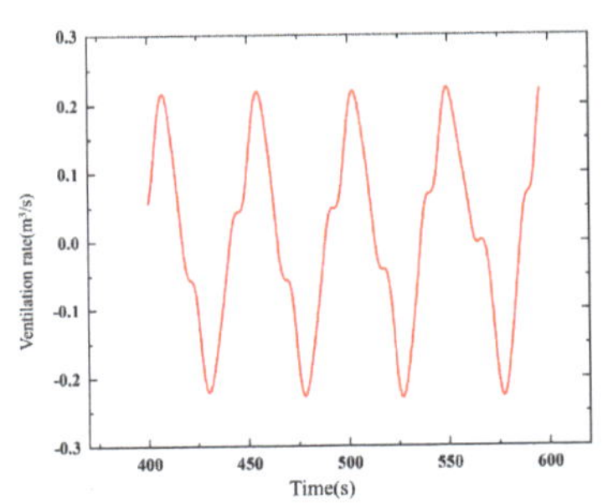

Figure 13. Time-range diagram of ventilation at $U_y = 1$ m/s, $s\prime = 0.25$.

3. Result and Discussion

3.1. *Pumping Airflow in a Single-Room Building*

In this part, the numerical simulation was used to examine the features of the pumping airflow with different incoming wind speed in a single-room building. The vortex-shedding frequency and the ventilation flow rates were analyzed.

3.1.1. Fluctuating Frequency

Vortex-Shedding Frequency

Four monitoring points were selected in order to analyze the vortex-shedding frequency, as shown in Figure 14. Point b was located at the center of the opening 1 with a height of 4.5 m. The time-domain and the frequency spectrum plots of the static pressure at point a with the incoming wind speed as 4 m/s are presented in Figure 15a,b, respectively. As shown in the time-domain plot, the static pressure changed periodically. From the frequency spectrum plot, it was possible to determine that the shedding frequency was 0.0825 Hz, at which the power spectral density achieves its maximum value.

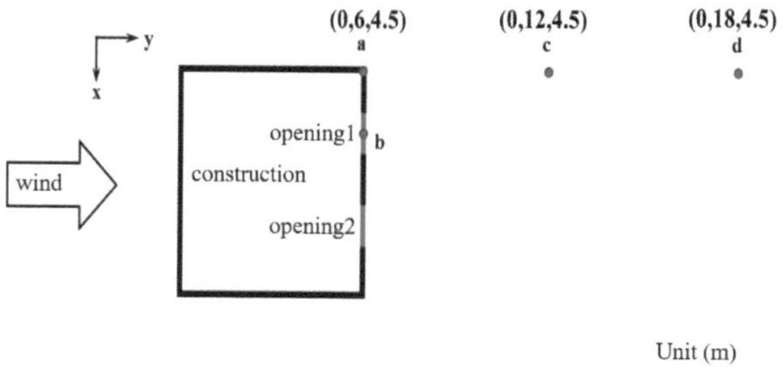

Figure 14. Schematic of the location of the different monitoring points.

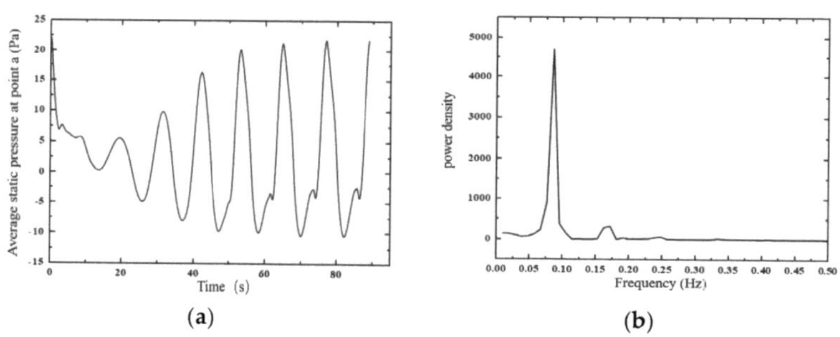

Figure 15. Time and frequency domain of static pressure at point a. (**a**) Time-domain spectrum. (**b**) Frequency spectrum.

The vortex-shedding frequencies of the four points with different wind speed are presented in Table 3. It was evident from the simulated data that with the same wind speed, the vortex-shedding frequencies essentially stayed constant at different places. The frequency at point b increased by 12.5% more than the other three places only at the case wind speed of 5 m/s. For various wind speeds, the frequency of vortex shedding at the same place increased as the wind speed rose.

Table 3. Frequency of vortex shedding at 4 points under different wind speeds.

Position Frequency (Hz) Velocity (m/s)	a	b	c	d
1	0.0224	0.0224	0.0224	0.0224
2	0.0400	0.0400	0.0400	0.0400
3	0.0619	0.0619	0.0619	0.0619
4	0.0825	0.0825	0.0825	0.0825
5	0.1000	0.1125	0.1000	0.1000

As a dimensionless number used to describe fluid flow, the Reynolds number (*Re*) represents the ratio of inertial forces to viscous or frictional forces acting on a moving object.

In the research of the wind field around buildings, the dimensionless number Strauhl number (St) was also utilized to investigate the flow frequencies.

$$Re = \rho dU/\mu \quad (8)$$

$$St = df/U \quad (9)$$

where d was the characteristic length of the object (m); μ was the dynamic viscosity ($N{\cdot}m/s^2$); U was the velocity (m/s); and f was the frequency (Hz).

Due to the fact that the relationship between the Reynolds number and Strauhl number in the vortex-shedding frequency research was not linear, Roshko proposed another dimensionless number, Roshko number (Ro) [21]:

$$Ro = \frac{fd^2}{\nu} \quad (10)$$

where ν was the kinematic viscosity (m^2/s). Previous studies had given the formulae for the corresponding Ro numbers as follows [10,22]:

$$150 < Re < 250 \quad Ro = 0.212Re - 4.7 \quad (11)$$

$$10^4 < Re < 4 \times 10^4 \quad Ro = 0.1477Re - 138 \quad (12)$$

where Equation (11) was obtained in the flow field across a cylinder with a low Re number, while Equation (12) was obtained for the 3D square cylinder model.

Based on the numerical simulation results and combined with the Fourier transform, f and St were obtained. Using Equations (8) and (10), the Ro number and Re number were determined, and the results are provided in Table 4. Daniel [10] discovered that the St values measured during vortex shedding in the 2D model case ranged from 0.07 to 0.21. In this study, the values of St in the quasi-3D model fell within the range of 0.12 to 0.14 and in a bandwidth of roughly 0.015. The result is relatively reasonable compared to the research of Daniel [10]. The following equation was developed by fitting the Re and Ro data using the least squares method, as shown in Figure 16.

$$4 \times 10^5 < Re < 2.1 \times 10^6 \quad Ro = 0.098Re + 4207.2 \quad (13)$$

Table 4. Parameter values with different wind speeds.

Parametric Velocity (m/s)	f (Hz)	St	Re	Ro
1	0.0224	0.1346	410,752	55,205
2	0.0400	0.1200	821,504	97,758
3	0.0619	0.1237	1,232,256	152,553
4	0.0825	0.1237	1,643,008	203,240
5	0.1000	0.1200	2,053,761	246,451

Therefore, it could be judged that there was also a good linear relationship between the Ro number and Re number in this study case, which was similar to that in other research.

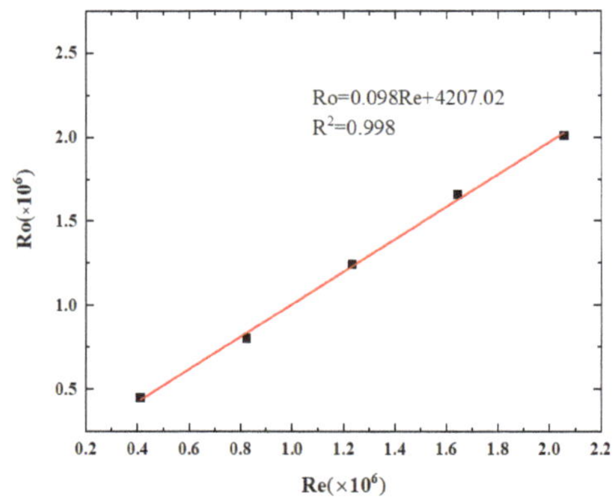

Figure 16. Linear relationship between *Ro* number and *Re* number.

Indoor Airflow Oscillation Frequency

During the simulations, the instantaneous ventilation flow rates of two openings were monitored. The results with wind speed as 1 m/s are shown in Figure 17, that positive ventilation values indicated the air flow into the room and negative values showed the air flow out. It could be seen that the ventilation flow rates of two openings were the same but the flow direction was opposite, and all of their sinusoidal properties changed with time.

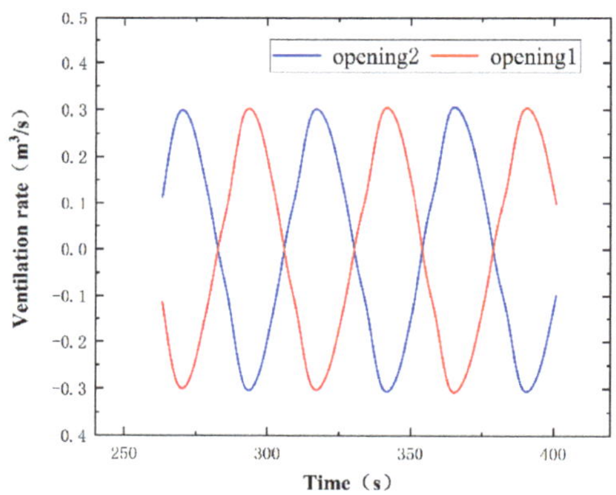

Figure 17. Time-domain plot of ventilation for two openings.

Figure 18 depicts the oscillation frequency of the airflow with various wind speeds, showing there was a significant association between the oscillation frequency of the airflow and the wind speed. As the wind speed increased, the frequency of the oscillations inside the building rose. The airflow oscillation frequency was basically the same as the vortex-shedding frequency around the building; all of them varied monotonically with the wind speed, as shown in Table 3 and Figure 18. Moreover, if the dimensionless opening distances s' changed from 0.25 to 0.5, the simulation results showed that the frequency was only different by about 5%. It could be concluded that the opening distance had almost no impact on the oscillation frequency of the indoor airflow.

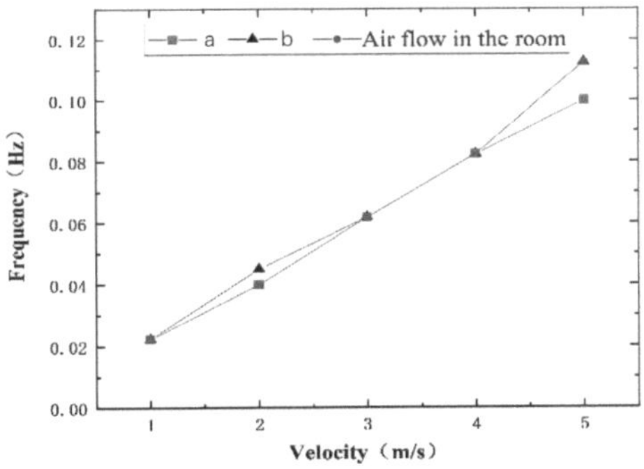

Figure 18. Frequency of airflow oscillations in different locations with different wind speeds.

3.1.2. Ventilation Flow Rates

In the analysis of natural ventilation, the most important parameter is the ventilation flow rates. In this study, the dimensionless ventilation flow rates could be obtained by the following equation.

$$Q\prime = \frac{Q}{U_R A_{ref}} \tag{14}$$

where U_R is the reference wind speed (m/s), which is considered equal to the incoming wind speed in this study; A_{ref} is the reference opening area (m^2).

The reference opening area is the effective opening area, which can be calculated by the formula:

$$A_{ref} = A_{eff} = \frac{A_1 A_2}{\left(A_1{}^2 + A_2{}^2\right)^{1/2}} \tag{15}$$

where A_1 and A_2 are the areas (m^2) of the two openings. In this study, $A_1 = A_2 = A$, so $A_{eff} = A/\sqrt{2}$.

According to the literature [23], the mean and fluctuating ventilation made up the instantaneous ventilation. The fluctuating ventilation rates can be predicted by the standard deviation as follows:

$$\sigma_Q = \left[\frac{1}{N_m}\sum\nolimits_1^{N_m}(Q - Q_m)^2\right]^{1/2} \tag{16}$$

where σ_Q is the fluctuating ventilation (m^3/s), N_m is the time step, Q is the monitored ventilation (m^3/s), and Q_m is the mean ventilation (m^3/s), which is the average of the sequence of instantaneous ventilation changes over time. In addition, $\overline{Q}\prime$ and $\sigma_Q{}'$ represent the dimensionless numbers of the mean and fluctuating ventilation, respectively.

The Influence of Different Wind Speeds

The ventilation flow rates with different wind speeds are shown in Table 5. From the results, it can be seen that the dimensionless mean ventilation $\overline{Q}\prime$ is larger than the dimensionless fluctuating ventilation $\sigma_Q{}'$ for all the five cases. It may indicate that although there are periodic changes in the direction of ventilation airflow, the average ventilation intensity is greater than the fluctuating ventilation in terms of the overall pumping ventilation effect on the room. Moreover, the simulated results show that almost all the ventilation flow rates increased with the wind speed except the dimensionless fluctuating air volume under the condition of wind speed as 1 m/s. The probable reason is that there was more airflow (which on the vortex scale was smaller than the opening size) entering or exiting the room from the openings at lower wind speeds, which increased the dimensionless fluctuating air volume.

Table 5. Dimensionless mean and fluctuating ventilation flow rates with different wind speeds.

Velocity (m/s)	1	2	3	4	5
$\overline{Q}'$	0.1917	0.2397	0.2414	0.2420	0.2534
$\sigma_Q{}'$	0.1350	0.1320	0.1333	0.1335	0.1345
Q (m^3/s)	0.1356	0.3389	0.5120	0.6840	0.8960

Based on the ventilation flow rates, incoming wind speed, and the effective opening area, the following formula can be obtained:

$$Q\prime = \frac{Q}{U_R A_{eff}} = 0.246 \quad (17)$$

The plot of the relationship after the data fitting and the Pearson's correlation coefficient are shown in Figure 19, and its Pearson's correlation coefficient (R^2) is 0.993. It indicates that there is a strong linear correlation between the ventilation flow rates and the wind speeds.

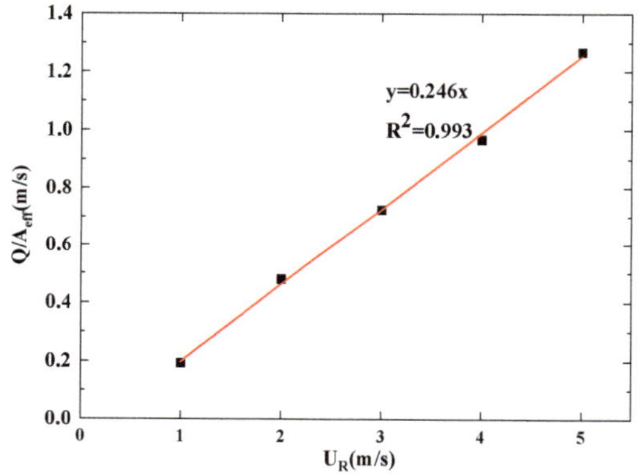

Figure 19. The relationship of ventilation flow rates and wind speed.

Warren [5] previously proposed a unilateral natural ventilation model, which was considered the most accurate model suitable for single-sided ventilation [5,24]. In addition, Daish [7] carried out special research about the pumping ventilation. Therefore, the two models were applied here to compare the simulated results. Using the simulation results as the base value, the relative error was calculated as:

$$\varepsilon = \frac{|Q_m - Q_n|}{Q_n} \times 100\% \quad (18)$$

where ε is the relative error (%); Q_m is the ventilation flow rates volume obtained by existing model (m^3/s); and Q_n is the numerically simulated ventilation flow rate (m^3/s).

The comparison results are shown in Table 6. The model of the Warren underestimated the ventilation flow rates to a greater extent with the average error about 70%. The results of Q_{Daish} were closer to the numerical data than the Warren model, with the relative error smaller than 15%, except for the wind speed of 1 m/s. It indicated that the Warren model was not suitable for the case of the pumping airflow. Comparing the results of Equation (17) and the model of Daish, as shown in Table 7, it could be found that the fitted function drawn in this study had a better ability to predict the pumping ventilation flow rates since there was only about a 10% relative error between the calculated results of the function obtained after fitting and the results measured by Daish [7].

Table 6. Ventilation flow rates obtained from different calculation models.

Wind Speed (m/s)	Q_n (m³/s)	Daish Q_{Daish} (m³/s)	Daish ε (%)	Warren Q_{Warren} (m³/s)	Warren ε (%)
1	0.1356	0.1928	42.19	0.0500	63.13
2	0.3389	0.3856	13.79	0.1000	70.49
3	0.5120	0.5784	12.98	0.1500	70.70
4	0.6840	0.7712	12.76	0.2000	70.76
5	0.8960	0.9640	7.600	0.2500	72.10

Table 7. Ventilation flow rates obtained by fitted equations.

Velocity (m/s)	Q' (m³/s)	Q_{Daish} (m³/s)	ε (%)
1	0.1739	0.1928	9.80
2	0.3479	0.3856	9.78
3	0.5218	0.5784	9.79
4	0.6958	0.7712	9.78
5	0.8697	0.9640	9.78

The aforementioned data showed that the model that only considered the windward or the parallel wind was no longer applicable for the leeward pumping flow, and it was possible to assume that the fitted function had a better ability to predict the pumping ventilation.

The Influence of Different Opening Spacings

Simulations were carried out for three cases with different opening spacings under the condition of a wind speed of 1 m/s. The dimensionless ventilation for the simulation results is given in Table 8 along with data from related studies in the literature [10,14]. The correlation between the two variables of the three sets of data was calculated to obtain the corresponding Pearson's correlation coefficient (R^2) to determine the pattern of change between the dimensionless ventilation and the dimensionless opening distance.

Table 8. Dimensionless ventilation flow rates with different opening distances.

	Method	Dimensionless Open Spacing (s')	Dimensionless Ventilation Flow Rates (Q')	R^2
1	2D simulation	0.12 0.36 0.6 0.84	0.1521 0.3596 0.4492 0.4900	0.84
2	wind tunnel test	0.25 0.5 0.75	0.067 0.081 0.095	0.99
3	quasi-3D simulation	0.25 0.5 0.75	0.1060 0.2460 0.3691	0.99

The results demonstrated that the data in this study and those in Danie [10] followed the same general pattern of law: namely, that the dimensionless ventilation Q' was linear with the dimensionless aperture distance s', and the Pearson's correlation coefficients (R^2) were both 0.99, which indicated a strong correlation. Additionally, Daish had demonstrated through an experiment that in contrast to the other three listed above, the dimensionless ventilation Q' was linearly proportional to the dimensionless aperture distance s'.

It should be mentioned that the simulation model or experimental subjects used in the table were different, which may account for the conclusions' discrepancy. Zhong [8] chose a two-dimensional model for the opening distance problem, whereas this paper used a quasi-3D model. The hanging window's opening angle was 14° in the study of Danie [10], whereas it was fully open in this work. However, all studies explained that the ventilation flow rates and the opening distances have a linear relationship.

Based on the simulation data, the following equation can be received.

$$Q\prime = \frac{Q}{U_R A_{eff}} = 0.4872s\prime = 0.4872\frac{S}{B} \tag{19}$$

where U_R is the reference wind speed (m/s); A_{eff} is the effective opening area (m^2); S is the opening spacing (m); and B is the building width (m).

The numerical simulation results were compared with the predicted values of the fitted function shown in Equation (19). The results showed that the formula can estimate the ventilation flow rates suitably under different opening sizes with the average relative error of 6.27%. Given a constant building size, the ventilation flow rates will increase with the opening sizes.

3.2. Pumping Airflow in Multi-Room Buildings

The pumping airflow of a multi-room building is analyzed in this section. The building model 2 is shown in Figure 20 with three rooms on each floor. There are a total of nine separate rooms, with a size of 18 m × 6 m × 9 m, and two openings on the leeward. The size of opening is 1 m × 1 m, and the distance between their center point is 3 m.

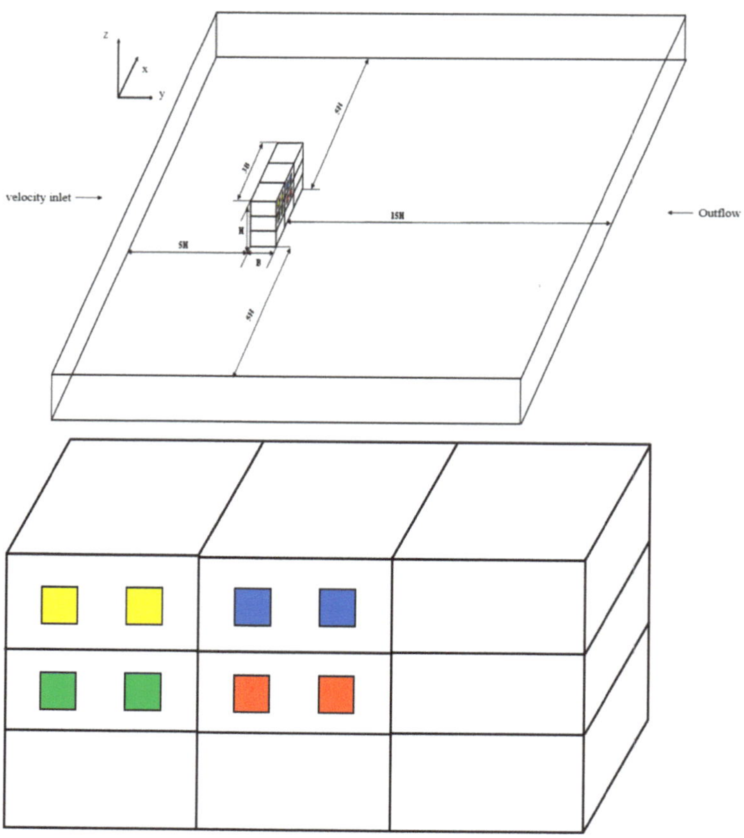

Figure 20. Schematic diagram and computational domain of the multi-room building.

The size of the external computational domain is 186 m × 108 m × 9 m. The gridding, boundary conditions, turbulence model, discrete formatting, and parameter settings for simulation were identical to those in the aforementioned sections. The wind speed is assumed to be 1 m/s.

Considering the symmetry of the building model, four rooms were analyzed. The openings of the rooms were colored yellow, green, red, and blue, respectively, and these are called room 1, room 2, room 3, and room 4.

3.2.1. The Flow Characteristics

The simulation calculation settings in this section are the same as those in Section 2. The pumping flow characteristics are the same as well. However, the simulation results showed that the indoor airflow oscillation frequency was smaller than that of building 1 (0.0224 Hz); their values were 0.0073 Hz, 0.0075 Hz, 0.0075 Hz, and 0.0079 Hz, respectively.

The exterior flow fields of the single-room and multi-room building are depicted in Figure 21. It was possible to see that with the same incoming wind speed, the building body's expansion caused a gradient increase in the velocity. In a result, a larger vortex formed and the velocity at the openings increased.

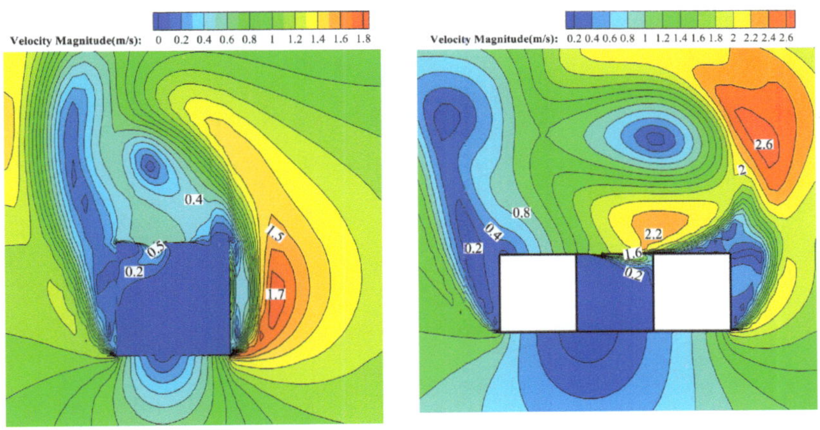

(a) (b)

Figure 21. Velocity distribution of outdoor flow field in different buildings. (**a**) Single-room buildings (model 1). (**b**) Multi-room buildings (model 2).

The indoor flow fields of the four rooms at t = 900 s are examined in Figure 22. It was discovered that for the side rooms (rooms 1 and 2), there was a greater difference in velocity magnitude at the openings, and the airflow entered the room at an angle to the windows; for the middle rooms 3 and 4, there was a smaller difference in velocity magnitude at the openings and the airflow flowed in and out of the windows almost parallel to the openings.

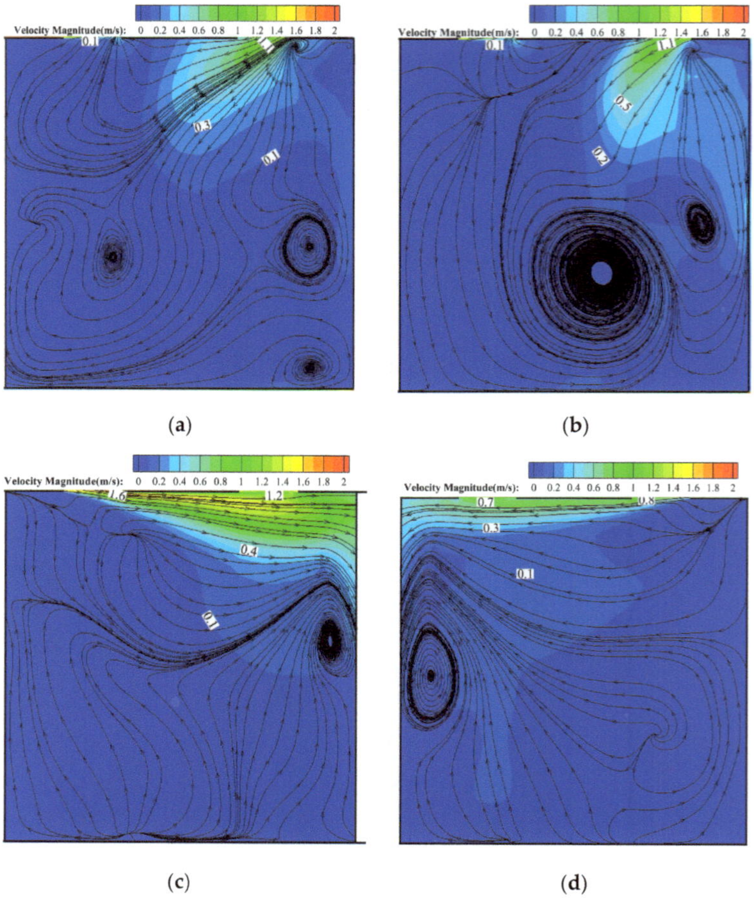

Figure 22. Velocity and streamline distribution of indoor flow field in different rooms in model 2. (**a**) Room 1. (**b**) Room 2. (**c**) Room 3. (**d**) Room 4.

3.2.2. Ventilation Flow Rates

Table 9 lists the ventilation flow rates in the various rooms. According to the table, room 4 was the sole instance in which the dimensionless fluctuation ventilation flow rates are larger than the dimensionless mean value with a 20% difference. While the dimensionless fluctuating ventilation flow rates in rooms 1 and 2 were much less than the average value with differences of 36% and 27%, respectively, and the dimensionless average ventilation flow rates in room 3 were comparable to the fluctuating value with a difference of roughly 7%. The greatest value of the dimensionless fluctuating ventilation rate (room 4) was 1.3 times higher than the minimum value (room 1) in these four rooms, and the maximum value of the dimensionless mean ventilation rate (room 4) was 39% different from the lowest value (room 1).

In summary, room 1 had the lowest ventilation flow rate, while room 4 had the greatest. The reason is the vortex produced by the side walls of the building. Moreover, it can be seen from Table 9 that the ventilation flow rates of single-room buildings are smaller than those of multi-room buildings under the condition of same incoming wind speed and same building height. The reason may still be the flow split caused by the sidewall.

Table 9. Ventilation flow rates in different rooms.

Ventilation	Room 1	Room 2	Room 3	Room 4
Dimensionless average ventilation $(\overline{Q}')$	0.2396	0.2479	0.3265	0.1917
Dimensionless fluctuations ventilation $(\sigma_Q{}')$	0.1756	0.1946	0.3035	0.1350
Ventilation (m^3/s)	0.1695	0.1753	0.2308	0.1356

4. Conclusions

In this study, the pumping airflows of the single-room and the multi-room buildings were numerically simulated with an SST k-ω turbulent model, which was verified by an existing experiment. Through the unsteady simulation, the qualitative and the quantitative characteristics of the pumping ventilation were analyzed alongside the influence of the wind speed and the opening size. The primary conclusions are as follows.

(1) For the pumping ventilation of the single-room buildings, the airflow direction at the leeward openings will change periodically. The ventilation flow rates also display the features of a regular fluctuation in a sine wave. Only the transient simulation can describe those flow characteristics. The vortex-shedding frequency is constant, which can be described in terms of the *Ro* number. For the case of this study, $Ro = 0.098Re + 4207.2$ ($4 \times 10^5 < Re < 2.1 \times 10^6$).

(2) The airflow oscillation frequency is basically the same as the vortex-shedding frequency around the building; all of them vary monotonically with wind speed. As the wind speed increases, the frequency of the oscillations inside the building rise. The variation in the dimensionless opening spacing has an insignificant impact on the room's airflow oscillation frequency.

(3) The ventilation flow rates of the Warren model are not suitable for the pumping ventilation. However, the ventilation flow rates are also proportional to the wind speed and the opening areas. By fitting the simulation data, the formula for calculating the amount of pumped ventilation in a single-room building was derived as $Q = 0.246U_R A_{eff}$. Additionally, a linear relationship was found between the dimensionless opening distance and the dimensionless ventilation flow rate as $Q = 0.4872U_R A_{eff} s\prime$.

(4) For multi-room buildings, the ventilation effects of various rooms are different. The room in the center of the building had a larger air change rate than the rooms on the sides of the structure. Moreover, the ventilation performance of any space in multi-room buildings is better than that in single-room buildings under the conditions of the same wind speed and the same building height.

The study can help engineers distinguish the ventilation effect of the pumping ventilation and even judge the different effects of the outdoor pollutants for various rooms. However, the research took the incoming wind speed as a constant value in every quasi-3D model. In a real project, the wind velocity will change with time, and the flow field will be affected by the ground. Moreover, the thermal stack effect also needs to be considered. The characteristics of the pumping ventilation will be more complex than those provided in this paper, all of which need to be further studied.

Author Contributions: Writing—original draft preparation, X.Q.; writing—review and editing, J.Z. (Junli Zhou), X.X., W.Z., J.Z. (Jiaji Zhang), S.G. and Y.P. All authors have read and agreed to the published version of the manuscript.

Funding: This work is supported by the Major R&D projects of the China Metallurgical group corporation(2022 No. 14).

Data Availability Statement: Data are contained within the article. The data presented in this study are available in the insert article.

Conflicts of Interest: Author Shuqiang Gui and Yangdong Peng were employed by the company WSGRI Engineering & Surveying Incorporation Limited. The remaining authors declare that the research was conducted in the absence of any commercial or financial relationships that could be construed as a potential conflict of interest. The authors declare that this study received funding from China Metallurgical group corporation. The funder was not involved in the study design, collection, analysis, interpretation of data, the writing of this article or the decision to submit it for publication.

References

1. Visagavel, K.; Srinivasan, P. Analysis of single side ventilated and cross ventilated rooms by varying the width of the window opening using CFD. *Sol. Energy* **2009**, *83*, 2–5. [CrossRef]
2. Van Hooff, T.; Blocken, B.; Tominaga, Y. On the accuracy of CFD simulations of cross-ventilation flows for a generic isolated building: Comparison of RANS, LES and experiments. *Build. Environ.* **2017**, *114*, 148–165. [CrossRef]
3. Tao, T. Research on Airflow Characteristics and Thermal Comfort of Thermopressurised Naturally Ventilated Rooms. Ph.D. Thesis, Yangzhou University, Jiangsu, China, 2015. (In Chinese).
4. Chu, C.R.; Chiu, Y.H.; Tsai, Y.T.; Wu, S.L. Wind-driven natural ventilation for buildings with two openings on the same external wall. *Energy Build.* **2015**, *108*, 365–372. [CrossRef]
5. Warren, P. The analysis of single sided ventilation measurements. *Air Infiltration Rev.* **1986**, *7*, 3–5.
6. Straw, M.P. Computation and Measurement of Wind Induced Ventilation. Ph.D. Thesis, University of Nottingham, Nottingham, UK, 2000.
7. Daish, N.C.; da Graça, G.C.; Linden, P.F.; Banks, D. Impact of aperture separation on wind-driven single-sided natural ventilation. *Build. Environ.* **2016**, *108*, 122–134. [CrossRef]
8. Zhong, H.Y.; Zhang, D.D.; Liu, D.; Zhao, F.Y.; Li, Y.; Wang, H.Q. Two-dimensional numerical simulation of wind driven ventilation across a building enclosure with two free apertures on the rear side: Vortex shedding and "pumping flow mechanism". *J. Wind. Eng. Ind. Aerodyn.* **2018**, *179*, 449–462. [CrossRef]
9. Zhong, H.Y.; Zhang, D.D.; Liu, Y.; Liu, D.; Zhao, F.Y.; Li, Y.; Wang, H.Q. Wind driven "pumping" fluid flow and turbulent mean oscillation across high-rise building enclosures with multiple naturally ventilated apertures. *Sustain. Cities Soc.* **2019**, *50*, 101619. [CrossRef]
10. Albuquerque, D.P.; Sandberg, M.; Linden, P.F.; da Graça, G.C. Experimental and numerical investigation of pumping ventilation on the leeward side of a cubic building. *Build. Environ.* **2020**, *179*, 106897. [CrossRef]
11. Zhong, H.Y.; Jing, Y.; Liu, Y.; Zhao, F.Y.; Liu, D.; Li, Y. CFD simulation of "pumping" flow mechanism of an urban building affected by an upstream building in high Reynolds flows. *Energy Build.* **2019**, *202*, 109330. [CrossRef]
12. Zhong, H.Y.; Jing, Y.; Sun, Y.; Kikumoto, H.; Zhao, F.Y.; Li, Y. Wind-driven pumping flow ventilation of highrise buildings: Effects of upstream building arrangements and opening area ratios. *Sci. Total Environ.* **2020**, *722*, 137924. [CrossRef]
13. Zhong, H.Y.; Lin, C.; Sun, Y.; Kikumoto, H.; Ooka, R.; Zhang, H.L.; Hu, H.; Zhao, F.Y.; Jimenez-Bescos, C. Boundary layer wind tunnel modeling experiments on pumping ventilation through a three-story reduce-scaled building with two openings. *Build. Environ.* **2021**, *202*, 108043. [CrossRef]
14. da Graca, G.C.; Albuquerque, D.P.; Sandberg, M.; Linden, P.F. Pumping ventilation of corner and single sided rooms with two openings. *Build. Environ.* **2021**, *205*, 108171. [CrossRef]
15. Kobayashi, T.; Sandberg, M.; Fujita, T.; Lim, E.; Umemiya, N. Simplified estimation of wind induced natural ventilation rate caused by turbulence for a room with minute wind pressure. *Proc.-Roomvent Vent.* **2018**, *2018*, 613–618.
16. Tominaga, Y.; Mochida, A.; Yoshie, R.; Kataoka, H.; Nozu, T.; Yoshikawa, M.; Shirasawa, T. AIJ guidelines for practical applications of CFD to pedestrian wind environment around buildings. *J. Wind. Eng. Ind. Aerodyn.* **2008**, *96*, 1749–1761. [CrossRef]
17. Yoshie, R.; Jiang, G.; Shirasawa, T.; Chung, J. CFD simulations of gas dispersion around high-rise building in non-isothermal boundary layer. *J. Wind. Eng. Ind. Aerodyn.* **2011**, *99*, 279–288. [CrossRef]
18. Liu, J.; Niu, J. CFD simulation of the wind environment around an isolated high-rise building: An evaluation of SRANS, LES and DES models. *Build. Environ.* **2016**, *96*, 91–106. [CrossRef]
19. Mou, B.; He, B.J.; Zhao, D.X.; Chau, K.W. Numerical simulation of the effects of building dimensional variation on wind pressure distribution. *Eng. Appl. Comput. Fluid Mech.* **2017**, *11*, 293–309. [CrossRef]
20. Jiang, Y.; Alexander, D.; Jenkins, H.; Arthur, R.; Chen, Q. Natural ventilation in buildings: Measurement in a wind tunnel and numerical simulation with large-eddy simulation. *J. Wind. Eng. Ind. Aerodyn.* **2003**, *91*, 331–353. [CrossRef]
21. Roshko, A. Experiments on the flow past a circular cylinder at very high Reynolds number. *J. Fluid Mech.* **1961**, *10*, 345–356. [CrossRef]
22. Monai, R. Wind tunnel experiments on vortex shedding frequency and end effects of square columns. *Exp. Mech.* **1994**, *9*, 6. (In Chinese)
23. Caciolo, M.; Stabat, P.; Marchio, D. Full scale experimental study of single-sided ventilation: Analysis of stack and wind effects. *Energy Build.* **2011**, *43*, 1765–1773. [CrossRef]
24. Ziarani, N.N.; Cook, M.J.; Freidooni, F.; O'Sullivan, P.D. The role of near-façade flow in wind-dominant single-sided natural ventilation for an isolated three-storey building: An LES study. *Build. Environ.* **2023**, *235*, 110210. [CrossRef]

MDPI AG
Grosspeteranlage 5
4052 Basel
Switzerland
Tel.: +41 61 683 77 34

Buildings Editorial Office
E-mail: buildings@mdpi.com
www.mdpi.com/journal/buildings

www.ingramcontent.com/pod-product-compliance
Lightning Source LLC
Chambersburg PA
CBHW041413210926
51850CB00013B/3265